2 Chemistry of Plant Protection

Degradation of Pesticides, Desiccation and Defoliation, ACh-Receptors as Targets

With contributions by
H. Bergmann, G. Engelhardt, D. Martin,
H.-J. Mengs, D. Otto, R. Richter,
U. Schoknecht, P. R. Wallnöfer

With 42 Figures and 47 Tables

Springer-Verlag
Berlin Heidelberg New York
London Paris Tokyo Hong Kong

Managing Editors:

Dr. G. Haug
Pflanzenschutzzentrum Monheim, Bayerwerk
D-5090 Leverkusen/FRG

Prof. Dr. H. Hoffmann
Leiter des Ressorts Chemische Forschung
Pflanzenschutzzentrum Monheim, Bayerwerk
D-5090 Leverkusen/FRG

This series continues the handbook
"Chemie der Pflanzenschutz- und Schädlingsbekämpfungsmittel"
edited by R. Wegler

ISBN-13:978-3-642-69795-1 e-ISBN-13:978-3-642-69793-7
DOI: 10.1007/978-3-642-69793-7

Library of Congress Cataloging-in-Publication Data
Degradation of pesticides, desiccation, and defoliation, ACh-receptors
of insects / Rainer Richter, Dieter Otto, H. J. Mengs, editors ; with
contributions by H. Bergmann ... [et al.].
p. cm. — (Chemistry of plant protection ; 2)
Includes index.
ISBN-13:978-3-642-69795-1
1. Pesticides. 2. Pesticides—Biodegradation. 3. Insecticides—
Physiological effect. 4. Insects—Physiology. I. Richter,
Rainer. II. Otto, Dieter, 1928– . III. Mengs, H. J. IV. Series.
SB951.D414 1989 632'.95—dc20 89-6378CIP

Softcover reprint of the hardcover 1st edition 1989

2152/3020-543210 — Printed on acid-free paper

Editorial Board

Editorial

The series "Chemistry of Plant Protection" is the successor to our handbook "Chemie der Pflanzenschutz- und Schädlingsbekämpfungsmittel" which was edited by R. Wegler.

Aims and Scope

Plant protection chemistry continues to develop rapidly. Important targets have been achieved and improvements with respect to selectivity, efficacy, mode of action, environmental and toxicological compatibility have been made. There have been new developments in the field of formulation and side-effect-research, and new classes of substances with high effectivity at very low dosages have been discovered.

These developments have been accelerated by new biological and biochemical discoveries as well as by advances in synthetic chemistry. In order to emphasize the interaction between these of different fields of crop protection chemistry, such as insecticide-, fungicide or herbicide research, has been abandoned. The following volumes will contain recent developments in research into new active chemical substances as well as reports on metabolism, residue analysis, biochemical mechanisms and other important innovations.

The volumes will be more or less topic-oriented. Each volume will have a "volume index" which approximately characterizes the contents. The aim of the publisher and the board of editors is to produce review articles of the highest quality by leading scientists in the field of plant protection. Suggestions for such contributions from all those involved in plant protection are, of course, always welcome.

Editorial Policy

The series publishes critical review articles in English from invited authors and covering new developments in this field. The topics should be covered comprehensively and the international literature evaluated. Ideally, contributions should be 40–80 typewritten pages long. Experimental details, except when not previously published, should be covered by citing appropriate references. — The responsible editor, that is the editor who invited the article, discusses the scope of the review with the author on the basis of a tentative outline which the author is asked to provide.

Editors Publisher

Table of Contents

Table of Contents of Volume 1

Microbial Degradation of Pesticides

Peter R. Wallnöfer and Gabriele Engelhardt
Bayerische Landesanstalt für Ernährung, Abt. Ernährung
Menzingerstr. 54, 8000 München 19/FRG

The fate of xenobiotic compounds e.g. pesticides in the environment is of great importance, since disappearance, persistence or transformation of such a compound determines its usefulness or its potentially hazardous effect. There are chemical as well as physical factors that influence the fate of pesticides in the environment. The soil which can be considered as the most complex microbial habitat is an area where pesticides are deposited either by direct application or by decaying foliage. It possesses many criteria for an efficient physical or chemical attack on a chemical, e.g. adsorption. leaching, volatization, microbial degradation, or photodecomposition which is restricted to the soil surface only. Since microbial degradation makes the greatest contribution to elimination of a xenobiotic compound from soil it is necessary to investigate the microbially induced metabolic reactions and to isolate and identify the microorganisms and degradation products in laboratory experiments.

Of the many aspects of pesticide behavior in the environment there has been a notable increase of knowledge on the microbial transformation of pesticides by pure cultures of microorganisms in recent years.

The present monograph presents information currently available on the microbial degradation of pesticides with particular respect to microorganisms responsible for the biochemical reactions involved, and to the main degradation products identified.

Chemistry of Plant Protection 2

1 Introduction

The behavior of pesticides in the environment is of great importance, since disappearance, persistence, or partial transformations of such a compound determines its usefulness or its potential effects.

The range of chemicals in this category includes insecticides and fungicides for controlling soil-borne pests and pathogens, systemic insecticides and fungicides intended for uptake by the underground parts of plants, herbicides, nematicides, and general soil sterilants.

There are chemical and physical factors that influence the fate of a pesticide, but the least predictable transformation is usually caused by microorganisms. It is important to know the actual biochemical reactions involved in pesticide metabolism, since these can give a basis for the understanding of their short or long persistence in a natural environment and can also contribute to the clarification of the relationship between chemical structure and susceptibility to probable microbial transformations. For this purpose it is necessary to investigate the metabolic reactions and to isolate and identify the resulting products in laboratory experiments.

2 Significance of Microorganisms in Degradation of Pesticides

2.1 Fate of Pesticides in Soil and Significance of Microbial Communities in Biodegradation

The reason for interest in the fate of pesticides in soil arises from concern in the possibilities of environmental contamination. The soil, for instance, which can be considered as the most complex microbial habitat and an area where pesticides or their derivatives are deposited either by direct application or by decaying foliage, possesses many characteristics for an efficient chemical, biochemical or physical attack on a molecule and for its removal by adsorption, leaching, volatilization, biodegradation or photodecomposition. Consequently, it is very difficult to distinguish in a soil environment between microbial, chemical, and physical factors contributing to the removal or transformation of a pesticidal molecule. It was also postulated that breakdown products of biological and physicochemical activity in soil are often similar because the reagents (oxygen, water, and nucleophiles) are the same in each instance. E.g., photolysis can cause the hydrolysis of esters and amides, dealkylation of amines, and other effects, and the same reactions can be initiated by enzymatic activity resulting in identical products. Biological transformation of xenobiotic compounds, for example pesticides, in soil, freshwater, or estuarine ecosystem, appears to be caused primarily by bacteria, actinomycetes, and fungi [1)].

Initial research on microbial pesticide metabolism was characterized mainly by isolating organisms capable of using a compound as their only source of both carbon and energy. Later it was realized that other mechanisms, such as cometabolic transformation, conjugation reactions, or the mere accumulation of a pesticide within a microbe, are important factors of microbial interference (see Chapt. 4). The complete degradation of a pesticidal molecule to inorganic compounds or its fragmentation into components which could be mineralized easily removes its potential toxicity completely from the environment. The mechanisms which cause only partial change or temporary removal do not eliminate the potential hazard of an applied chemical or its transformation product in nature [1].

In many cases the transformation of a pesticide does not occur by only one type of mechanism during its exposure to one organism or to a whole microflora under natural conditions. In addition, a specific compound can be metabolized by various pathways in the environment.

All microbial transformations are caused by enzymes, and since most of the applied pesticides are xenobiotic compounds possessing a molecular configuration that may normally not occur in nature, it becomes evident that many of the enzymes catalyzing certain reactions are induced which often causes an initial lag period until metabolic activity can be observed. Although many enzymes are induced, the transformations they catalyze are usually reactions also encountered in the metabolism of natural substances. However, it is difficult to predict which molecular change can be expected by a specific microorganism, since each group of organisms, even various strains of one genus, can metabolize a selected molecule in a different manner.

2.1.1 Pesticides as Substrate for Growth and Energy

The complete microbial breakdown of an organic molecule to inorganic components would be the desired mechanism if one is interested in avoiding the persistence of a potentially hazardous compound in the environment. As figured out in laboratory experiments, some organic pesticides can serve as the sole carbon or energy source for growth and proliferation of certain microorganisms. If a pesticide can be used in such a way, it is degraded and fragmented to compounds that can be transferred into the known metabolic pathways which are necessary for providing energy.

In order to determine whether a pesticide can serve as the only carbon source needed for growth, the general experimental approach proceeds by an enrichment culture technique. After isolation of the surviving microorganisms, the pesticide is added to a basal salt medium and further observations are usually performed with a pure culture.

It appears to be an obvious conclusion that in a natural ecosystem like the soil, with an abundance of various microorganisms, there is an even greater possibility for the single or combined transformation and complete use of a specific pesticide by the microbial population.

2.1.2 Cometabolism

The phenomenon that a microorganism can transform a chemical without deriving energy to support its growth is a relatively recent observation [2] and its detection and significance is related to the modern use of xenobiotic compounds in various environments. The potential importance of cometabolism for the transformation of pesticides was first pointed out by Alexander [3], and the findings of many investigations can now be explained by this process [4]. Cometabolism generally does not result in extensive degradation of a pesticidal molecule, but it can cause a reduction, elimination or possibly increase of toxicity in the environment. However, it also was demonstrated that different microorganisms can degrade a certain pesticide considerably by subsequent cometabolic attack; for instance, the herbicides 2,4,5-trichlorophenoxyacetate and 2,3,6-trichlorobenzoate are converted by a cometabolic oxidation to 3,5-dichlorocatechol by a *Brevibacterium* sp. [5,6], whereas an *Achromobacter* sp. was capable of cooxidizing the resulting 3,5-dichlorocatechol to 3,5-dichloro-2-hydroxymuconic semialdehyde [7]. It was also speculated that cometabolism could account for complete mineralization of a chemical if a carbon and energy source is supplied to mixed microbial populations in the form of a biodegradable analog of the chemical under investigation [4].

The process of cometabolism is effected by bacteria as well as actinomycetes and fungi, and therefore it can be assumed that its occurrence is widespread in natural ecosystems. Many observations of microbial transformations that could not be understood, since the microorganisms did not derive any energy or nutritional use from it, are now interpretable by this mode of metabolism. The process of cometabolism, especially as a factor in the microbial transformation of pesticides, requires still further intensive study for a clear understanding of its actual cause, its transformation capabilities as related to the structure of chemicals, and the extent to which such a transformation forms a more or less persistent compound. Although it is difficult to demonstrate unequivocally that microbial cometabolism also occurs under natural conditions, there is little doubt that it takes place, and therefore the ecological importance of this reaction has to be fully explored [1].

2.2 Pesticide Effects on Soil Microorganisms

Very often it has been claimed that effects of pesticides on non-target organisms, e.g. soil microorganisms, were mainly the result of high rates of pesticide application and values of 100 kg a.i. ha^{-1} were frequently cited [8]. It is now apparent that effects of one kind or another are being obtained with normal field rates which are often well below 5 kg a.i. ha^{-1}, and occasionally effects are obtained with sub-normal field rates. This is undoubtedly due to the considerable advances that have been made in pesticide technology and, in particular, to the higher chemical reactivity of modern pesticidal chemicals [8].

Bacterial numbers and varieties in soil are not generally adversely affected by any of the four groups of pesticides whereas nitrification is more sensitive to fungicides, insecticides and other pesticides than it is to herbicides. Rhizobia and legume nodulation appear more sensitive to insecticides than to fungicides or herbicides, although both of the latter two groups may tend to be slightly more inhibitory than either stimulatory or without effect.

Soil fungi and actinomycetes are not as susceptible to herbicides or insecticides as they are to fungicides and others, many of which are fumigant fungicides. This generalized result, like the effect ratio for herbicides on algae, is not unexpected. What is unexpected is the very high positive (non-inhibitory) effect of fungicides in general on pathogenic microorganisms and their antagonists. It is necessary to emphasize that these findings are not concerned with the effects of fungicides on pathogens known to be controlled by particular fungicides, but on other pathogenic fungi (and their antagonists) which are not the intended target for the particular fungicide. Herbicides, on the other hand, appear to be effective in controlling the growth of many pathogens and antagonists of pathogens.

Cellulolytic activity and organic matter degradation are not adversely affected by the majority of herbicides and insecticides. Other pesticides, however, many of which are fumigant fungicides, appear to be predominantly inhibitory to these organisms. Since many fungi and actinomycetes are very actively involved in organic matter degradation, it is interesting to note that the effect ratios for herbicides and insecticides on these organisms follow the same positive (non-inhibitory) trend as the effect ratios for cellulolytic activity and organic matter degradation. Soil respiratory activity, on the other hand, which is ment to be indicative of organic matter degradation, does not correlate well with effect ratios for the latter, especially in the case of other pesticides. Other soil enzymic activities appear more susceptible to fungicides and other pesticides than to either herbicides or insecticides, while ammonification seems to benefit from the application of all groups of pesticides to soil.

In most cases the effects noted are transient, lasting a few days or a few weeks, but in other cases extended periods of effects have been recorded. Doubtless the more transient the particular effect the better, but based upon even the most theoretical average for microbial propagation time in soil, it is difficult to judge how transient an effect should be to qualify it as unimportant. The inhibition of nitrification for one month may be regarded as unimportant in agronomic terms but the inhibition of a natural antagonist of a plant pathogen for one week may be sufficient for the pathogen to infect a crop.

Some effects of pesticides, applied at normal field rates, are disturbing in their practical implication. The increase, for example, in plant pathogenic infestations of crops following pesticide treatments, the impact of some pesticides on plant rhizosphere populations or the profound changes in species composition in pesticide-treated soils, indicate a very real necessity for microbial ecological monitoring programmes [8)].

With regard to those parameters which are most frequently studied in relation to pesticide effects it is at least reassuring to know that, in general saprophytic soil microorganisms are very resilient and that re-colonization of pesticide-decimated soil will occur.

3 Methods for Isolation of Pesticide Degrading Microorganisms from the Soil Environment

3.1 Isolation Techniques and Cultivation of Soil Microorganisms

Laboratory studies of the microbial metabolism of pesticides utilize microorganisms either obtained from culture collections or isolated from soil, aquatic environments, etc. Using soil as source of microorganisms first the evidence of the microorganisms for pesticide degradation has to be proved.

Evidence for microbiological degradation of pesticides can be obtained by observing the effects of sterilization on the rate of breakdown. In numerous cases autoclaving, irradiation, or treating the soil with chemical sterilants such as sodium azide and sodium ethylmercurithiosalicylate greatly retards degradation. Unfortunately all of these methods have effects on soil other than simply killing microorganisms. Autoclaving is particularly drastic. On the other hand it can not be assumed that complete sterilization has been achieved or that reinvasion has been prevented. The quantitative effects of sterilization cannot therefore be interpreted precisely and related simply to the contribution that microbial processes make to the overall degradation rate [3].

Enrichment and isolation of soil microorganisms is often carried out under field or laboratory simulated field conditions. Some workers, however, have favoured the use of soil suspension or perfusion techniques [9,10,11]. This technique involves continuous circulation of an aerated solution of pesticide through a soil column, a process which gradually enhances the capacity of the microbial population to degrade the pesticide. The soil is then said to be "enriched". The mechanism of enrichment has not been established with certainty. It could be an "ecological" effect, the presence of the chemical favouring the growth of those components of the microbial population, possibly including mutants, which can utilize it as a source of energy. However, the effects seem to be too reproducible and consistent for such a mechanism and many workers therefore favour the alternative theory that enzymes capable of more effective degradation can be induced by the presence of the chemical; in other words microorganisms can adapt their metabolic processes to cope with the new material. The adapted organism would then multiply rapidly, favoured by the supply of substrate and lack of competition from unadapted species [12]. Whatever the mechanism, enrichment makes it much easier to isolate organisms responsible for degradation, after which they can be grown in pure culture using the pesticide as the sole source of carbon [13].

The ultimate proof of the involvement of soil microorganisms in the removal of a pesticide from soil is the isolation of the responsible organism, and demonstration of its ability to inactive the pesticide in culture solutions. The isolation of cultures of these organisms is mostly carried out by dilution and plating techniques [14] on appropriate synthetic media. The methods used rarely isolate more than a small proportion of the viable microbial population and in particular are likely to miss auxotrophic organisms, many non-sporing fungi and organisms with highly specific nutrient or physical requirements. Furthermore, enrichment techniques tend to favour organisms which can metabolize the pesticide as a carbon and energy source and may exclude those capable of co-metabolic transformation.

3.2 Degradation Studies with Pure Cultures

During or following isolation, those microorganisms are selected that are capable of metabolizing the pesticide in solid or liquid culture media. The method of assessment differs widely from one pesticide to another. Pesticide degradation studies generally are following the traditional methodology: the organisms are incubated in a variety of media containing the pesticide in the presence or absence of other sources of carbon, nitrogen and energy. Special growth requirements of some microorganisms can be accommodated by the inclusion of, for example, amino acids and growth factors. The detection of pesticide metabolites is an essential criterion for confirming transformation or degradation, since losses of unchanged pesticide from the medium can occur by volatilization, intracellular accumulation, and by adsorption to cell surfaces. The use of a ^{14}C-labeled substrate is often recommended, particularly in order to collect volatized products evolved from the culture medium into the air, when closed incubation systems are used [15].

An exact proof of microbial pesticide metabolism is also obtained by isolation and, where possible, by purification of the active enzyme systems.

3.3 Maintenance of Pesticide Degrading Microbial Cultures

The maintenance of laboratory cultures of microorganisms able to metabolize pesticides usually is performed by traditional methods, although it has not always proved easy. Sometimes a progressive loss of metabolic ability of microorganisms is noted, when the microorganisms are maintained by regular subculture and growth on or in nutrient media, even when the medium contains the pesticide. It is suggested that such instability might be due to loss of the genetic elements essential for processes such as pesticide metabolism during the growth phase [16]. In order to avoid or reduce these losses of activity in pesticide degradation which commonly occurs with some soil organisms maintained on laboratory nutrient media [17,18] microbial cultures can be stored either lyophilized (freeze-dried) or at extremely low temperatures (e.g. under liquid nitrogen; at —196 °C).

The use of lyophilization is well established and has been found to preserve the viability of many biological specimens almost indefinitely [19]. Disadvantages may arise by the fact that many mycelial forms are difficult to lyophilize and many fungi prepared from spore suspensions fail to survive. Within the last two decades ultra low temperature storage of cultures, has attracted considerable attention [17,18,20]. The technique requires relatively little storage space and although some cells may suffer death the majority of cultures show a reasonable proportion of survivors. It has been reported that often a higher proportion of cells survived after freezing than after lyophilization [21] and no indications of genetic changes in bacteria subjected to the freezing, storage and thawing procedures are found [22].

3.4 Taxonomy of Pesticide Degrading Microorganisms

The present state of knowledge of the microbial breakdown of xenobiotic compounds e.g. pesticides mostly consisting of aromatic and heteroaromatic structures has already been summarized in a number of reviews [23,24,25]. Eubacteria, yeasts, and higher fungi from the soil environment are capable of breaking down or transforming pesticides. Up to now, the species, listed in Table 1 have been reported to carry out these reactions.

4 Biochemical Reactions Involved in Pesticide Degradation by Microorganisms

Metabolic degradation or transformation of pesticides by soil microorganisms is predominantly an aerobic process. However, a variety of chemicals are decomposed also in the absence of molecular oxygen by reactions involving hydrolysis, dechlorinations or reductions, or hydrolytic ring cleavage.

Table 1. Microorganisms responsible for the degradation of pesticides. (Names used formerly are in brackets)

Eubacteriales (Gram negative)		
1. *Aerobic, microaerophil, motile:*		
	Azospirillum	*lipoferum*
	Hafnia	sp.
2. *Aerobic rods and cocci:*		
Pseudomonadaceae:	*Pseudomonas*	*alcaligens*
		arvilla
		acidovorans
		aeruginosa
		cepacia
		convexa
		fluorescens
		multivorans
		putida
		striata
		testosteroni
		(= *Comomonas terrigeri*)
	Hydrogenomonas	(untertain affiliation)
	Xanthomonas	sp.
Azotobacteriaceae	*Azotobacter*	sp.
Rhizobiaceae	*Rhizobium*	*meliloti*
		(= *japonicum meliloti*)
	Agrobacterium	sp.
Neisseriaceae	*Moraxella*	sp.
Other genera of group 2:	*Flavobacterium*	*meninginosepticum*
		pelegrinum
		sp.
	Paracoccus	sp.
	Alcaligenes	*faecalis*
3. *Facultative anaerobic rods:*		
Enterobacteriaceae	*Escherichia*	*coli*
	Proteus	*vulgaris*
	Enterobacter	sp.
	Klebsiella	*aerogenes*
		pneumoniae
	(= *Aerobacter aerogenes*)	
	Klebsiella	*pneumoniae* ssp. *ozeanae*
	Serratia	*marcescens*
Eubacteriales (Gram positive)		
1. *Cocci:*		
Micrococcaceae	*Micrococcus*	sp.
		denitrificans
2. *Endospore-forming rods and cocci:*		
	Bacillus	*cereus*
		coagulans
		firmus
		sphaericus
		sp.
	Clostridium	sp.

Table 1. (continued)

3. *Non-endospore-forming rods:*		
(irregular, swollen, and branched cells)	*Corynebacterium*	sp.
	Arthrobacter	*simplex*
		sp. DSM 20389
		sp.
	Brevibacterium	sp.
4. *Mycobacteria:*		
Mycobacteriaceae	*Mycobacterium*	sp.
5. *Nocardioforms:*		
	Nocardia (uncertain affiliation,	
	probably belonging to	*Rhodococcus*
		opaca
		coeliaca
		restricta
		sp. DSM 43251
	Rhodococcus	*equi*
		(= *Nocardia restricta*)
		rubropertinctus
		(*R. corallinus*)
	Nocardiopsis?	
6. *Actionomycetes:*	*Actinomyces*	sp.
	Streptomyces	*aureofaciens*
		griseus
		sp.
Cyanophyta	*Anacystis*	*nidulans*
	Microcystis	*aeruginosa*
Mycota		
1. Zygomycetes	*Mucor*	*plumbeus*
	Mortierella	*ramanniana*
	(= *Mucor ramanniani*)	
	Rhizopus	*arrhizus*
		oryzae
	(= *Rhizopus japonicus*; *Rh. peka*)	
		stolonifer
	(= *Rh. nigricans*)	
	Cunninghamella	*echinulata*
2. Ascomycetes		
Saccharomycetaceae	*Saccharomyces*	*pastorianus*
	Hansenula	*anomala*
		saturnus
	Lipomyces	*starkeyi*
Sordariaceae	*Neurospora*	*crassa*
		sp.
Hypocreaceae	*Hypocrea*	*cf. pilulifera*
	Gliocladium	*roseum*
		catenulatum
	Fusarium	*moniliforme*
	Trichoderma	*harzianum*
		lignorum
		virgatum
		viride
		sp.

Table 1. (continued)

Mycota		
Polystigmataceae	*Glomerella*	*cingulata*
Chaetomiaceae	*Chaetomium*	*globosum*
Capnodiaceae	*Aureobasidium* (= *Pullularia*)	*pullulans*
Trichocomaceae	*Talaromyces*	*wortmanni*
3. Hyphomycetes		
Aspergillaceae	*Aspergillus*	*candidus*
		carneus
		fumigatus
		nidulans
		niger
		oryzae
		sp.
		terreus
	Penicillium	*notatum*
	Fusarium	*oxysporum* Schlecht
		(oxysporum f. sp.)
		lycopersici
		culmorum (= *F. roseum*)
		solani
		sp.
	Rhodotorula	*gracilis*
	Paecilomyces	sp.
	Alternaria	*mali*
		sp.
	Pyricularia	*oryzae*
	Geotrichum	*candidum*
		sp.
	Cladosporium	*cucumerinum*
	Cephalosporum	*gramineum*
4. Agonomycetes	*Rhizoctonia*	*practicola*
		solani
5. Basidiomycetes		
Polyporaceae	*Trametes*	*versicolor*
Chlorophyta		
Chlorococcales	*Chlorella*	*vulgaris*

Enzymatic reactions in the decomposition of pesticides to smaller molecules are energy dependent. Various organic chemicals can serve as source of energy to at least some microorganisms. However, degradation of a large variety of pesticides as e.g. chlorinated aromatic pesticides do not go as far as to complete mineralization and are seldom energy yielding. Furthermore besides intact microbial cells also enzymes which have been isolated from soil microorganisms and demonstrated to catalyze the cleavage of specific pesticides are suggested to contribute to pesticide degradation as free soil enzymes. Finally, compounds of intermediary metabolism as e.g. acetic and other organic acids or the methyl group derived from S-adenosylmethionine can be transferred to the pesticidal molecule or a metabolite derived from it thus forming conjugated pesticidal derivatives. Aromatic pesticidal metabolites can also be incor-

Table 2. Summary of main reactions involved in microbial pesticide metabolism

Process		Type of transformation	Type of compounds attacked
Oxidative reactions	β-Oxidation	$R{-}CH_2{-}CH_2{-}COOH \rightarrow R{-}COOH + CH_3COOH$	Phenoxyalkanoic acid herbicides
	Phosphorothioate Oxidation	$-P{=}S \rightarrow -P{=}O$	Thioorganophosphorus insecticides
	Oxidation of sulfur compounds	$>S \rightarrow SO;\ >SO \rightarrow >SO_2-$	Insecticides, containing a thioether or sulfinyl group Oxime carbamate nematicides with thioether groups
	Epoxidation	$R{-}C{=}C{-}R \rightarrow R{-}C(O)C{-}R$ (epoxide)	Cyclodiene insecticides
Oxidative dealkylations	O-Dealkylation	$Ar{-}O{-}CH_2{-}R \rightarrow Ar{-}OH + CHO{-}R$	Phenoxyalkanoic acid herbicides
	N-Dealkylation	$R_3{-}N(R_1)(R_2) \rightarrow R_3{-}N(R_1)(H) \rightarrow R_3{-}N(H)(H)$	Substituted urea herbicides, triazine herbicides, amide herbicides
	C-Dealkylations	$R{-}(CH_2)_n{-}CH_3 \rightarrow R\ (CH_2)_n{-}COOH \rightarrow R{-}(CH_2)_nH + CO_2$	Various alkylsubstituted aromatic pesticides
Hydroxylations of alkyl side chains		$Ar{-}CH(R_1)(R_2) \rightarrow Ar{-}COH(R_1)(R_2)$	Various alkylsubstituted aromatic pesticides
Hydrolysis	Ester hydrolysis	$R_1{-}COOR_2 \rightarrow R_1{-}COOH + R_2{-}OH$	Benzoic acid ester herbicides
	P-Ester hydrolysis	$R{-}OP\ X \rightarrow R{-}OH + HOP\ X$	Organophosphorus insecticides
	Amide hydrolysis	$R{-}CO{-}NH_2 \rightarrow R_1{-}COOH + NH_3$	Amide herbicides
	Amide hydrolysis	$R_1{-}NHCOO{-}R_2 \rightarrow R_1{-}NH_2 + CO_2 + R_2{-}OH$	Carbamate insecticides and herbicides
	Amide hydrolysis	$Ar{-}NH{-}CO{-}N(OCH_3)(CH_3) \rightarrow Ar{-}NH_2 + CO_2 + HN(OCH_3)(CH_3)$	Methoxysubstituted phenylurea herbicides
	Nitrile hydrolysis	$R{-}CN \rightarrow R{-}CONH_2 \rightarrow R{-}COOH$	Nitrile herbicides

Nitroreduction		$R{-}NO_2 \longrightarrow R{-}NH_2$	Various nitrosubstituted pesticides
Transformations of amino groups	Hydroxyl replacement	$R{-}NH_2 + H_2O \longrightarrow ROH + NH_3$	*s*-Triazine herbicides, picloram (herbicide)
	Deamination by benzene ring dioxygenation	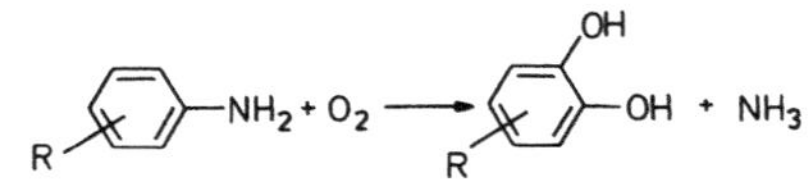	Pesticide derived anilines
	Reductive deamination	$R{-}NH_2 \longrightarrow R{-}H + NH_3$	*s*- and *as*-Triazine herbicides
Dehalogenations	Dehydrohalogenation	$R{-}CH_2CCl_3 \longrightarrow R{-}CH{=}CCl_2 + HCl$	Organohalogen pesticides containing aliphatic halide groups (e.g. DDT)
	Reductive dehalogenation	$R{-}CCl_3 \longrightarrow R{-}CHCl_2 + HCl$ $Ar{-}Cl \longrightarrow Ar{-}H + HCl$	Organohalogen pesticides containing aliphatic halide groups (e.g. DDT) PCP
	Hydroxyl replacement	$R{-}CH_2Cl + H_2O \longrightarrow RCH_2OH + HCl$ $Ar{-}Cl + H_2O \longrightarrow Ar{-}OH + HCl$	Chloro-*s*-triazine herbicides, organochlorine insecticides and fungicides, haloalkanoic acid herbicides
	Halogenid elimination after benzene ring fission, elim. of halogen from aliphatic acids	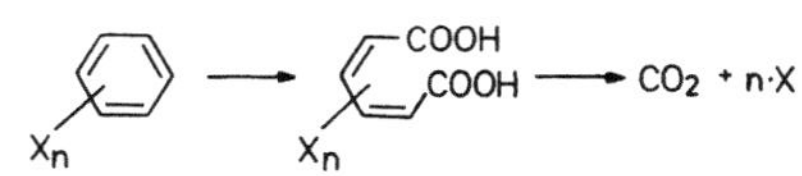	Haloalkanoic acid herbicides, halogenated aliphatic herbicides, halogenated phenol and aniline based pesticides etc.
Aerobic aromatic ring hydroxylation and ring cleavage		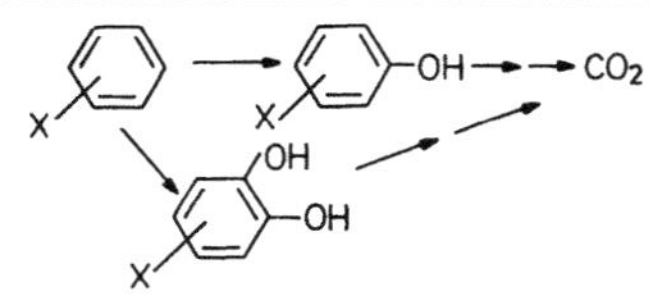	Phenoxyalkanoic acid herbicides, pyridazinone herbicides, phenolic pesticides; aniline based pesticides etc.
Anaerobic aromatic ring cleavage		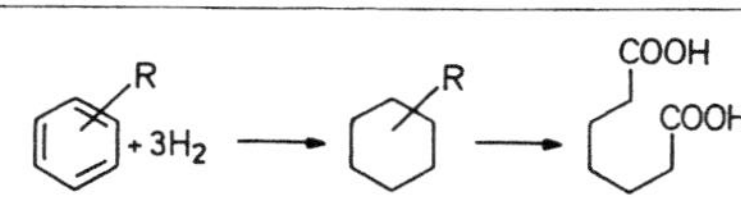	Benzoate and phenol based pesticides

Table 2. (continued)

Process		Type of transformation	Type of compounds attacked
Heterocyclic ring cleavage		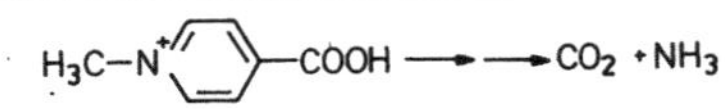	Bipyridyl herbicides
		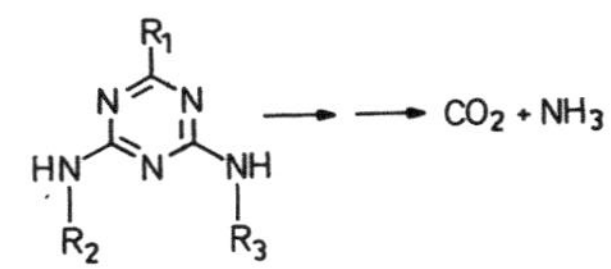	*s*-Triazine herbicides
		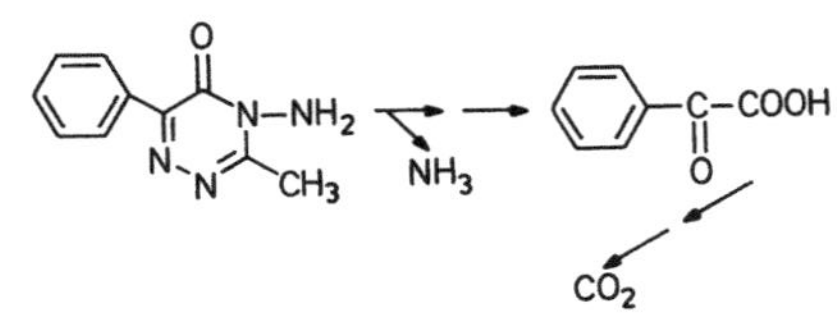	*as*-Triazine herbicide metamitron
		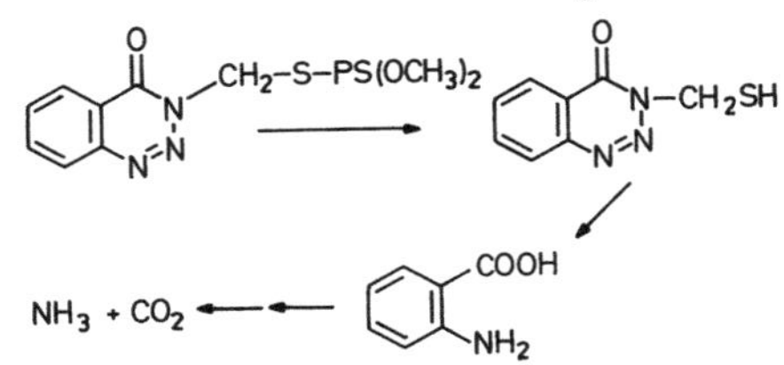	Organophosphorus insecticide azinphos-methyl
Conjugations	Acylation	$R_1-NH_2 + HOOC-R_2 \longrightarrow R_1-NH-CO-R_2$	Aniline based pesticides
	Alkylation	$R-OH + Alk-S-Adenosin \longrightarrow R-O-Alk + HSAd.$	Phenolic pesticides
Condensations		$n \cdot R \longrightarrow (R)_n \quad n=2,3,(4)$ $(R_1-OH)_n + (R_2-OH)_n \longrightarrow (R_1-O-R_2-O)_n$	Aniline based pesticides Phenolic pesticides and metabolites

porated into humic compounds via polymerization, oxidation, and reduction reactions. Enzymatic reactions involved in pesticide decomposition and transformation are therefore classified in four major groups:

a) direct degradation in the central metabolism of microorganisms in which they serve as energy sources to supply growth (catabolism);
b) incidental transformation by peripherical metabolic processes in which the microorganisms cannot derive energy for growth (co-metabolism);
c) incidental transformation by extracellular enzymes;
d) conjugation of the entire pesticidal molecule or a metabolite derived from it with compounds of intermediary metabolism or condensation with naturally occurring substances giving more complex new products.

Although details of many enzymatic reactions with pesticides are still lacking, the volume of knowledge is steadily increasing. Various data have been obtained from laboratory studies with isolated microbial strains grown under controlled culture conditions or from studies with crude cell extracts or purified enzymes obtained from these microorganisms.

The main biochemical reactions found till now to be involved in microbial pesticide metabolism include oxidative reactions as e.g. β-oxidation of aliphatic compounds or alkyl side chains of aromatics, hydroxylation and oxidative dealkylations of side chains, thioether or phosphorothioate oxidations and epoxidation of carbon-carbon double bonds, hydrolysis of amide, ester or nitrile groups, dehalogenations, transformations of nitro and amino groups, conjugate formation and condensation reactions and finally hydroxylation and cleavage of aromatic and heterocyclic rings (Table 2).

The number of reactions is few as compared to the number of compounds metabolized. Various correlations between the chemical structures of different compounds and their routes of metabolism have been noted.

On the other hand it strongly depends on the group of organisms or even on a specific strain of one genus which reactions are employed for the metabolism of a selected pesticide molecule. On the basis of the degradation pathways known till now for a large variety of xenobiotic compounds of pharmaceutical or environmental importance a certain mode of biochemical attack can be anticipated for the breakdown or transformation of a specific chemical from its chemical structure.

In the following the main enzymatic reactions involved in the transformation or cleavage of the major part of pesticides and other xenobiotic compounds in soil microorganisms are classified according to the respective chemical structure attacked.

4.1 Non Aromatic Compounds and Side Chains of Aromatics

4.1.1 β-Oxidation

Many aromatic pesticides have fatty acid side chains or substituents which are readily converted into the latter. Microbial metabolism of these fatty acid moieties is often performed by β-oxidation, a stepwise cleavage of C-2 fragments requiring ATP and coenzyme A (Fig. 1).

Side chain degradation by β-oxidation has been found as main inactivation reaction in the ω-phenoxyalkanoate herbicides with side chains consisting of more than two carbon atoms both by soil bacteria such as *Nocardia*, *Pseudomonas*, and *Micrococcus*

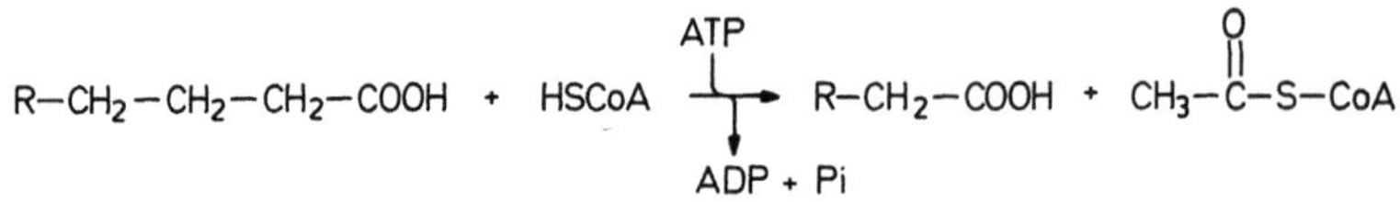

Fig. 1. β-Oxidation of fatty acid side chains

and the fungus *Aspergillus niger* [24, 26]. Alkanoic acid units with large even numbers of carbon atoms are cleaved by β-oxidation to products with smaller even numbers, and valerate or propionate derivatives are formed from side chains with odd numbers of carbons (Table 14). Further oxidations of the propionate derivative gives an unstable phenoxyformate which results in the release of CO_2 and the corresponding phenol (Fig. 2). The biological activity of a pesticide is not necessarily destroyed by β-oxidation, in some cases it may even operate as an activation mechanism [24].

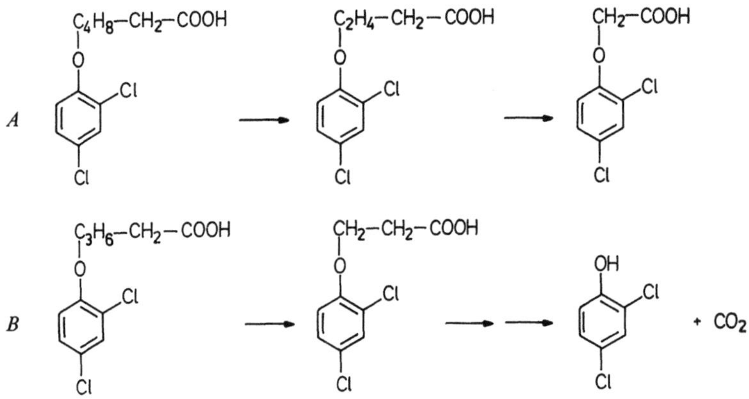

Fig. 2. β-Oxidation of 2,4-dichlorophenoxyalkanoic acids. (A) Formation of 2,4-D from acids with an even number of carbons in the side chain and (B) formation of 2,4-dichlorophenol from acids with an odd number of carbons in the side chain (after Loos [26])

4.1.2 Oxidative Dealkylation

Alkyl groups of pesticides are commonly attached to nitrogen and oxygen atoms, less frequently to the carbons and only rarely to sulphur and other atoms. Although N- and O-linked alkyl groups are readily dealkylated by microorganisms it would appear that those directly bonded to a carbon skeleton are often more resistant to metabolism [24]. Methylthio- and methylsulfinyl groups are not removed or dealkylated before ring cleavage to form the respective muconic semialdehyde derivatives as shown for the metabolism of 4-(methylthio)- and 4-(methylsulfinyl)substituted phenols by a *Nocardia* species (Table 13) [27, 28]. The limited number of observations of S-dealkylation may be due to the greater tendency for oxidation of the sulfur to the sulfone and sulfoxide.

The dealkylation reactions are mostly non-specific oxidations requiring mixed-function oxidases and a hydrogen donor such as NADPH and proceed via the formation of hydroxyalkyl intermediates [24].

4.1.2.1 N-Dealkulation, Hydroxylations of N-Alkyl Groups

N-Dealkylation is one of the most common reactions in the transformation of the side chains of a series of pesticides such as the phenylureas, dinitroanilines, *s*-triazines or bipyridyls (Fig. 3). It is catalyzed by the activity of mixed function oxidases requiring ferrous ions and NADPH and has been shown to proceed stepwise via a hydroxyalkyl and carboxy derivate as demonstrated for the demethylation of N-methyl groups (Fig. 4). For some N-dialkylated pesticides it was suggested that the monodesalkyl derivative is formed much more readily than the didesalkyl compound [24].

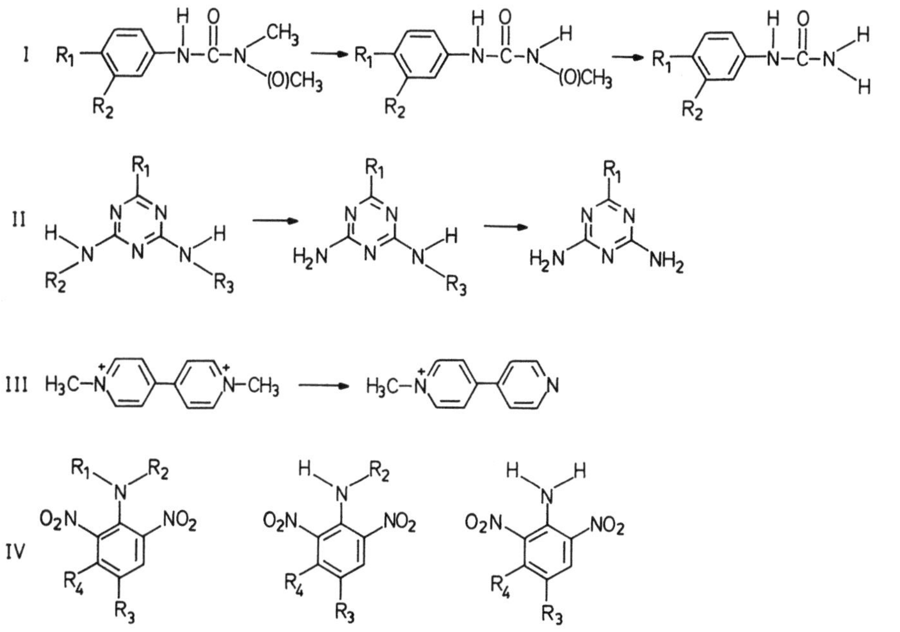

Fig. 3. Typical microbial N-dealkylation reactions of phenylureas (I), s-triazines (II), bipyridylium compounds (III), and of dinitroanilines (IV)

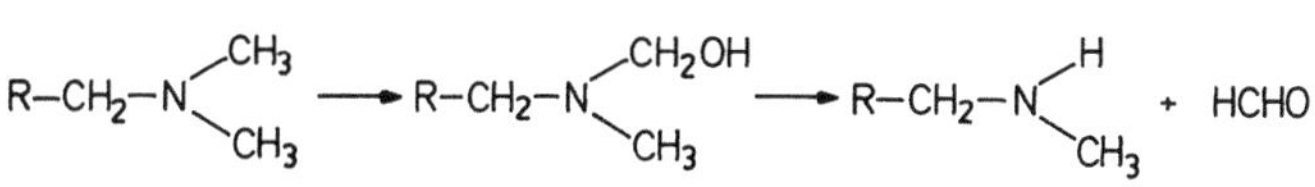

Fig. 4. Reaction sequences during N-demethylation

N-Dealkylation represents the major pathway of urea herbicide metabolism and has been demonstrated for a number of dialkyl and alkoxyalkyl substituted phenylureas [29]. The methyl groups of the quarternary bipyridyl herbicides diquat and paraquat are also dealkylated by microbial mixed function oxidases and in the microbial inactivation of *s*-triazine herbicides the oxidative removal of N-alkyl groups is a main detoxification process in microorganisms. During degradation of simazine by *Aspergillus fumigatus*, e.g., 2-chloro-4-amino-6-ethylamino-*s*-triazine was identified as main product and both 2-chloro-4-amino-6-isopropylamino- and 2-chloro-4-amino-

6-ethylamino-*s*-triazine were identified as degradation products of atrazine by various soil fungi (Table 9). Similarly several representatives of the chloro-, methoxy-, and methylthio-*s*-triazines were dealkylated by *A. fumigatus* Fres. [30].

Dealkylation reactions have been attributed mainly to the activity of soil fungi, however, recently it was shown that soil bacteria as different *Pseudomonas* sp. [31] and a *Nocardia* species [32] are also able to dealkylate atrazine (Fig. 26).

As observed in plants, the ease of alkyl group cleavage by microorganism and soil decreases in the sequence ethyl, isopropyl to larger or more branched alkyl groups [30]. Oxidation of N-alkyl side chains must not always go as far as to removal of the complete alkyl moiety thus leading to the release of metabolites hydroxylated in the N-alkyl side chain which are not oxidized further by the producing microorganism. N-Hydroxylalkyl derivatives have been found after transformation of the acylanilide herbicides karsil and dicryl by *Rhizopus japonicus* [33,34] or of the herbicide metolachlor by *Actinomyces* sp. [35] (Table 5).

4.1.2.2 O-Dealkylation

Alkoxy substituents generally are cleaved from aromatic compounds to give the parent phenol, with the concomitant liberation of the alkyl moiety as an aldehyde. Vanillic acid (3-methoxy-4-hydroxybenzoic acid), for example, is converted to protocatechuic acid by loss of its methoxy group in the form of formaldehyde. As in the hydroxylation of methyl groups, these reactions are catalyzed by monooxygenases which introduce hydroxyl groups forming semi-acetals. Hydrolysis of the latter (probably a rapid non-enzymic process) yields the corresponding phenol plus an aldehyde. Higher ethers such as *p*-ethoxybenzoate and *p*-n-propoxybenzoate can be similarly metabolized.

In the microbial degradation of organophosphorus insecticides and phenoxyalkanoic acid herbicides O-dealkylation is responsible for the loss of biological activity. Besides stepwise removal of C-2 fragments by β-oxidation degradation of phenoxyalkanoic acid herbicides as 4-chloro-2-methylphenoxyacetic acid and 2,4-dichlorophenoxyacetic acid by bacteria can also be initiated by such an oxidative ether cleavage to give the substituted phenol and glyoxylic acid (Fig. 5).

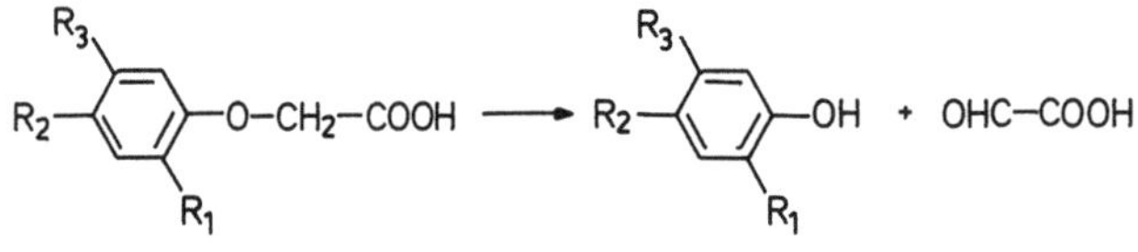

Fig. 5. Oxidative dealkylation of phenoxyalkanoic acids

This reaction has been described during soil degradation studies and also for a variety of isolated soil microorganisms, their crude extracts and purified enzyme preparations. Pesticides with more than one alkoxy substituent can lose only one or more alkoxy substituents, as found for the dealkylation of the insecticide bromophos (O-(4-bromo-2,5-dichlorophenyl)O,O-dimethylphosphorothioate) by several soil fungi [24].

Dealkylation of methoxy groups of aromatic compounds, however, does not always occur before ring cleavage. *Pseudomonas putida* degrades syringic acid (3,5-dimethoxy-4-hydroxybenzoic acid) via 3-0-methylgallic acid, which then undergoes ring cleavage [36]. Similarly, in the conversion of phenylacetic acid to homogentisic acid (2,5-dihydroxyphenylacetic acid) and 4-hydroxyphenylacetic acid to homoprotocatechuic acid(3,4-dihydroxyphenylacetic acid) the carboxymethyl substituent remains intact before ring cleavage occurs.

4.1.2.3 C-Dealkylation, Hydroxylation of Alkyl Groups

Alkyl substituents of aromatic and heterocyclic xenobiotics may either be oxidized or remain intact during hydroxylation and subsequent ring cleavage.

Thus, *m*- and *p*-cresol or *p*-cymene can be transformed to hydroxymethyl derivatives by an oxygen-dependent hydroxylation and by successive dehydrogenations forming the carboxylic acid via the aldehyde. However, *o*-, *m*-, and *p*-cresol can also be directly hydroxylated to give 3-methyl- and 4-methylcatechol intermediates, respectively [36, 37]. Similar pathways are described for the degradation of toluene. Some *Pseudomonas* and *Achromobacter* sp. oxidize toluene to 3-methylcatechol, whereas a strain of *Pseudomonas aeruginosa* oxidizes the methyl group of toluene to benzylalcohol and further to benzaldehyde and benzoic acid [38]. The oxidative metabolism of ring-substituted methyl groups has been recorded for many aromatic pesticides in biological systems, the benzylic alcohol intermediate in many cases having been isolated. In microorganisms, there have been few such observations reported. Thus, the methyl substituents of the fungicide furacarbanil(2,5-dimethyl-3-furancarboxanilide) were oxidized to hydroxy-methyl groups in fungal culture (Fig. 6) and the organochlorine insecticide methoxychlor(1,1,1-trichloro-2,2-di-4-methoxyphenyl ethane) was dealkylated by soil microorganisms. Conversely, the methyl group of MCPA (4-chloro-2-methyl-phenoxyacetic acid) was retained past the muconic acid stage when the herbicide was metabolized by a soil bacterium [24].

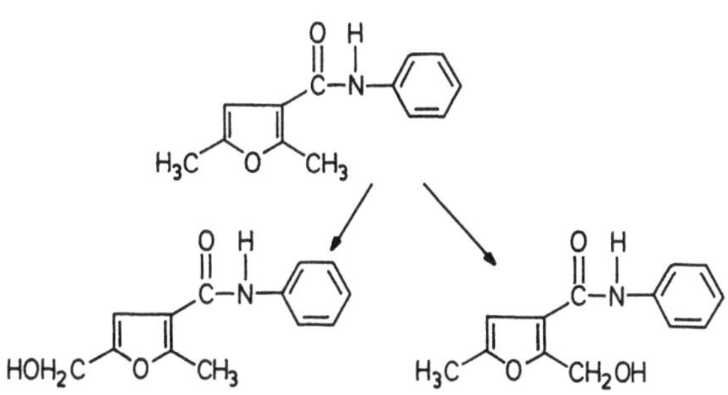

Fig. 6. Hydroxylation of the alkyl side chains of furacarbanil by *Rhizopus japonicus*

Larger alkyl side chains my also remain intact during the formation of the corresponding catechol or undergo oxidation. Carboxylic acids are then formed by oxidation of the terminal methyl group which is followed by a β-oxidation pathway. A number of alkylbenzenes or alkylbenzene sulfonates are oxidized in this manner [36].

Similarly, oxidation of large alkyl groups attached directly to the aromatic rings of pesticides has been recorded. Both bufencarb (3-(1-methylbutyl)phenylmethyl-carbamate and 3-(1-ethylpropyl)methylcarbamate) and diazinon (O,O-diethyl-O-(2-isopropyl-6-methyl-4-pyrimidinyl)phosphorothioate) are oxidized at the alkyl side chain, in soil and in microbial culture [24].

Although generally remaining unaltered, carboxyl groups of aromatic rings can also be eliminated. This may either occur via oxidation as in the conversion of anthranilic acid, salicylic acid, and benzoic acid to catechol, or nonoxidatively as in the conversion of 4,5-dihydroxyphthalic acid to protocatechuic acid in *Pseudomonas* and *Nocardia* sp., and of protocatechuic acid to catechol in *Klebsiella aerogenes* [36,39].

4.1.3 Epoxidation

Oxidation of unsaturated carbon-carbon double bonds of aromatic and aliphatic structures by microbial monoxigenases leads to the formation of an epoxide derivative. Thus, in fungi grown in presence of a second carbon source double hydroxylation of aromatic compounds results in the formation of *trans*-dihydrodiols via the formation of arene oxide (epoxide) intermediates [40,41].

Formation of epoxides during aerobic degradation of cyclodiene insecticides in soil was attributed at least in part to the activity of soil microorganisms (Table 3). In bacterial and fungal cultures, however, aldrin was transformed to the epoxide dieldrin and heptachlor to heptachlor epoxide and chlordene epoxide. A mono-oxygenase was isolated from a *Pseudomonas* species catalyzing the oxidation of alkenes to their corresponding epoxides in the presence of NADH and molecular oxygen [24,42].

4.1.4 Phosphorothioate Oxidation

The oxidation of phosphorothioates to the respective organophosphates is commonly regarded as the main reaction occurring in animal metabolism of organophosphorus insecticides in contrast to the hydrolytic and reductive reactions representing the major metabolic transformations of these pesticides in microorganisms [24].

Oxidation products of phosphorothioates have been rarely detected in microbial cultures: paraoxon was formed from parathion in a mixed bacterial culture adapted to growth on parathion as sole carbon source [43] and O-ethyl-S-phenyl ethanephosphonothioate was found as a product of fonophos in cultures of different soil fungi [44] (Table 12). Whether this oxidation was a chemical or a biochemical reaction, however, was not determined.

4.1.5 Hydrolysis

One of the reactions of predominating importance in the transformation of pesticides by soil microorganisms is hydrolysis, which is catalyzed by a large group of commonly occuring enzymes as esterases, amidases, nitrilases and phosphatases. It is the main inactivation step in the decomposition of phenylamide herbicides and fungicides, carbamate insecticides and herbicides, nitrile herbicides, organophosphorus insecticides and phenoxyacetic and benzoic acid ester herbicides (Table 2). Hydrolysis of the amide or ester bond by various bacteria has been shown for the phenylcarbamate herbicides IPC, CIPC and CEPC with the liberation of the substituted aniline, CO_2

and the respective alcohol (Table 10). The hydrolase catalyzing the hydrolysis of a variety compounds related to CIPC has been isolated and characterized. Similarly, carbamate insecticides such as carbaryl, carbofuran or aldicarb are detoxified by hydrolysis by numerous fungal and bacterial cultures from soil having either esterase or amidase activity [45] (Table 10). Acylanilide herbicides and fungicides as propanil, dicryl, karsil, solan and furacarbanil, pyracarbolid or carboxin were hydrolysed by different microbial amidases to give the respective substituted aniline and organic acid (Table 5). During studies on the microbial degradation of linuron and other phenylurea herbicides a phenylamide inducible aryl acylamidase (E.C.3.5.1) was isolated from a *Bacillus sphaericus* strain which catalyses the hydrolysis of a variety of phenylamide pesticides including the phenylcarbamates, acylanilides and the methoxysubstituted phenylureas (Fig. 7). Dimethyl substituted urea herbicides, however, were not cleaved by this enzyme [46].

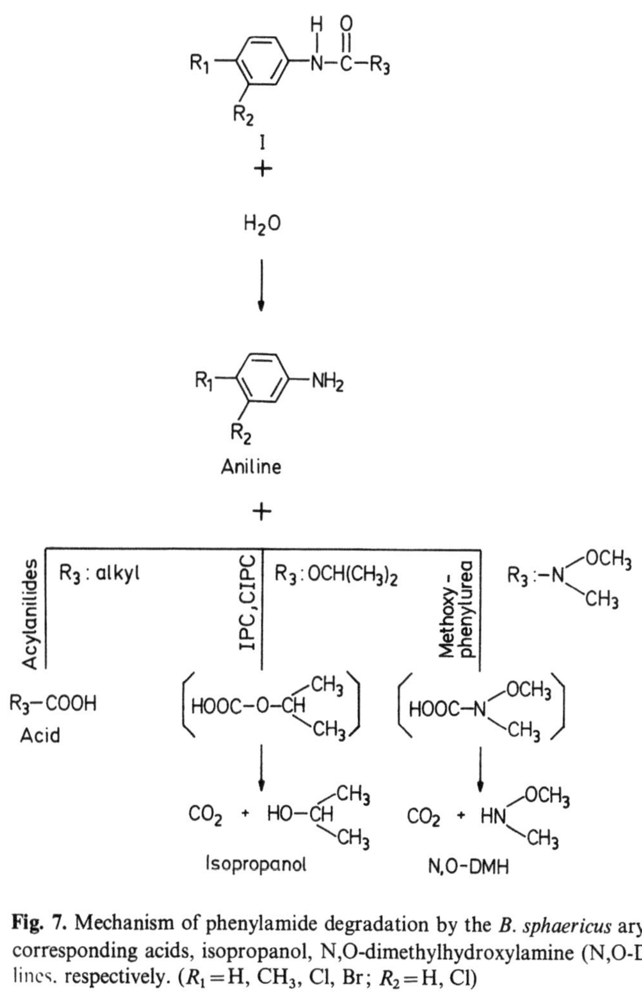

Fig. 7. Mechanism of phenylamide degradation by the *B. sphaericus* aryl acylamidase leading to the corresponding acids, isopropanol, N,O-dimethylhydroxylamine (N,O-DMH), and halogenated anilines, respectively. (R_1 = H, CH_3, Cl, Br; R_2 = H, Cl)

Ester hydrolysis is the first reaction in the microbial metabolism of the benzoic and phenoxyalkanoic acid ester herbicides and of the organophosphorus, thio-, and dithio-organophosphorus insecticides and fungicides (Fig. 8). Plasmid involvement has been demonstrated for parathion hydrolysis by *Pseudomonas diminuta* [47]. From a series of microorganisms the amidases and esterases active in pesticide cleavage have been partially purified and characterized with particular emphasis on the substrate range of the individual enzyme. Application of cell-free hydrolases capable of detoxifying pesticide wastes such as residues in pesticide containers or in soil-spill clean up has been discussed. For instance, parathion hydrolase isolated from an enrichment culture grown on parathion hydrolyzes a wide variety of organophosphorus pesticides as methylparathion, paraoxon, diazinon, Dursban, EPN, triazophos, fenitrothion, cyanophos, propetamphos and quinalphos and was found to be suitable for industrial application [45].

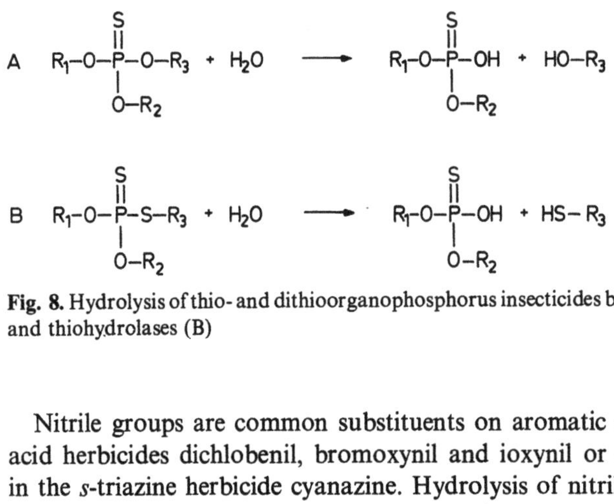

Fig. 8. Hydrolysis of thio- and dithioorganophosphorus insecticides by microbial phosphoesterases (A) and thiohydrolases (B)

Nitrile groups are common substituents on aromatic rings as e.g. in the benzoic acid herbicides dichlobenil, bromoxynil and ioxynil or on aliphatic side chains as in the *s*-triazine herbicide cyanazine. Hydrolysis of nitriles is commonly a two step hydrolytic reaction catalysed by a group of enzymes called nitrilases (E.C. 3.5.51) having the acylamide as an intermediate product as shown in cyanazine degradation in soil. In the metabolism of bromoxynil by *Klebsiella pneumoniae* subsp. *ozeanae*, isolated from soil, however, the bromoxynil hydrolyzing enzyme could not catalyze the release of ammonia from 3,5-dibromo-4-hydroxybenzamide, and no 3,5-dibromo-4-hydroxybenzamide could be detected in bromoxynil containing cultures of this organism. It was therefore suggested that in this strain the nitrilase catalyzed reaction is a single step reaction [48] (Fig. 9).

Since amidases, esterases, and nitrilases are very common in soil microorganisms hydrolysis of the respective chemical bonds occurs very rapidly in soil and the com-

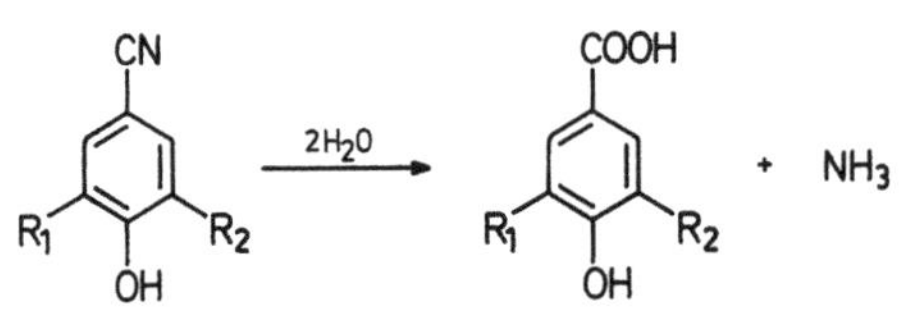

Fig. 9. Hydrolysis of benzonitriles by microorganisms R_1, R_2 = J: Ioxynil; R_1, R_2 = Br: Bromoxynil

pounds possessing these structures are frequently classified among the "biodegradable" pesticides.

In the case of the phenylcarbamates, it is difficult to determine whether the carbamate is cleaved by an esterase or amidase, since both enzymes would yield essentially the same products, i.e. aniline, CO_2 and an alcohol. The rate of the reaction, however, is strongly influenced by inductive, steric, and resonance effects, as well as molecular size. The rate of hydrolysis of several structurally related phenylcarbamate herbicides was significantly altered by inductive effects exerted by *meta* substitution of electron-withdrawing groups and steric effects imposed by increasing the size of the alcohol group. As the alcohol increased in size from the ethyl, through propyl and pentyl to benzyl the rate of hydrolysis decreased. Likewise, as the inductive effect of various meta substituents decreased in the order $NO_2 > CH_3 - CO > Cl > CH_3O > H$, the rate of hydrolysis decreased proportionally [49].

4.1.6 Transformation of Nitro and Amino Groups

Although nitro groups are frequently reduced to amines, probably via nitroso- and hydroxylamino intermediates, these are occasionally found as byproducts rather than as obligatory intermediates. Thus, aminobenzoic acids are present in cultures of bacteria growing at the expense of different nitrobenzoic acids. The latter acids, however, are not degraded by reduction but by replacement of the nitro substituents by hydroxyl groups with the concomitant formation of nitrite. Similar pathways operate for the degradation of 3,5-dinitro-*o*-cresol by *Arthrobacter simplex*, of 2- and 4-nitrophenol by *Pseudomonas*, *Moraxella*, and *Bacillus* sp. [36, 50, 51], of 3-nitrophenol by the same *Bacillus* sp., and of 2,4-dinitrophenol by a *Pseudomonas* sp. [51]. Nitro substituents of aromatic compounds have been shown to be reduced to amino groups in anaerobic environments as in the presence of oxygen. Thus, a strain of *Pseudomonas aeruginosa* reduces 2,4,6-trinitrophenol (picric acid) to the mutagen 2-amino-4,6-dinitrophenol under anaerobic conditions [52].

Reduction of nitro to amine in nitro-containing insecticides is also efficiently carried out by various microorganisms. For example, the major metabolic route of parathion in the presence of *Bacillus subtilis* is reduction to aminoparathion. In the case of fenitrothion, the major metabolite, accounting for 65 per cent of the added insecticide, was aminofenitrothion. Methyl parathion was metabolized in the same manner, but twice as quickly. After adding parathion to an algal culture of *Chlorella pyrenoidosa* the principal compound found within the cells was parathion, whereas the major one in the medium was aminoparathion. It was suggested that aminoparathion is first formed within the cells, and, once formed, it is rapidly released. Parathion was also reduced by two species of *Rhizobium*, *R. japonicum* and *R. meliloti*. Under low oxygen conditions, parathion was reduced via a reductive pathway to *p*-aminoparathion and subsequently hydrolyzed to *p*-aminophenol and diethylthiophosphoric acid. The mixed cultures used in the studies consisted of at least nine isolates, five of which were fluorescent pseudomonads, and the remaining four were classified as species of *Brevibacterium*, *Azotomonas*, *Xanthomonas* and one unknown. Reduction of parathion to aminoparathion was accomplished only when mixed culture suspensions were grown in batch culture without aeration. From studies with the bacterium *P. putida* and the algae *Microcystis aeruginosa* (IND 1036) and

Anacystis nidulans (TX-20) a flavoprotein has been partially isolated which requires an anaerobic incubation condition and a flavin cofactor for full catalytic activity [53].

In the case of the dinitrosubstituted aromatic herbicides as e.g. dinoseb, DNOC, trifluralin, oryzalin or isopropalin, nitralin, and dinitramine, nitroreduction leads to a series of mono- and diaminosubstituted metabolites and is performed by a large variety of aerobic and anaerobic soil and rumen microorganisms (see Tab. 7). The nitro group can also remain unchanged, however, as in the formation of 4-nitrocatechol from 4-nitrophenol by *Pseudomonas* sp. ATCC 29358 [54] or during dealkylation of amino groups as in the dinitramine and fluochloralin transformation in soil and by soil fungi (Tab. 7).

Arylamines are main intermediates in the microbial degradation of phenylamide type pesticides (Sect. 4.1.5) as well as in the catabolism of 3,5-dinitro-*o*-cresol by *Pseudomonas* species. In these cases it is the amino group that can be replaced by hydroxyl as in the conversion of *p*-aminobenzoic acid to *p*-hydroxybenzoic acid and of 5-hydroxyanthranilic acid to gentisic acid [36]. Similary the amino-group of the herbicide pichloram and that of *s*-triazines such as atrazine and simazine formed after dealkylation is replaced by a hydroxyl with the concomitant liberation of ammonia (see Tab. 9). In the microbial metabolism of aniline by *Pseudomonas* and *Nocardia* species, of 2-, 3-, and 4-chloroaniline by *Alcaligenes faecalis* and a *Pseudomonas multivorans* strain, and of 3,4-dichloroaniline by *A. faecalis* and *P. putida*, however, the deamination step is shown to occur by dioxygenation forming catechol [55, 56], 4-chlorocatechol [57, 58, 59] (Table 6), or 3,4-dichlorocatechol [60, 61], respectively. Recently, Latorre et al. [62] isolated 2-, 3- and 4-chloraniline degrading bacteria by natural genetic exchange between an aniline or toluidine degrading *Pseudomonas* strain and the chlorocatechol assimilating *Pseudomonas* B 13. Dioxygenation of 2- and 3-chloroanilines was shown to be highly regioselective. 3-Chlorocatechol was formed from 2-chloroaniline through 1,6-dioxygenation and from 3-chloroaniline through 1,2-dioxygenation. By contrast, only 4-chlorochatechol was generated from 4-chloroaniline.

Contrarily, the amino substituent of an aromatic substrate may also remain unchanged during ring cleavage. Thus, *o*-aminophenol is cleaved by the intradiol dioxygenase, pyrocatechase, of a *Pseudomonas arvilla* strain, in an extradiol manner to give picolinic acid as the major product [63].

Amino groups of heterocyclic pesticides can be replaced by a hydrogen thus giving deaminated products. This desamination occurs in the transformation of the herbicide metamitron by soil bacteria (see Fig. 28) and to a much greater extent by mucoraceae as e.g. *Rhizopus japonicus* or *Cunninghamella echinulata* Thaxter [25]. Similarly, the *as*-triazine herbizide metribuzin is deaminated by *C. echinulata* giving desaminometribuzin [64] and in the degradation of the *s*-triazine herbicide atrazine a deamination step follows dealkylation to give 4-amino-2-chloro-1,3,5-triazine in a *Nocardia* strain (Table 9).

4.1.7 Transformation of Thioether, Sulfinyl and Sulfonyl Side Chains

Thioether and sulfinyl substituents commonly are not removed or dealkylated from aromatic structures before ring cleavage occurs. As shown for the metabolism of 4-(methylthio)- and 4-(methylsulfinyl)-substituted phenols by a *Nocardia* species the respective catechol and muconic semialdehyde derivatives were formed (Table 13).

The limited number of S-dealkylations observed may be due to the greater tendency for oxidations of the sulfur to the sulfoxide and sulfone. Conversively, the sulfinyl substituent can be reduced by soil bacteria to the sulfide under oxygen-limited conditions as shown for the reduction of fensulfothion and its hydrolysis product 4-methylsulfinyl phenol to fensulfothion sulfide and 4-methylthiophenol, respectively, by *Klebsiella pneumoniae* [65]. A variety of sulfoxide and sulfone derivatives from phosphorodithioate, methylthiosubstituted and thiocarbamate pesticides have been detected in soil, however, very few thioether oxidations have been conclusively attributed to microbial activity [24, 66]. Examples for thioether oxidation are the formation of "aldicarbsulfoxide" from the thiocarbamate aldicarb by soil fungi [67], the detection of 2-(ethylsulfonyl)ethanethiole and the corresponding disulfide formed during degradation of metasystox R by various soil bacteria, or the release of fenthion sulfoxide and fenoxon sulfoxide in cultures of *Rhizopus japonicus* (Table 12). Disulfide compounds have also been found during metabolism studies of the organophosphorus type. It appears that a mixed-function oxydase forms catechol and simultaneously insecticides azinphos-methyl by the soil bacterium *P. fluorescens* DSM 1976 and by soil [66], and fonofos by different soil fungi [44] (Table 12).

The sulfoxidation products of pesticides still are biologically active as the parent pesticide and may also possess increased hydrolytic stability [68, 69].

There appear to be at least two different ways in which sulfonic acid substituents of aromatic compounds are metabolized by bacteria. Benzenesulfonate, e.g., is oxidized by a *Bacillus* species that is able to degrade detergents of the alkylbenzene-sulfonate releases the sulfonic acid substituent as sulfite. However, results with another *Pseudomonas* sp. strongly suggested that the sulfonic acid group of *p*-toluenesulfonate is not removed as a sulfite but as a sulfate [36].

4.1.8 Dehalogenation Reactions

Elimination of halogen from aromatic compounds is of special interest because of its relevance to the widely used chloroaromatic compounds which are known to decompose very slowly.

Direct dechlorination of aromatic compounds is very rarely described in the literature. One mechanism of such dechlorinations from the aromatic ring catalyzed by microbial monoxygenases is suggested to follow the so-called "NIH-shift" in which the *p*-substituent migrates to the *m*-position or is eliminated (Fig. 10) [24, 70].

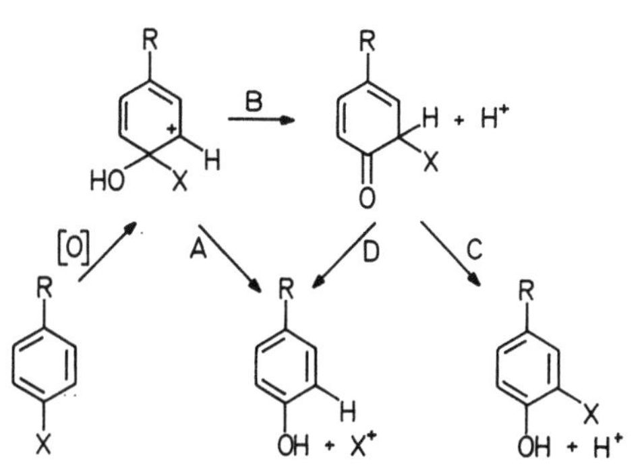

Fig. 10. Halogen migration (C) or elimination (A, D) during NIH-shift-reactions [70]

Recently, another mechanism of direct oxidative dehalogenation of various halogen-substituted benzenes, phenols, or anilines has been demonstrated [71], which is catalyzed by extremely unspecific microbial dioxygenases from *Pseudomonas*, *Alcaligenes*, and *Moraxella* species.

By this reaction molecular oxygen is introduced into the substrate molecule leading to an ortho-dihydroxylated product and halogenid ions.

Much more common among microorganisms than direct elimination of chlorine from the aromatic ring, however, is the *dechlorination of aliphatic intermediates* formed after ring cleavage. Ring cleavage of halocatechols is considered to be the crucial step for the total degradation of these compounds because of steric hindrance and inductive and mesomeric effects of the substituents. Dehalogenation of halocatechols always occurs after ring cleavage during one of the isomerization steps between muconic acid and the tricarboxylic acid cycle intermediates. Thus, 2,4-dichlorophenoxyacetic acid (2,4-D) is degraded by an *Arthrobacter* sp. via 2,4-dichlorophenol to 3,5-dichlorocatechol which undergoes intradiol ring cleavage to give the dichlorosubstituted muconic acid. Elimination of the *p*-chlorosubstituent occurs during lactonization of this muconic acid, whereas it is suggested that the *o*-chlorosubstituent is eliminated from the chlorosuccinic acid formed after cleavage of the halogenated 3-oxo-adipic acid [70,72] (Fig. 11).

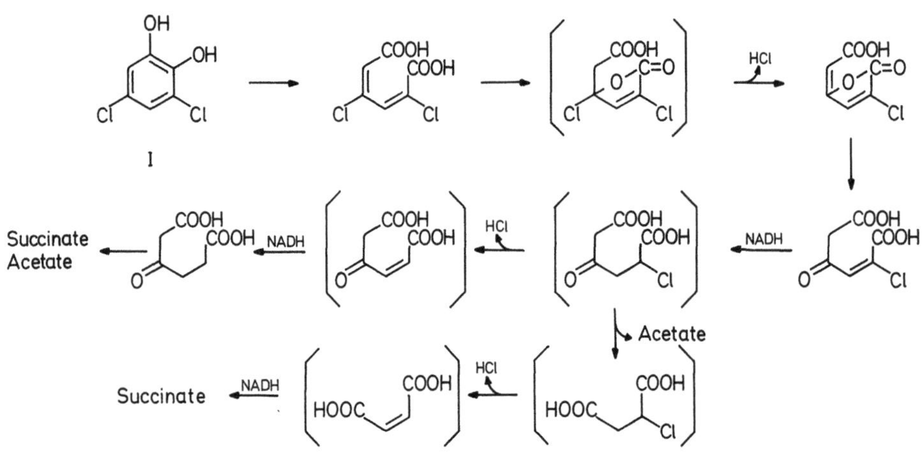

Fig. 11. Elimination of chloro substituents during microbial metabolism of 3,5-dichlorocatechol derived from 2,4-D (after Knackmuss [72])

Fluorine can also be eliminated from the aromatic ring during initial dioxygenation [72]. Catechol is formed, if the 2-position is involved in the formation of the intermediate cis-diol; fluorine is, however, retained in the product 3-fluorocatechol, if the 6-position is occupied. Ring cleavage of 3-fluorocatechol is recognized as a critical step in 2- and 3-fluorobenzoic acid degradation forming 2-fluoro-*cis,cis*-muconic acid as the final metabolite [73]. The failure of many fluorosubstituted aromatic compounds to support growth has frequently been attributed to their breakdown to fluoroacetic acid with subsequent poisoning of the tricarboxylic acid cycle.

Dechlorination of aromatic substances can also occur by *direct hydroxyl replacement* as shown in the transformation of 4-chlorobenzoic acid to 4-hydroxybenzoic acid by an *Arthrobacter* species [70]. Hydroxyl replacement is, however, very rare in the metabolism of aromatic compounds and the mechanism of this reaction is not clear till now. Since in the *s*-triazine ring the nitrogen withdraws electrons from the adjacent carbons, hydrolytic removal of halogen substituents is easily performed and represents a common reaction especially in plant metabolism. The dehalogenation by hydrolases was found e.g. during degradation of atrazine by *Pseudomonas* species [31] or of deethylsimazine by partially purified enzyme fractions of *Rhodococcus corallinus* [74].

Inducible halidohydrolases (E.C. 3.8.1 a) play a major role in the microbial dechlorination of halogenated aliphatic compounds as shown from studies on the metabolism of 2-halogenoalkanoic acids. Specific halidohydrolases have been isolated from different *Pseudomonas* strains which can catalyze the hydrolytic dechlorination of chloro-, fluoro-, or iodoacetic acid, respectively [70], the optimum pH of the reaction being in the strong alkaline region of pH 9.3 to 9.5.

Dehydrohalogenation has been observed with aliphatic halides. It involves the removal of adjacent halogen and hydrogen atoms and the formation of a C,C double bond. The formation of DDE from DDT and of γ-PCCH from γ-BHC are typical examples (Fig. 12A).

Dechlorination of dalapon (2,2-dichloropropionic acid) by an *Arthrobacter* strain leads to pyruvate and chloride ions [75]. The partially purified enzyme responsible for the liberation of chloride ions from dalapon was shown to exhibit a pH-optimum of pH 8.0 and a significant substrate specificity for other α-chloroalkanoic acids. Whereas this enzyme was shown to attack only α-chloroalkanoic acids, Bollag and Alexander [76] described a strain of *Micrococcus denitrificans* dehalogenating exclusively β-chloroalkanoic acids. The mechanism of the reaction was suggested to occur via dehydrohalogenation with the concomitant formation of HCl and a double bond [7]. Since dehalogenations of short chain aliphatic carboxylic acids, however, have been reported to be catalyzed by enzymes of the amino acid metabolism it is still unclear if the dehalogenating enzymes described are specific "dehalogenases" [70].

Reductive dehalogenation is a major degradation route, favored by anaerobic conditions and like dehydrohalogenation, has been reported for many of the chlorinated pesticides. It probably involves carriers such as reduced cytochrome oxidase and

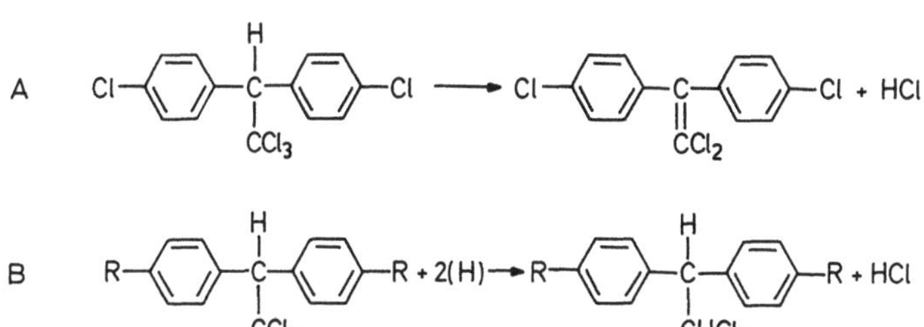

Fig. 12. (A) Dehydrochlorination of DDT forming DDE; (B) Reductive dechlorination of DDT forming DDD and of methoxychlor forming 2,2-bis-(4-methoxyphenyl)-1,1-dichloroethane) (R = Cl: DDT; R = OCH_3: methoxychlor)

FAD[24,53]. The reaction proceeds by replacing a chlorine atom with a hydrogen atom. Conversion of DDT to DDD(2,2'-bis-(*p*-chlorophenyl)-1,1-dichloroethane) is a classical example of this reaction (Fig. 12B) and has been shown to occur in anaerobic soil, lake water, anaerobic sludge, and by a wide variety of intestinal and soil microorganisms[24,77]. Furthermore, a strain of *Hydrogenomonas* is able to degrade DDT and its analogs reductively to DDD (bis(*p*-chlorophenyl)methane) (Fig. 12B). Similarly, degradation of methoxychlor to 2,2-bis(*p*-methoxyphenyl)-1,1-dichloroethane (Fig. 12B) and of the insecticide γ-BHC to γ-TCCH by different soil bacteria is accomplished by reductive dechlorination[77,78].

Reductive dechlorination has been postulated to be one of the main steps in pentachlorophenol degradation in soil and by a series of isolated soil bacteria liberating the chlorine as anion[79]. A variety of different dehalogenation products were identified including tetra-, tri-, and dichlorosubstituted phenols, hydroquinones, and benzoquinones (Fig. 13).

Recently, anaerobically digested municipal sewage sludge which had been acclimated to monochlorophenol degradation for more than 2 years was shown to degrade pentachlorophenol (PCP)[80]. Di-, tri-, and tetrachlorophenols accumulated when PCP was added to the individual acclimated sludges. When the 2-chlorophenol-(2-CP),

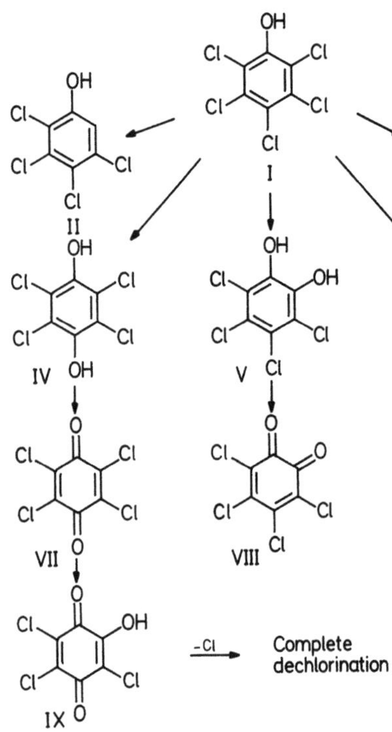

Fig. 13. Dechlorination of pentachlorophenol by microorganisms (I = pentachlorophenol; II = 2,3,4,5-tetrachlorophenol, III = 2,3,4,6-tetrachlorophenol; IV = 2,3,5,6-tetrachlorohydroquinone; V = 3,4,5,6-tetrachlorocatechol; VI = 2,4,5,6-tetrachlororesorcinol; VII = 2,3,5,6-tetrachloro-1,4-benzoquinone; VIII = 3,4,5,6-tetrachloro-1,2-benzoquinone; IX = 3,5,6-trichloro-2-hydroxy-1,4-benzoquinone)

3-CP-, and 4-CP-acclimated sludges were mixed in equal volumes, PCP was completely dechlorinated. The same results were obtained in sludge acclimated to the three monochlorophenol isomers simultaneously. With repeated PCP additions, 3,4,5-trichlorophenol, 3,5-dichlorophenol, and 3-CP accumulated in less than stoichiometric amounts. Incubations with ^{14}C-PCP resulted in 66% of the added ^{14}C being mineralized to $^{14}CO_2$ and $^{14}CH_4$. Pentabromophenol was also rapidly degraded by the mixture of acclimated sludges. These results clearly show the complete reductive dechlorination of PCP by the combined activities of three chlorophenol-degrading populations.

Similarly, reductive biodegradation of a variety of haloaromatic substrates by sewage sludge microorganisms under methanogenic conditions was reported [81]. The microorganisms dehalogenated different chlorophenols and degraded 2,4-dichloro- and 2,4,5-trichlorophenoxyacetic acid by first converting them to the corresponding chlorophenols (see Table 14). The dehalogenation reaction was partially inhibited by high sulfate levels.

4.1.9 Isomerization Reactions

Isomerization reactions are common reactions found in the metabolism of organochlorine insecticides. Various cyclodiene compounds are transformed to more stable products by isomerization catalyzed by microbial enzymes. Examples are the conversion of γ-BHC to α-BHC [82] or of endrin to δ-ketoendrin [83] and dieldrin to photodieldrin [84] (Table 3).

4.2 Aromatic Compounds

4.2.1 Aerobic Cleavage of the Aromatic Nucleus

In the microbial metabolism of aromatic compounds by ring fission the aerobic ring cleavage is predominating. However, in recent years knowledge on the cleavage of a variety of aromatic pesticides and other xenobiotics in the absence of molecular oxygen is increasing, demonstrating that aromatic ring cleavage is obtained by reaction sequences in which the benzene nucleus is first reduced and then cleaved by hydrolysis.

Depending on the number of rings and especially on the type of substituents aromatic compounds are either totally or partly degraded by microorganisms. The basic reaction sequences involved include entry into the cell, modification of side chains and substituents, formation of ring fission substrates, ring fission, and transformation of the products of ring cleavage into common compounds of intermediary metabolism. Entry of aromatic compounds into the cells has often been assumed to occur by diffusion, but there is evidence that specific transport mechanisms exist [85,86,87]. Transformation of side chains and substituents before ring cleavage occurs by various reactions common for non-aromatic compounds as summarized in Chapt. 4.1.

4.2.1.1 Reactions Converting Aromatic Compounds into Ring Fission Substrates

All aromatic compounds must be transformed into *ortho*- or *para*-dihydroxybenzenes before ring cleavage can occur. Converging metabolic pathways are used to transform the substrates into relatively few of those di- or trihydric phenolic key intermediates of which the most important are catechol, protocatechuic acid, homogentisic acid, homoprotocatechuic acid, gentisic acid, and gallic acid[88,89,90]. Catechol, for example, is formed from a variety of mono- and 1,2-disubstituted aromatic compounds such as benzoic acid, anthranilic acid, salicylic acid, and phenol (Fig. 14).

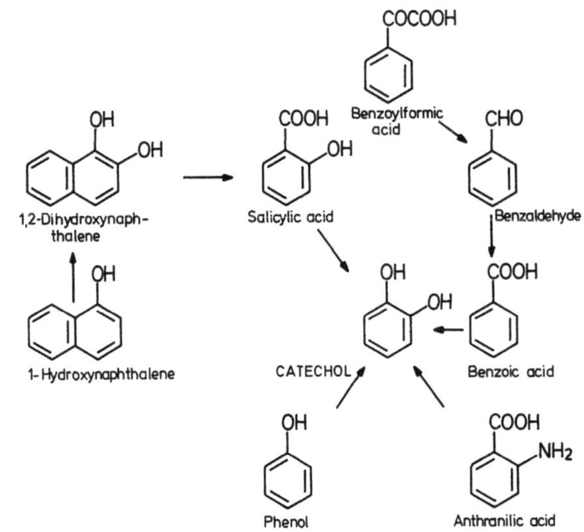

Fig. 14. Catechol as central metabolite in the microbial degradation of benzenoid compounds

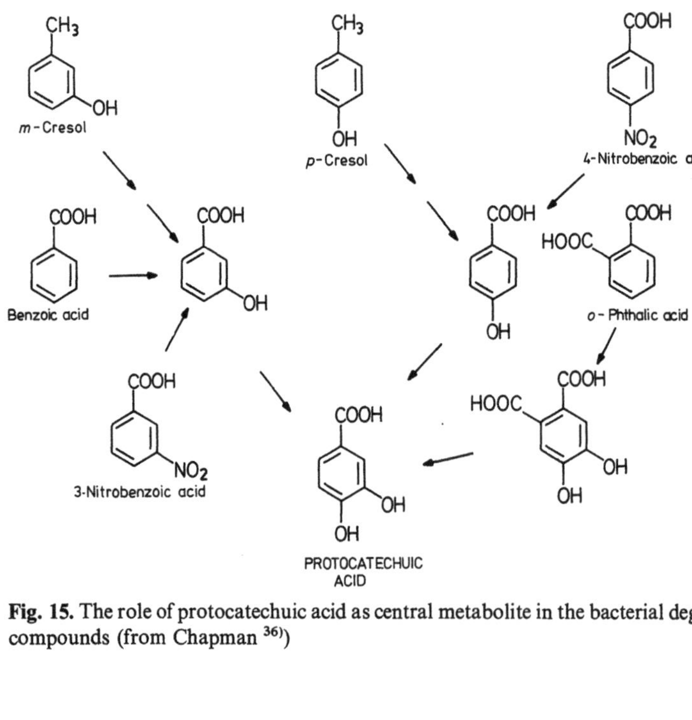

Fig. 15. The role of protocatechuic acid as central metabolite in the bacterial degradation of benzenoid compounds (from Chapman[36])

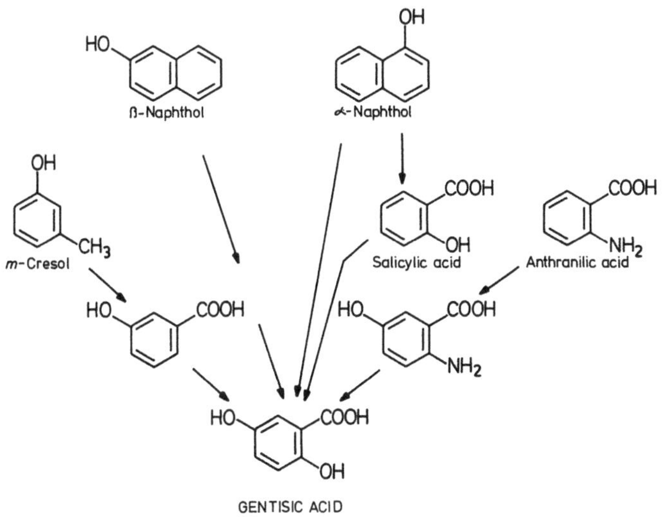

Fig. 16. The role of gentisic acid as a central metabolite in the degradation of benzenoid compounds (from Chapman [36])

Protocatechuic acid is formed by different reactions from 1,2-, 1,3-, and 1,4-substituted aromatic compounds, as e.g., 3- and 4-hydroxybenzoic acid, phthalic acid, 3- and 4-nitrobenzoic acid, and *m*- and *p*-cresol (Fig. 15).

Among the dihydroxy benzenes which are substrates for ring cleavage gentisic acid is the most prominent one. It is formed from anthranilic acid, salicylic acid, *m*-cresol, and α- and β-naphthol [36,91] (Fig. 16).

Trihydroxy benzenes may result from the bacterial oxidation of *meta*-dihydroxy benzenes such as orcinol, resorcinol, or thymol [36]. Since much has been published in this field the reader may refer to reviews containing more detailed information on the various reactions leading to substrates for ring fission enzymes [36,89,40,92], and to all the orginal work.

Microbial hydroxylation of the benzene nucleus involves the incorporation of molecular oxygen into the substrate and is accomplished by two different mechanisms. In the first, the introduction of a single hydroxyl group is catalyzed by a monoxygenase (hydroxylase) according to the following reaction type in Eq. (1):

$$R{-}H + NADH + H^+ + O_2 \rightarrow R{-}OH + NAD^+ + H_2O \quad (1)$$

Many of the enzymes are flavoproteins, although in a few cases the first hydroxylation involves a prosthetic group other than flavin. For instance, a cytochrome P-450 is used by a species of *Pseudomonas* for hydroxylation of D-camphor and both mandelate 4-hydroxylase from *Pseudomonas convexa*, and benzoate 4-hydroxylase from *Aspergillus niger* have an absolute requirement for tetrahydropteridine [89]. Characteristic of the microbial monooxygenases is their stringent substrate specificity as compared to the broad substrate specificity of, e.g., liver cytochrome P-450 hydroxylase. Thus, bacterial hydroxylases direct the catabolism of different aromatic substrates to the specific degradative sequences. Reactions of the monooxygenase type are exemplified by the bacterial oxidation of a number of phenols and phenolic acids.

The second mechanism of microbial hydroxylation forms *ortho*-dihydroxy benzenes. In bacteria it is accomplished by the simultaneous incorporation of both atoms of molecular oxygen with the intermediate formation of a dihydrodiol which is invariably in the *cis*-configuration and requires the participation of both a dioxygenase and a dehydrogenase. These enzymes which are suggested to exist as a complex [93] are involved in the bacterial oxidation of benzoic acid, anthranilic acid, phthalic acid, and some mononuclear and polynuclear hydrocarbons. *Trans*-dihydrodiols of aromatic compounds can also be formed in bacteria, but probably only as anabolic intermediates.

Another mechanism of double hydroxylation of aromatic compounds is carried out by fungi when grown in the presence of a second carbon source. The reaction leads via the formation of arene oxide (epoxide) intermediates which can be hydrated to a *trans*-dihydrodiol as in other eukaryotes [40,41].

4.2.1.2 Pathways of Aromatic Ring Cleavage and Subsequent Reactions

a. Ortho-Fission Pathways of Catechol and Protocatechuic Acid

Ring cleavage is achieved by reactions in which both atoms of molecular oxygen are introduced into the substrate molecule by dioxygenases. In the case of the *ortho*-dihydric phenols catechol and protocatechuic acid, intradiol ("*ortho*") cleavage (reactions A and C in Fig. 17) or extradiol ("*meta*") cleavage (reactions B, D, F in Fig. 17) may occur.

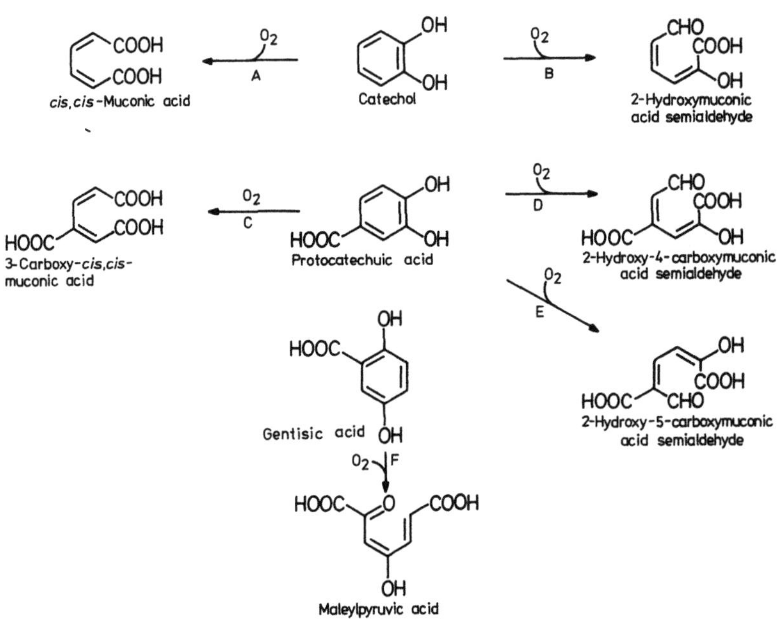

Fig. 17. Main enzymatic reactions involved in cleavage of dihydroxy benzenes (from Fewson [23])

Extradiol fission may be distal (reaction D in Fig. 17) or proximal (reaction E in Fig. 17) to the primary substituent. Cleavage of catechol by catechol 1,2-dioxygenase and of protocatechuic acid by protocatechuate 3,4-dioxygenase leads to *cis,cis*-muconate (*cis,cis*-2,4-hexadienedioic acid) or its 3-carboxy derivative. Alternatively, the products of the extradiol mode of ring fission by catechol 2,3-dioxygenase or protocatechuate 4,5-dioxygenase are 2-hydroxymuconic semialdehyde (2-hydroxy-6-oxo-2,4-hexadienoic acid) and 2-hydroxy-4-carboxymuconic semialdehyde (4-carboxy-2-hydroxy-6-oxo-2,4-hexadienoic acid). The *para*-dihydric phenol gentisic acid (reaction F in Fig. 17) is split by gentisate 1,2-dioxygenase forming maleylpyruvic acid (3-hydroxy-1-oxo-2,4-pentadienedióic acid) as ring fission products. Homogentisic acid and homoprotocatechuic acid are cleaved analogously [36].

Many of the dioxygenases have been crystallized and extensively characterized (reviews in Hayaishi [94]; Boyer [95]). They have a fairly strict but not an absolute specificity for one substrate: thus, catechol 1,2-dioxygenases from various organisms will tolerate the presence of methyl and ether alkyl substituents at C-3 or C-4, the rate of attack depending upon the source of the enzyme. This is particularly important in the metabolism of xenobiotics. In addition, catechol 1,2-dioxygenase can catalyze a limited degree of extradiol cleavage [96].

The mode of ring-fission strongly depends on the structure of the 1,2-dihydric phenol, on the bacterial species, and on the substrate used for growth. Similarly, the kind of substituent on the benzene nucleus influences the method of ring fission used. In *Nocardia* sp. DSM 43251, for example, phenols substituted with electron-

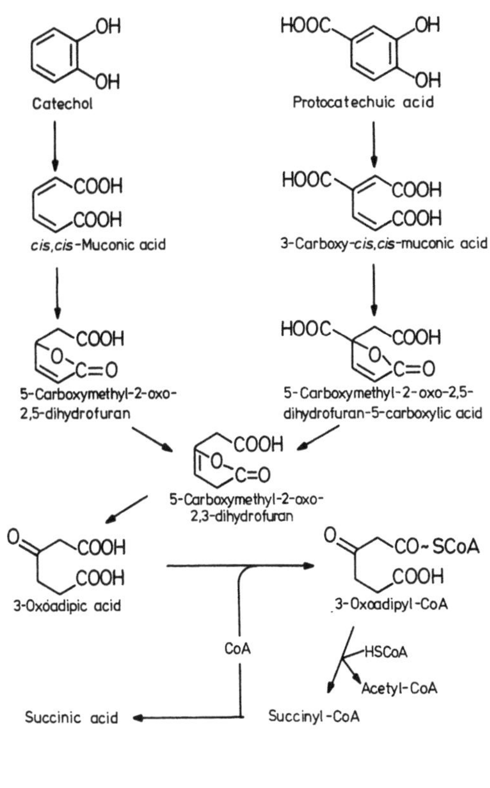

Fig. 18. The *ortho*-fission pathways of catechol and protocatechuic acid in bacteria

donating substituents as dimethyl, methoxy, and methylthio groups elicit the induction of the enzymes of the 2,3-cleavage pathway of catechol. When grown with unsubstituted phenols or phenols substituted by electron-withdrawing groups, however, the enzymes of the 1,2-cleavage pathway of catechol are induced [97].

After ring fission the resulting open chemical structures enter the main channels of metabolism, such as the tricarboxylic acid cycle. The reactions involved include specific hydrations, hydrolyses, and aldolase fissions. Thus, the products of intradiol cleavage of catechol and protocatechuic acid are converted to succinic acid and acetyl-CoA by the 3-oxoadipate pathway. The pathways of both compounds converge at 5-carboxymethyl-2-oxo-2,3-dihydrofuran (formerly 3-oxoadipate enol lactone) beyond which the reaction steps are common to both pathways (Fig. 18).

b. Meta-Fission Pathways of Catechol and Protocatechuic Acid

By extradiol fission of 1,2-dihydric phenols an entirely different sequence of reactions is initated (reviews by Dagley [88,89,90]). The final products are amphibolic intermediates such as pyruvic acid and fumaric acid, or closely related compounds such as acetaldehyde, succinic semialdehyde, propionaldehyde, or acetoacetic acid. Therefore, catechol and its analogs are cleaved by a 2,3-dioxygenase to 2-hydroxymuconic semialdehyde or its derivatives which may then either undergo hydrolysis or NAD^+-dependent dehydrogenation. The reactions from 2-oxo-pent-4-enoic acid are common to both pathways [36] (Fig. 19). In 3-methylcatechol only the hydrolytic route exits, since after ring cleavage an oxo acid and not a semialdehyde is formed.

The bacterial metabolism of protocatechuic acid via *meta*-fission is a distinctive property of the non-fluorescent species of *Pseudomonas*, namely *P. acidovorans* and *P. testosteroni* [98], and of the micrococci [99]. Analogous to the hydrolytic cleavage of catechol, after ring fission at the 4,5-position both hydrolytic elimination of formic acid and NAD^+-dependent dehydrogenation occur, followed by hydration to the 2-oxo-4-hydroxy-4-methylglutaric acid and by aldolase cleavage to pyruvate. The

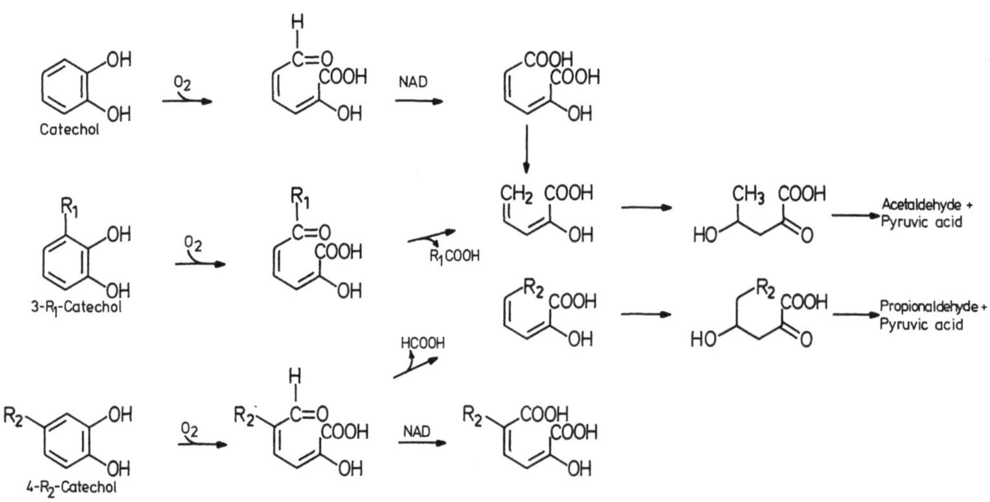

Fig. 19. The *meta*-fission pathways of catechol and its homologs in bacteria (R_1 = H, CH_3, Cl; R_2 = H, S—CH_3, CH_3, Cl)

alternative pathway via dehydrogenation, however, is different from the catechol *meta*-cleavage pathway since instead of the decarboxylation step the tricarboxylic acid is reduced to the hydroxy oxo acid followed by aldolase cleavage to pyruvic acid and oxaloacetic acid (Fig. 20) [36].

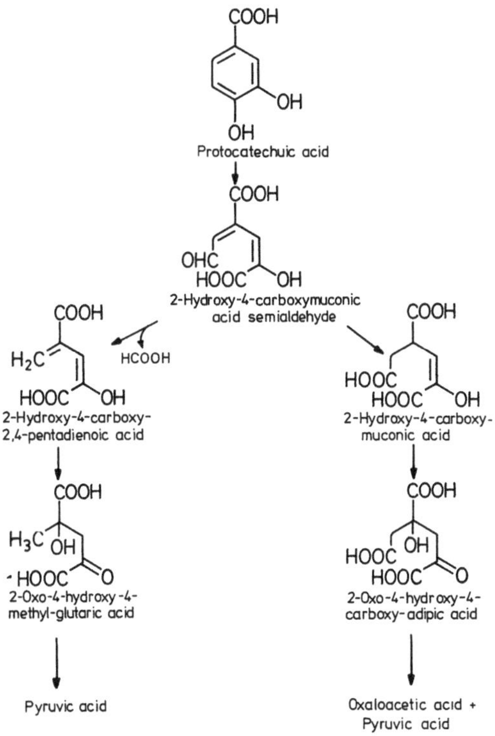

Fig. 20. The *meta*-fission pathways of protocatechuic acid metabolism (from Chapman [36])

c. Degradation of Gentisic Acid

Gentisic acid is cleaved by a dioxygenase at the carbon-carbon bond between the carboxyl group and the neighboring hydroxyl group to give maleylpyruvic acid followed by a glutathione-dependent isomerization to fumarylpyruvic acid which is then hydrolyzed to fumaric acid and pyruvic acid (Fig. 21). In some organisms,

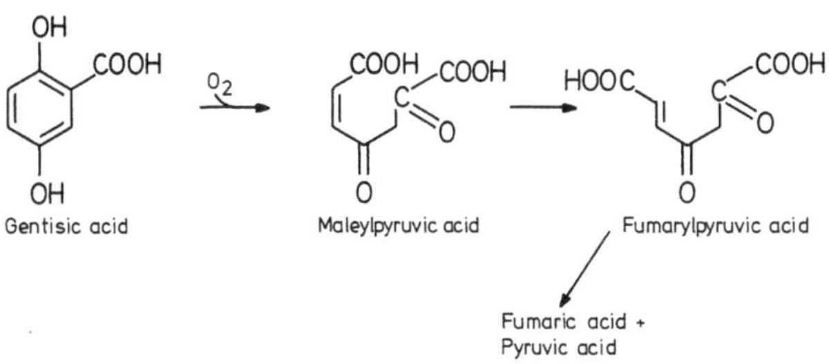

Fig. 21. The pathway of gentisic acid metabolism in bacteria

however, hydrolysis occurs without isomerization, as e.g., in the degradation of xylenols by *Pseudomonas* sp. The pathway of homogentisic acid metabolism is analogous [36].

d. Degradation of Trihydroxy Benzenes

Metabolism of 1,2,4-trihydroxybenzene involves *ortho*-fission yielding maleylacetic acid.

The two trihydroxy benzenes 2,3,5-trihydroxytoluene and 3-hydroxythymoquinol can be considered as substituted catechols which undergo *meta*-cleavage forming 2,4,6-triketo acids. Similarly, alkylcatechols are substrates for *meta*-fission while alkylsubstituted quinols are not [36].

4.2.2 Anaerobic Degradation of Aromatic Compounds

Degradation of aromatic compounds during methane fermentation was reported as early as 1934 by Tarvin and Buswell [100] but it has been shown only fairly recently how the benzene nucleus is opened and degraded by bacteria in the absence of molecular oxygen; this may be accomplished by photometabolism, nitrate respiration, and methanogenic fermentation (review by Evans [101]). In these pathways which are also used for the catabolism of xenobiotics, the benzene ring is first reduced and then cleaved by hydrolysis to give aliphatic acids for cell growth.

a. Anaerobic Photometabolism

Different species of the purple non-sulfur bacteria, the *Rhodospirillaceae*, can use simple aromatic carboxylic acids such as benzoic acid and the hydroxybenzoic acids as the sole source of carbon both aerobically by respiration as well as anaerobically in the presence of light [101]. Anaerobically grown cells were devoid of the enzymes of the aerobic pathways and produced none of the known intermediates of aerobic aromatic metabolism. A new method of aromatic ring metabolism which includes successive reduction steps of benzoic acid (or a derivative) to give cyclohexanecarboxylic acid followed by a coenzyme-A mediated β-oxidation phase (pathway A in Fig. 22) was proposed for the photometabolism of benzoic acid by Dutton and Evans [102] and Guyer and Hegeman [103]. Using a washed chromatophore suspension of *Rhodopseudomonas palustris*. Evans [101] obtained strong evidence for the presence of a light-dependent membrane-bound proton-translocating redox system. The low potential reductant was suggested to be a ferredoxin. Furthermore, an appropriate β-oxidation suite of enzymes exists in these cells which is responsible for the sequence of β-oxidations resulting in ring cleavage.

Besides the simple aromatic acids which are metabolized photosynthetically by all species of the *Athiorhodaceae*, some can also degrade phloroglucinol by photometabolism [101].

b. Metabolism of Aromatic Compounds Through Anaerobic Nitrate Respiration

Anaerobic metabolism of benzoic and hydroxybenzoic acids in the presence of nitrate is reported for different soil microorganisms [104,105]. A reductive pathway is shown to be operating in the decomposition of benzoic acid to adipic acid by a

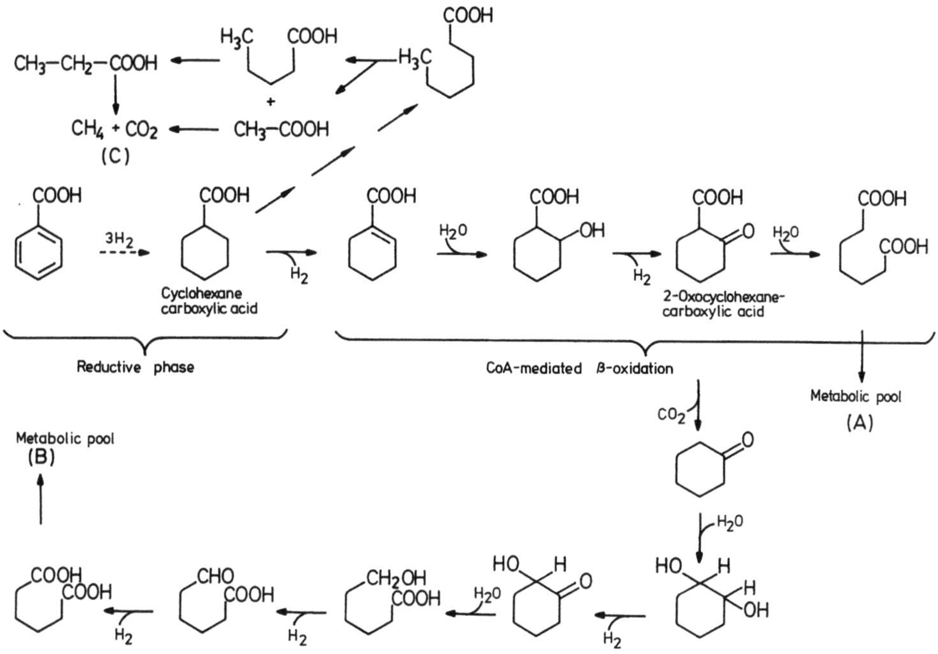

Fig. 22. Pathways of the anaerobic metabolism of benzoic acid in photometabolism by *Rhodopseudomonas palustris* (A), nitrate respiration by a *Moraxella* sp. (B) (after Evans [101)]), and methanogenic fermentation by a microbial consortium (C)

Moraxella sp. as well as in the anaerobic decomposition of phenol and other aromatic compounds by a mixed bacterial population in the presence of nitrate [106,107)]. Concomitant with the disappearance of the aromatic substrate, nitrate is reduced mainly to nitrogen gas. Cyclohexanecarboxylic acid, cyclohex-1-enecarboxylic acid, 2-hydroxycyclohexanecarboxylic acid, and adipic acid are identified in the nitrate-dependent metabolism of benzoic acid by the *Moraxella* sp. as well as by a *Pseudomonas* sp. strain PN1. The production of adipic instead of pimelic acid is explained by a divergence at the 2-oxocyclohexanecarboxylic acid stage, with its decarboxylation to cyclohexanone followed by alicyclic ring cleavage by an unknown mechanism (pathway B in Fig. 22). In the mixed culture growing on ring U-^{14}C phenol $^{14}CO_2$-labeled n-caproic acid and acetic acid are detected. Recently, evidence was obtained that phthalic acids, toluene, or 1,3-dimethylbenzene are also biodegradable by denitrifying, mixed cultures of bacteria present in sediments, sludge or groundwater infiltration zones [108,109)].

A probable interpretation of the events in the anaerobic nitrate-dependent catabolism of aromatic compounds is given by Evans [101)]: the reductive phase is accomplished by a ferredoxin-type reductant followed by the β-oxidation sequence and ring cleavage to aliphatic acids. Since these have to serve as the carbon and energy source, a part must be oxidized and the resulting reduced coenzymes re-oxidized via a membrane bound proton-translocating redox system which is coupled by the electron transport chain to nitrate through nitrate reductase.

c. Anaerobic Degradation of Aromatic Compounds During Methane Fermentation

The formation of methane and CO_2 from simple aromatic compounds occurs in the absence of nitrate, sulfate, and light by a consortium of microorganisms. The composition of the fermentation gases is shown to be in good accordance with the following equation [101]:

$$4\,{}^{14}C_6H_5COOH + 18\,H_2O \rightarrow 15\,{}^{14}CH_4 + 9\,{}^{14}CO_2 + 4\,CO_2 \qquad (2)$$

Using isotopic trapping experiments with benzoate-methanogenic cultures cyclohexanecarboxylic acid, cyclohex-1-enecarboxylic acid, heptanoic acid, valeric acid, butyric acid, propionic acid, and acetic acid were identified as intermediates [110]. However, only cyclohexanecarboxylic acid, cyclohex-1-enecarboxylic acid, propionic acid, and acetic acid are detected during methanogenic fermentation of benzoic acid by both rumen and sewage sludge cultures, from which the latter also contain adipic acid [111].

It is unlikely that the microbial communities studied by different workers are identical in the composition of their bacterial species, although they produce the same intermediates during fermentation of benzoic acid. In all these cultures the aromatic substrate is first reduced with the intermediate formation of cyclohexanecarboxylic acid and cyclohex-1-enecarboxylic acid and then cleaved to form volatile aliphatic acids (heptanoic, valeric, butyric, propanoic, and acetic acid) which are degraded further to substrates for methane bacteria. The electrons generated are assumed to be excreted as H_2 which is necessary for the reduction of CO_2 to CH_4. During the anaerobic degradation of benzoic acid C-1 is converted to methane and C-4 and C-7 primarily to CO_2. The carboxyl carbon of propanoic acid, a possible intermediate, is derived from C-4 of benzoic acid [112]. The formation of heptanoic acid necessitates a break between the annular C-1 and C-6, and the formation of acetic acid from C-1 and C-7 necessitates a break between the annular C-1 [113]. Previously breaks have been indicated between C-1 and C-6 and between C-3 and C-4 [112]. Including these breaks and the identified intermediates a possible pathway is proposed for the methanogenic fermentation of benzoic acid by a microbial consortium [113] (Fig. 22 pathway C). The acetic acid is suggested to arise by β-oxidation of heptanoic and valeric acid. The reactions leading to butyric acid, however, could not yet be determined.

Besides benzoic acid different ligno-aromatic compounds are biodegradable to methane, as shown for eleven simple aromatic lignin derivatives [114] and recently, degradation of a variety of halogenated phenols and phenoxyacetic acids by methanogenic communities from sewage sludge was described. After reductive dehalogenation and side chain cleavage mineralization of the resulting phenol forming CH_4 and CO_2 occurs [80,81] (see 4.1.7).

Microbial communities acclimated to a particular aromatic substrate are simultaneously adapted to other selected aromatic substrates. From carbon balance measurements with vanillic and ferulic acid evidence was obtained that the aromatic ring is cleaved and that the amounts of methane produced closely agree with the calculated values. Thus, nearly stoichiometric amounts of the substrate are converted to CO_2 and CH_4 which indicates that half or more of the organic carbon in aromatic ring derivatives can be potentially converted to methane gas. These results also suggest that aromatic carbon derivatives can be mineralized to CO_2 and CH_4 in highly anaerobic environments.

4.3 Common Reactions in the Microbial Metabolism of Heterocyclic Compounds

More than one third of all known organic compounds have heterocyclic structures. They include many biologically important compounds as metabolites, coenzymes, or components of macromolecules and other natural products as well as substances with pharmaceutical or pesticidal properties. On the other hand, many heterocyclic compounds are present in industrial effluents and before discharge have to be removed or degraded to harmless products to avoid pollution.

The hetero atom(s) can be the nutritional source for biosynthetic purposes. This is especially the case with many heterocyclic compounds containing nitrogen or sulfur. Microorganisms can use these compounds as sole sources of carbon and energy as well as sole nitrogen or sulfur sources [115]. Though occasionally excreted as methylamine, nitrogen is nearly always eliminated as ammonia. Sulfur finally yields sulfate or hydrogen sulfide or mercaptans of low molecular weight. Oxygen is finally incorporated into hydroxyl groups, frequently the —OH in water, or in carbon dioxide.

Heterocyclic compounds vary considerably in stability. Some can be easily cleaved through hydrolysis. Thus, 5-valerolactone and 6-capronolactone are hydrolyzed rapidly to 5-hydroxyvaleric acid and 6-hydroxycaproic acid, respectively [116]. Similar to the benzene derivatives other heterocyclic compounds must first be transformed to hydroxyl derivatives by introduction of one or more hydroxyl groups before ring cleavage can occur.

These quite stable ring systems, however, are hydroxylated without the participation of oxygen and without the expenditure of reducing power. The sequences shown in

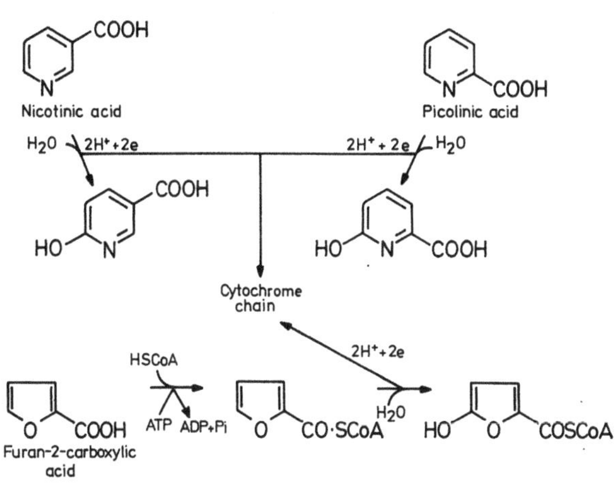

Fig. 23. Reaction sequences in the hydroxylation of nicotinic acid, picolinic acid, and furan-2-carboxylic acid (from Dagley [117])

Fig. 23 demonstrate that the oxygen atom of water is incorporated into the pyridine rings of nicotinic acid and picolinic acid. Similarly, a species of *Pseudomonas* uses water to hydroxylate the furan ring of furan-2-carboxylic acid, converting the substrate first into its coenzyme A ester. This initial activation involves an investment of energy, but the actual hydroxylation reaction does not. In each of these systems, in contrast to hydroxylation in the benzene ring, electrons are not supplied to the enzymes but are removed through the cytochrome chain.

The conversion of nicotinic acid to 6-hydroxynicotinic acid has also been demonstrated for strictly anaerobic systems. Thus, a FAD-containing (and non-heme iron) enzyme from a *Clostridium* species inserts oxygen derived from water into nicotinic acid, and, in this case, NADP serves as the necessary acceptor of electrons. 6-Hydroxynicotinic acid is therefore the initial product formed from nicotinic acid by both aerobic and anaerobic bacteria.

The ability to use the oxygen of water for hydroxylation may be due to the fact that the nitrogen withdraws electrons from the adjacent carbons [117]. In the following the degradation pathways of some structurally simpler heterocyclic compounds of both biological and ecotoxicological importance are presented.

a. Five Membered Ring Systems

Many biologically and ecotoxicologically important compounds possess one or more five membered heterocyclic ring systems or one such ring condensed to a benzene ring or another heterocyclic one.

In the heterocyclic rings of pyrrol, furan, and thiophene the heteroatoms can each contribute a lone pair of electrons into the unsaturated ring resulting in an aromatic character of these substances.

Furan-2-carboxylic acid (formerly called furoic or pyromuconic acid) is metabolized to 2-oxoglutaric acid which enters the tricarboxylic acid cycle [116]. The first reaction involved was shown to be the "activation" of furan-2-carboxylic acid with coenzyme A and ATP giving the coenzyme A derivative which is then hydroxylated as described above. Since the introduction of the hydroxyl group results at least in the formation of a lactone these compounds are easily cleaved by hydrolysis as demonstrated earlier for 5-valerolactone and 6-caprolactone. Thiophene-2-carboxylic acid seems to be degraded by an analogous reaction sequence as was shown for furan-2-carboxylic acid [116].

Thiophene and benzothiophene which is a contaminant of crude oils and coaltar distillates together with other condensed thiophenes are degraded by an anaerobic bacterium under formation of hydrogen sulfide [116]. The imidazole ring which contains two nitrogen atoms seems to be easily opened by bacteria. Hence, L-histidine is degraded by a reaction sequence very similar to that employed in the degradation of furan- and thiophene-2-carboxylic acid. The imidazole ring is transformed to give an imidazolone which is then cleaved hydrolytically [116].

The microbial degradation of nitrogeneous compounds containing a five membered saturated heterocyclic ring system has been investigated by several authors. 2-Pyrrolidone is hydrolyzed by *Pseudomonas aeruginosa* into γ-aminobutyric acid, and pyrrolidine is shown to be degraded by *P. fluorescens* and an *Arthrobacter* sp. to succinic acid via γ-aminobutyric acid [118].

b. Six Membered Heterocyclic Compounds

Among the heterocyclic compounds the pyridine ring is considered to be the most important from the view of environmental and industrial chemicals. Pyridine and its alkyl derivatives are widely used as industrial solvents. They have to be removed from industrial effluents before such effluents can be safely discharged. In addition, pyridine compounds find their way into the environment, e.g., by application of the pyridine herbicides diquat and paraquat. Pyridine is degraded by comparatively few bacteria, actinomycetes being predominating, when the compound is supplied as the sole source of carbon, nitrogen, and energy. In sewage, pyridine is rapidly removed by activated sludge. Degradation of pyridine by a *Bacillus* and a *Nocardia* sp. has been shown to occur via two distinct metabolic routes [119]. Both organisms grow rapidly on this compound as the sole C, N, and energy source. The monohydroxypyridines, tetrahydropyridine, piperidine, and some other analogs are not utilized for growth by these bacteria. In both organisms an initial reduction of pyridine produces 1,4-dihydropyridine which, in the case of *Nocardia* Z1 (route A, Fig. 24) undergoes a hydrolytic N-C-2 ring cleavage and subsequent deamination to glutaraldehyde followed by successive oxidations to glutaric semialdehyde, glutaric acid, and glutaryl-CoA. In the *Bacillus* sp. (route B), on the other hand, hydrolysis or a dioxygenase attack at the C-2, C-3 double bond forms 4-(formylamino)-but-3-en-1-al which undergoes immediate oxidation to the corresponding semialdehyde, 4-(formylamino)-but-3-enoic acid. Hydrolysis of the C-6-N bond liberates formamide and succinic semialdehyde, the former of which is attacked by a specific formamide amidohydrolase to produce formate and NH_3, whereas succinic semialdehyde is further oxidized to succinic acid by an equally specific dehydrogenase (Fig. 24) [119].

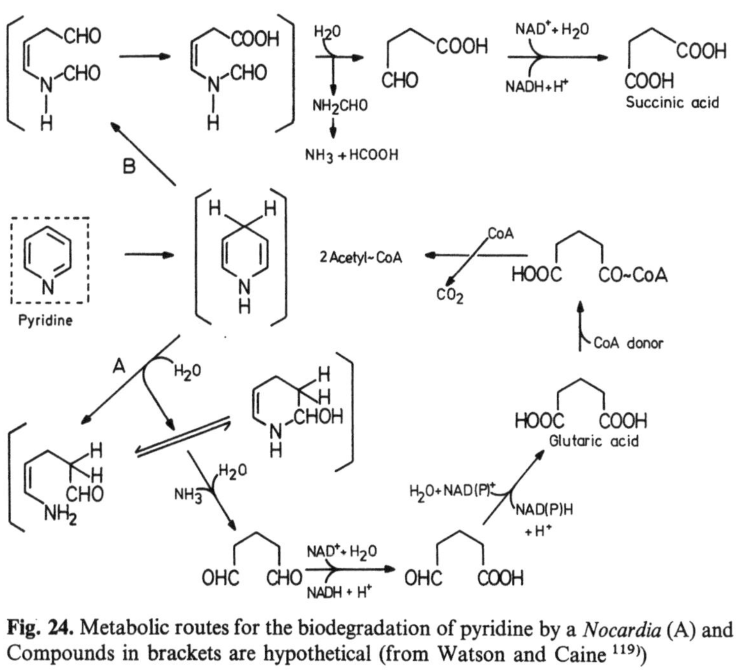

Fig. 24. Metabolic routes for the biodegradation of pyridine by a *Nocardia* (A) and a *Bacillus* sp. (B). Compounds in brackets are hypothetical (from Watson and Caine [119])

Nicotinic acid is degraded by pseudomonads via 6-hydroxynicotinic acid via two successive hydroxylations and decarboxylation to give the *para*-diol 2,5-dihydroxypyridine. This is then cleaved by an oxygenase to formate and maleamic acid from which maleic and finally fumaric acid is formed. Alternatively, a *Bacillus* species produces 2,6-dihydroxynicotinic acid from the 6-hydroxy compound. It is suggested to be decomposed via pyridine 2,3,6-triol yielding maleamic acid and formic acid.

In these two pathways and those describing the breakdown of other pyridine-ring compounds (also route B of pyridine degradation) the overall strategy is broadly the same i.e., the pyridine ring is normally cleaved between C-2 and C-3 to give a C—N—C—C—C—C open-chain compound. The isolated carbon is disposed of (usually as formic acid) and then the amino group is lost, leaving a four carbon compound which is converted to a tricarboxylic acid cycle intermediate.

As shown with monohydroxy pyridines the pyridine *para*-diol is not always the substrate for ring cleavage. Thus, the 2- and 3-hydroxy derivatives are both converted to the *para*-diol 2,5-dihydroxypyridine, whereas the *ortho*-diol 3,4-dihydroxypyridine is formed from the 4-hydroxy isomer. This pyridine *ortho*-diol is analogous to the *ortho*-dihydroxy compounds commonly encountered as ring-cleavage substrates in the degradative pathways of benzene-ring compounds. This *ortho*-diol pathway, however, is suggested to be not as common as the *para*-diol route for pyridine degradation [116]. The fermentation of nicotinic acid by a *Clostridium* sp. has been reported, the end products being propionic acid, acetic acid, carbon dioxide, and ammonia [116].

1-Methyl-4-carboxypyridinium ion, the photolytic and microbial decomposition product of the herbicide paraquat (Tab. 9) can also be degraded by soil bacteria via two different degradation pathways (Fig. 25). In the first which is performed by an *Achromobacter* species N-methylisonicotinic acid is hydroxylated on C-2, followed by demethylation of the ring nitrogen atom. The compound is then hydroxylated further to the 2,6-dihydroxy derivative before ring cleavage. In the alternative pathway in an *Arthrobacter* strain the N-methyl group is released as methylamine very late

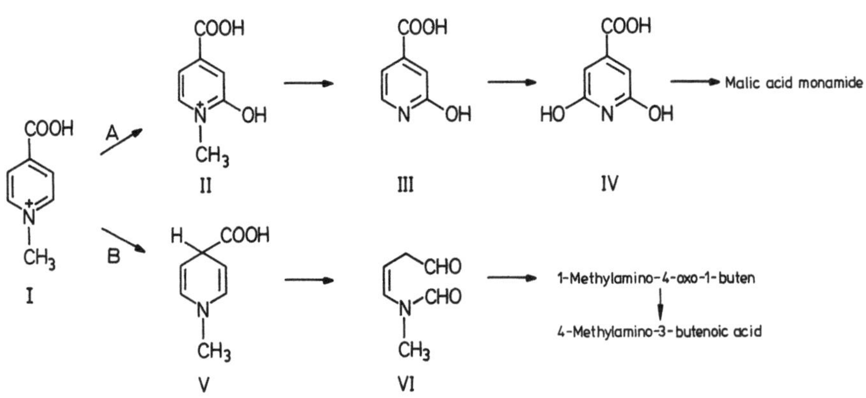

Fig. 25. Initial steps in the catabolism of N-methyl-4-carboxylpyridinium ion (I). A = Diol pathway; B = pathway involving direct oxygenative ring cleavage (after Callely [116]). (II = 2-hydroxy-1-methyl-4-carboxylpyridinium ion; III = 2-hydroxy-4-pyridinecarboxylic acid; IV = 2,6-dihydroxy-4-pyridinecarboxylic acid; V = 1-methyl-1,4-dihydro-4-pyridinecarboxylic acid; VI = 1-(N-formyl-N-methyl-amino-4-oxo-1-butene)

in the metabolic sequence after ring fission has occurred. In the latter pathway no hydroxylated heterocylic compounds are involved. In contrast, the pyridine ring is first reduced followed by a direct oxygenative cleavage of the link between carbon atoms C-2 and C-3 analogous to route B of pyridine degradation performed by a *Nocardia* sp. (Fig. 25). This direct oxygenase cleavage of a double bond without previous hydroxylation represents a new mode of ring fission performed by micro-organism [116].

Six membered hetrocyclic rings containing two nitrogen atoms are very common in nature. Four types of reactions on the pyrimidine molecule are known: hydroxylation, ribosidation, ring cleavage, and hydrogenation [92]. Ring cleavage of these compounds occurs by similar mechanisms as described for pyridine. Thus, uracil is first hydroxylated to give barbituric acid which is then cleaved hydrolytically yielding urea and malonic acid. The pyrimidine herbicides bromacil and terbacil, the insecticides diazinon, pirimicarb, pirimiphos-ethyl and -methyl, and the fungicide ethirimol undergo microbially catalyzed ring cleavage in soil, however, no ring fission intermediates could be identified [24]. Studies on the microbial degradation of the pyridazine herbicide pyrazon showed that the aromatic moiety of the compound is mineralized by dihydroxylation and *ortho* cleavage with concomitant release of the pyridazine ring [120] (Table 9) which was not metabolized further.

Six membered heterocyclic rings containing three nitrogen atoms include a group of very important pesticides, as e.g. the *s*- and the *as*- triazine herbizides or the insecticide azinphos-methyl.

Since the electronic configuration of the *s*-triazine ring is to a certain extent very similar to that of the benzene ring, the *s*-triazine compounds are considerable resistant to biodegradation. Similar to the aromatic nucleus the *s*-triazine ring is stabilized by delocalization of its π-electrons. Because of the greater electronegativity of the nitrogen atoms as compared to the carbon atoms, however, the π-electrons are localized in the vicinity of the nitrogen atoms rather than evenly distributed over the whole ring. As a result of the relative electron deficiency of the ring carbon atoms, the stability of the *s*-triazine ring is greatly influenced by the inductive and mesomeric effects of substituents at the carbon atoms; the nucleophilic attack being facilitated by electron-withdrawing substituents, as e.g. chlorine, and being inhibited by electron-supplying substituents such as amino groups.

Because of the worldwide application of large amounts of *s*-triazine herbicides microbial degradation of *s*-triazines has been studied over years by several authors [115]. The most rapid degradation was observed with cyanuric acid and ammeline, whereas decomposition of N-alkyl substituted *s*-triazines could not be proved for a long time. Some *Pseudomonas* and *Klebsiella pneumoniae* strains were later shown to utilize several of the *s*-triazine compounds melamine, ammeline, ammelide, N-ethylammeline, N-isopropylammelide and cyanuric acid (Fig. 26), respectively, as sole nitrogen sources [115]. Quantitative yields of ammonium ion were obtained from some of those *s*-triazines and ring carbon atoms were released as carbon dioxide as detected by radiorespirometry of $^{14}CO_2$ yields from ring-U-^{14}C-labeled *s*-triazines, but no intermediate products of heterocyclic ring cleavage could be detected.

A bacterium identified as *Rhodococcus corallinus* was isolated from soil enrichment cultures and shown to utilize desethylsimazine (6-chloro-N-ethyl-*s*-triazine-2,4-diamine) as a nitrogen source [115]. The compound was converted to 6-(ethylamino)-*s*-

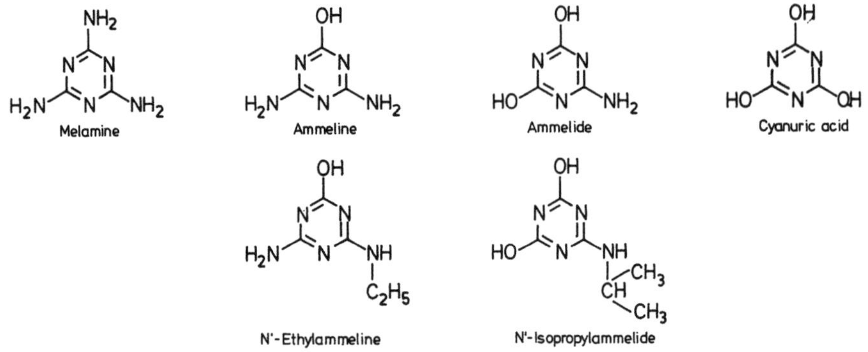

Fig. 26. Structures and names of some *s*-triazines degraded

triazine-2,4(1H,3H)-dione, 1 mol of chloride ion and 1 mol of ammonium ion, 4 amino-6-(ethylamino)-*s*-triazin-2 (1H)- one being tentatively identified as transient intermediate in the degradation. Growth of the *Rhodococcus* strain together with a *Pseudomonas* sp. able to utilize 6-(ethylamino)-*s*-triazine-2,4(1H,3H)-dione resulted in complete conversion of desethylsimazine to cell material.

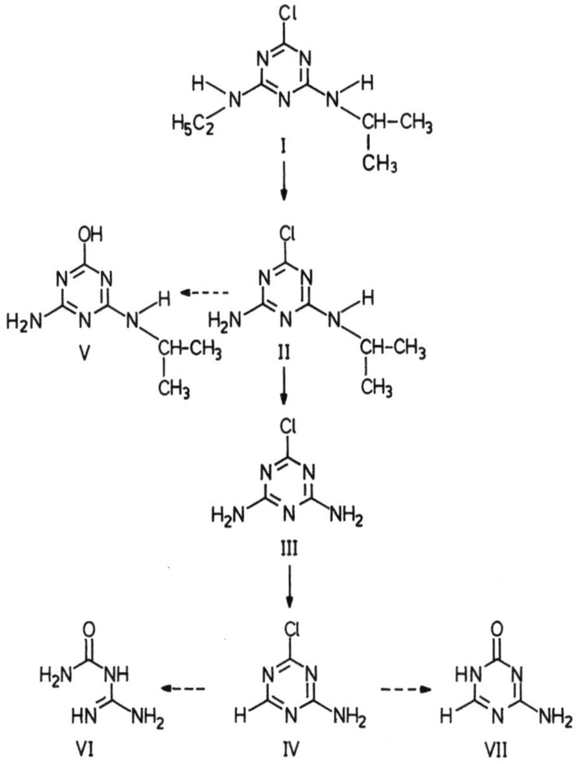

Fig. 27. Microbial (solid-line arrows) and chemical (broken-line arrows) degradation of atrazine (I) (from Giardi et al. [32]). II = 4-Amino-2-chloro-6-isopropylamino-*s*-triazine; III = 2-chloro-4,6-diamino-*s*-triazine; IV = 4-amino-2-chloro-*s*-triazine; V = 4-amino-2-hydroxy-6-isopropylamino-*s*-triazine; VI = dicyanodiamide; VII = 4-amino-1,2-dihydro-*s*-triazin-2-(1H)-one

Recently, a degradation scheme, differentiating the microbial degradation route from chemical reactions was proposed for atrazine metabolism in a *Nocardia* sp. [32]. In this strain atrazine is decomposed via 2-chloro-4,6-diamino-*s*-triazine (III), 4-amino-2-hydroxy-6-isopropylamino-*s*-triazine (V), 4-amino-1,2-dihydro-*s*-triazin-2-one (VII) and dicyanodiamide (VI) (Fig. 27). The formation of (III) was attributed to a microbial N-dealkylation, while the detection of (VII) and (VI) was attributed to a chemical degradation of the intermediate 4-amino-2-chloro-*s*-triazine (IV). Compound (IV) undergoes rapid hydrolysis and does not accumulate in the culture medium.

By contrast to the *s*-triazines, *as*-triazine derivatives seem to be easily metabolized by soil microorganisms. Cleavage of the benzotriazinone ring of the organophosphorus insecticide azinphos-methyl has been demonstrated recently [121, 66]. The compound is degraded in soil to a great extent. Liberation of two benzoic acid derivatives (benzamide, and salicylic acid), and of $^{14}CO_2$ from [carbonyl-^{14}C] and [ring U-^{14}C]-azinphos-methyl indicates that the 1,2,3-benzotriazinone ring is cleaved in soil. Of the 17 metabolites identified, 3-(mercaptomethyl)-1,2,3-benzotriazin-4-(3H)-one (III), 3-(methylthio)methyl-1,2,3-benzotriazin-4(3H)-one (IV) and the corresponding sulfinyl (V) and sulfonyl (VI) derivatives represent key intermediates in azinphos-methyl degradation in soil. Azinphos-methyl and the compounds IV, V, and VI are

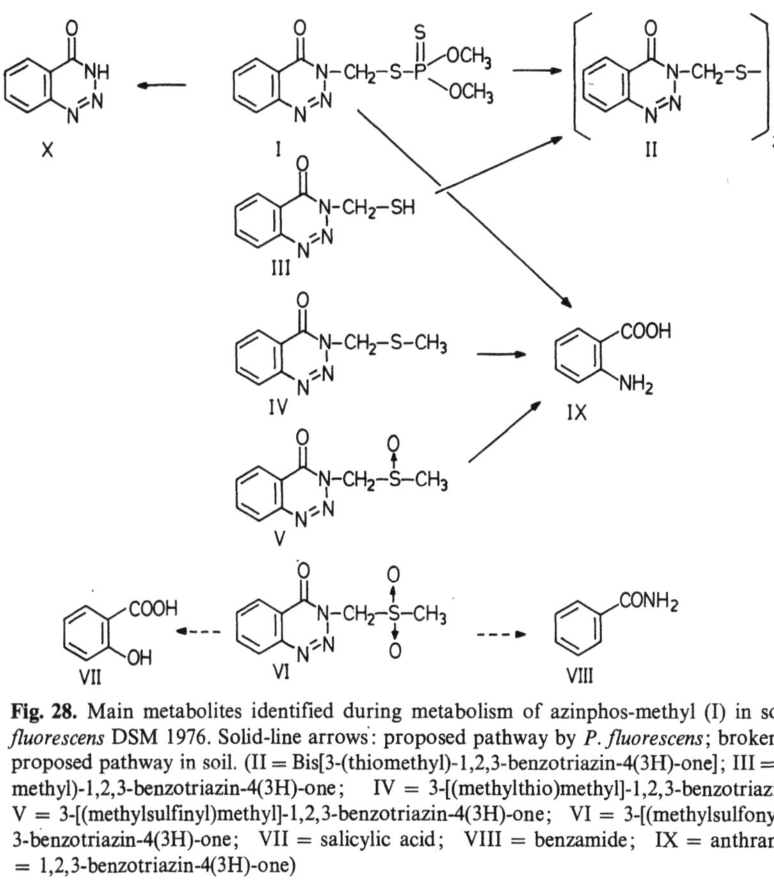

Fig. 28. Main metabolites identified during metabolism of azinphos-methyl (I) in soil and by *P. fluorescens* DSM 1976. Solid-line arrows: proposed pathway by *P. fluorescens*; broken-line arrows: proposed pathway in soil. (II = Bis[3-(thiomethyl)-1,2,3-benzotriazin-4(3H)-one]; III = 3-(mercapto-methyl)-1,2,3-benzotriazin-4(3H)-one; IV = 3-[(methylthio)methyl]-1,2,3-benzotriazin-4(3H)-one; V = 3-[(methylsulfinyl)methyl]-1,2,3-benzotriazin-4(3H)-one; VI = 3-[(methylsulfonyl)methyl]-1,2,3-benzotriazin-4(3H)-one; VII = salicylic acid; VIII = benzamide; IX = anthranilic acid; X = 1,2,3-benzotriazin-4(3H)-one)

cleaved by the *Pseudomonas* isolate forming anthranilic acid as the main metabolite, thus demonstrating that the insecticide can be metabolized in soil via easily decomposable benzoic acid derivatives (Fig. 28).

From the liberation of anthranilic acid from azinphos-methyl and from the methylthiomethyl-, methylsulfinylmethyl-, and methylsulfonylmethyl-substituted benzotriazinone derivatives found as soil degradation products, a hydrolytic cleavage of the heterocyclic ring between the carboxy group and N-1 is suggested.

A similar hydrolytic cleavage of the amide bond in an *as*-triazinone ring has been demonstrated for the decompositon of the herbicide metamitron by *Arthrobacter* sp. DSM 20389 [122] (Fig. 29). Benzoylformic acid acetylhydrazone (II) and benzoylformic acid (III) could be identified as ring fission products. Small amounts of desaminometamitron (V) and 2-(carboxy)-3-methyl-2,3-dihydro-1,2,4,5-tetrazine (VI) were also found. Besides these compounds, 3-methyl-6-phenyl-1,2,4,5-tetrazine (IV), its 4′-hydroxy derivative and 4-(benzalimino)-3-methyl-6-phenyl-1,2,4-triazin-5(4H)-one, 4-(benzalimino)-3-methyl-5-phenyl-1,2,4-triazole and dibenzalhydrazine were determined as products of combined microbial and abiotic reaction sequences. The hypothetic molecule (Fig. 29) could not be isolated under the given conditions. Since both the triazinone ring of azinphos-methyl and of metamitron are opened by a similar

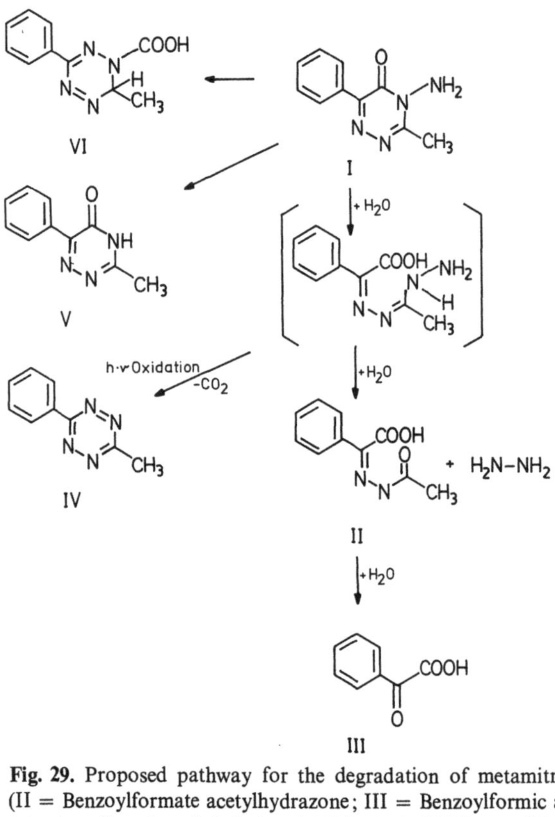

Fig. 29. Proposed pathway for the degradation of metamitron (I) by *Arthrobacter* DSM 20389. (II = Benzoylformate acetylhydrazone; III = Benzoylformic acid; IV = 3-methyl-6-phenyl-1,2,4,5-tetrazine; V = 3-methyl-6-phenyl-1,2,4-triazin-5(4H)-one; VI = 2-(carboxy)-3-methyl-6-phenyl-2,3-dihydrotetrazine)

hydrolytic attack it was of interest, if the bacterial strain, able to cleave the 1,2,3-benzotriazinone ring of azinphos-methyl also degrades the 1,2,4-triazinone ring of metamitron and vice versa. The results obtained showed that *Pseudomonas* sp. DSM 1976 as well as a variety of other *Pseudomonas* and *Arthrobacter* species isolated from soil cleaved the metamitron ring [123] and that the azinphos-methyl molecule is also split by the metamitron degrading *A.* sp. 20389 [124]. This indicates that the hydrolytic cleavage of such triazinone derivatives studied represents a reaction type which is very common among soil bacteria.

4.4 Formation of Pesticide Conjugates

Products formed by the endogeneous reaction of pesticides and their metabolites with compounds of intermediary metabolism are referred to as pesticide conjugates. Generally conjugates are bioalteration products resulting from the reaction of an exogeneous chemical or its metabolite (exocon) with an endogeneous compound (endocon). Such endogenous reagents include amino acids, alkylating agents as acetyl-CoA or malonyl-CoA, and S-adenosylmethionine. Since the majority of conjugation reactions involve the linkage of xenobiotics with sugars, sulfate ions, amino acids, and glutathione, the resulting products are more water-soluble than the original compounds and are rapidly eliminated. Therefore conjugation is frequently considered as a detoxification process. Lipophilic conjugates, however, are formed by the linking of xenobiotic molecules to alkyl moieties or e.g. to long chain fatty acids. By these processes the excretion of xenobiotics is delayed.

Although conjugative reactions are more commonly found in xenobiotic metabolism in plants and animals, — especially glucosidation, glucuronidation and amino-acid conjugation —, a series of conjugated metabolites have also been described during studies on the microbial metabolism of pesticides.

Direct conjugation with an amino acid — presumably alanine, serine, or phosphoserine — was found for the conversion of amitrole (3-amino-1,2,4-triazole) to 3-amino-1,2,4-triazolylalanine in *Escherichia coli*; the mechanism of the reaction being unknown (Table 9). Since the product has a structure analogous to the amino acid histidine it is incorporated into cellular protein [125].

Acylation is a reaction found very commonly among microorganisms. Substituted anilines formed after hydrolysis from phenylamide type herbicides and fungicides can be acylated resulting in formylated, acetylated or malonic acid conjugated derivatives by soil fungi, bacteria and cyanobacteria [126, 127, 128, 129, 130, 131] and also by the microflora of silage during the ensilage process [132] (Table 6). Acetylation and successive methylation of pentachlorophenol and its dechlorinated phenolic metabolites (Fig. 13) by various soil microorganism leads to pentachlorophenylacetate and 3,4,5,6-tetrachlorocatechol diacetate and a series of anisole derivatives as 2,3,4,5,6-pentachloroanisole, 2,3,5,6-tetrachloroanisole or 2,3,5,6-tetrachloro-4-methoxy-, and -1,4-dimethoxy-benzene; S-adenosylmethionine being the donor of the methyl groups [79].

Similarly, methylation of several other phenolic pesticide metabolites has been reported. Thus, 2,4-dichloroanisole was isolated from a culture of an *Arthrobacter* sp. metabolizing the herbicide 2,4-D [133] and pentachlorothioanisole was detected in cultures of different species of *Fusarium oxysporum* degrading pentachloronitrobenzene [134].

4.5 Condensations and Polymerizations

In addition to acylated products a variety of substituted azobenzenes formed by condensation reactions have been described as microbial transformation products from the substituted anilines derived from phenylamide type pesticides (Table 6). For example, 3,4-dichloroaniline is transformed in soil to 3,3′,4,4′-tetrachloroazobenzene, 3,3′,4′-trichloro-4-(3,4-dichloroanilino)azobenzene, and 1,3-bis(3,4-dichlorophenyl)-triazene [135]. At high aniline concentrations in soil the formation of complex oligomers like azobenzenes appears to predominate, whereas at lower concentrations simpler degradation products as e.g. acylated compounds and oxidation products have been found [136].

No azobenzenes were found from aniline and trichloroaniline whereas e.g. monochloroanilines and 2,3-, 2,4-, 3,4-, and 3,5-dichloroanilines are oxidized in soil forming azobenzenes. Addition of different substituted anilines led to the formation of asymmetrical azobenzenes in addition to the expected symmetrical compounds [137]. The formation of azo compounds has been demonstrated to occur through the oxidation of the aniline by peroxidase enzymes in soil and by soil microorganisms.

From the accumulation of 3,3′,4,4′-tetrachloroazoxybenzene, -azobenzene or -triazene in cultures of *Fusarium oxysporum* Schlecht growing in vitro in the presence of 3,4-dichloroaniline it was suggested that halogenoaniline metabolism proceeds

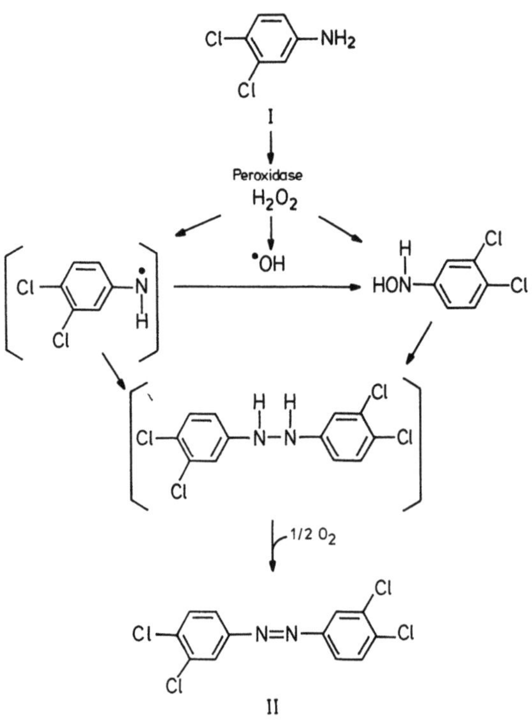

Fig. 30. Reaction scheme proposed for the peroxidase catalyzed formation of 3,3′,4,4′-tetrachloroazobenzene (II) from 3,4-dichloroaniline (I) (from Bordeleau et al. [140])

by oxidation of the amino group via the intermediate formation of phenylhydroxylamine, nitroso-, and nitrobenzene [138]. Furthermore when incubated with 4-chloroaniline in the cultures of the same fungus the chlorophenylhydroxylamine, nitroso- and nitrobenzenes as well as the expected azo and azoxybenzenes were detected [139].

In soil azobenzene formation was found to be correlated with peroxidase activity and in vitro incubation of 3,4-dichloroaniline with peroxidase and hydrogen peroxide resulted in the formation of 3,3′,4,4′-tetrachloroazobenzene. A reaction scheme proposed for the peroxidase catalyzed condensation of 3,4-dichloroaniline is presented in Fig. 30.

Finally, a fungus isolated from soil, *Geotrichum candidum*, possesses a high peroxidase activity and in addition an enzyme, called aniline oxidase. Upon incubation of these enzymes with a large number of substituted anilines a variety of reaction products were identified; the structural configuration of these products allowing predictions about the environmental fate of pesticide derived aniline residues [140].

A pathway quite different from the peroxidase system described by Bordeleau and Bartha [140, 141, 142] has been suggested by Corke et al. [135] and Minard et al. [136] in which a bis-substituted triazene is synthesised in soil when a diazonium ion formed by the reaction of nitrite with aniline, reacts with residual aniline. With bacteria containing a nitrate reductase conversion of 3,4-dichloroaniline into 3,3′,4,4′-tetrachloroazobenzene, 1,3- bis (3,4-dichlorophenyl)triazene, and chlorinated biphenyls [135] and of 4-chloroaniline into 1,3-bis(*p*-chlorophenyl)triazene [136] was described. The rate of azobenzene formation was examined in correlation with the nitrate activity of the bacteria and a pathway was proposed for the conversion of 3,4-dichloroaniline which involves the relatively rapid chemical reactions of diazotation and coupling (Fig. 31).

From studies on 4-chloroaniline metabolism in a *Paracoccus* sp. Minard et al. [136] outlined the probable course of the chemical transformation during nitrosation and diazotation as shown in the following equations:

$$NO_3^- \xrightarrow[\text{slow}]{\textit{Paracoccus}\text{ sp.}} NO_2^- \quad \text{biologically mediated} \tag{3}$$

$$H^+ + NO_2^- \overset{\text{fast}}{\rightleftharpoons} HONO \quad \text{affected by pH} \tag{4}$$

$$H^+ + HONO \overset{\text{fast}}{\rightleftharpoons} H_2ONO^+ \quad \text{affected by pH} \tag{5}$$

$$H_2ONO^+ + NO_2^- \overset{\text{slow}}{\rightleftharpoons} NONO_2 + H_2O \tag{6}$$

$$NONO_2 + ArNH_2 \xrightarrow{\text{fast}} ArNH_2NO^+ + NO_2^- \tag{7}$$

$$ArNH_2NO^+ \xrightarrow{\text{fast}} ArN_2^+ + H_2O \tag{8}$$

$$ArN_2^+ + ArNH_2 \rightarrow ArN{=}NNHAr + H^+ \tag{9}$$

Recently, the laccase of the fungus *Trametes versicolor* was shown to polymerize various halogen-, alkyl-, and alkoxysubstituted anilines [143] showing a substrate specificity similar to that of horseradish peroxidase. The transformation of 4-chloroaniline gave rise to a number of oligomers.

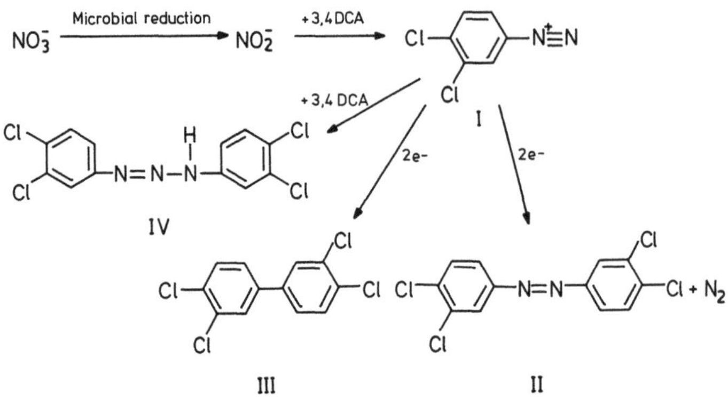

Fig. 31. Proposed sequence of conversions of 3,4-dichloroaniline (3,4-DCA) by *Escherichia coli* via the 3,4-dichlorodiazonium ion (I) to 3,3',4,4'-tetrachloroazobenzene (II), 3,3',4,4'-tetrachloro-triazene (IV), and 3,3',4,4'-tetrachlorobiphenyl (III) (from Corke [135])

Alternatively, when primarily hydroxylated ortho to the amino group condensation of two molecules of 4-chloroaniline leads to the formation of 7-chloro-2-amino-3-H-phenoxazine-3-one, which has been isolated as a product of aniline metabolism of a number of microorganisms [144, 129] (Table 6). At field application levels, however, only a small portion of the chloroanilines liberated is polymerized by microbial oxidase and peroxidases. The bulk of aniline moieties, are bound to the soil organic matter [145].

Polymerization has also been described as a main reaction for the transformation of pesticide derived phenolic metabolites by an extracellular laccase of *Rhizoctonia practicola*. The enzyme catalyzes the formation of oligomeric products from halogenated phenolic intermediates of phenoxyalkanoate herbicides and from naphtholic products arising from carbamate insecticides [146]. Similarly, phenolic pesticides as

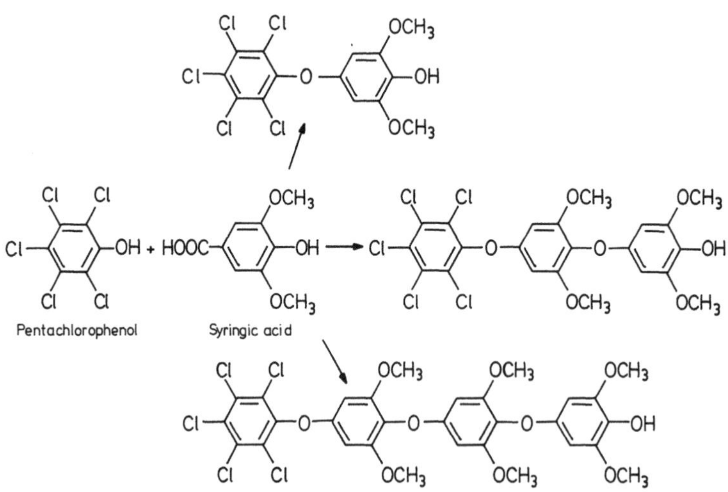

Fig. 32. Proposed structures of hybrid products of syringic acid and pentachlorophenol (from Bollag and Liu [147])

e.g. pentachlorophenol (PCP) or phenolic intermediates of microbial pesticide metabolism, especially chlorophenols, can be co-polymerized during the humification process with phenolic intermediates of lignin degradation by the action of microbial laccases and peroxidases forming bound residues.

The copolymerization of halogenated phenols and syringic acid, a representative of numerous humus oligomers, was demonstrated recently for the laccase of *Rhizoctonia practicola* [147]. Two types of hybrid products were found: (1) phenols covalently bound to an orthoquinone product of syringic acid giving rise to quinoid oligomers, and (2) phenols covalently bound to decarboxylated products of syringic acid resulting in the formation of phenolic oligomers. The structural configurations proposed by Bollag and Liu [147] for the PCP-syringic acid hybrid oligomers are presented in Fig. 32. Due to the strong activity of peroxidases in the oxidation and coupling of aromatic amines it is suggested that anilinic pesticide residues are bound to the soil organic matter by reactions similar to that found for the phenolic pesticide residues.

4.6 Miscellaneous Reactions

Some reactions are described only for few pesticides. In these transformations it has not yet been proved with certainty, whether they are due to biochemical rather than chemical processes.

4.6.1 Dehydrogenations Leading to C,C-Double Bonds

In the metabolism of the alicyclic insecticide lindane (hexachlorocyclohexane; γ-HCH) in submerged soil and by soil microorganisms under anaerobic conditions the removal of electrons from the alicyclic ring leads to pentachloro-1-cyclohexene (Table 3) and further to an aromatization of the molecule, various penta-, tetra- and trichlorobenzenes and -chlorophenols being described as products of combined dechlorination and dehydrogenation reactions [148,149]. Whether this aromatization occurs through chemical or biochemical reactions, however, remains to be elucidated.

4.6.2 Cyclisations

Microbial degradation of the side chains of a pesticide can lead to reactive groups giving rise to a cyclisation of the side chains if there is a suitable substituent at the adjacent aromatic carbon atom. Such reactions are described e.g. for the transformation of the herbicides alachlor and antor by *Chaetomium globosum*, a dihydroindol derivative (Table 5) being the product of the reaction. Similarly, during the metabolism of the herbicide dinitramine in soil or by soil fungi, respectively, cyclisation of the side chains leads to a benzimidazole derivative (Table 7).

5 The Pesticides and Their Microbial Degradation Products

General Comments for Using the Tables

In the following tables, the pesticides are classified and grouped according to chemical characteristics. Biological activity, if not the same for all compounds in a particular table, is given in parentheses after the name, i.e. (I) insecticide, (H) herbicide, (F) fungicide, (A) acaricide.

Table 3. Alicyclic insecticides

Pesticide	Metabolite (s)	Microorganism or Source	Refs.
Lindane 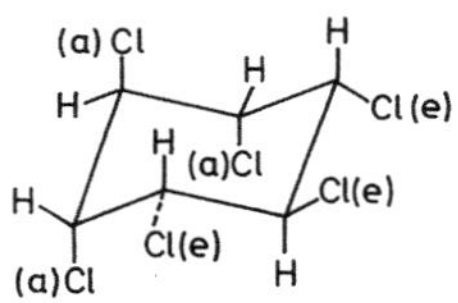	1,3,4,5,6-Pentachloro-1-cyclohexene 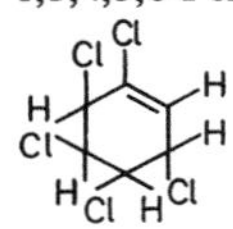	Soil	150)
	γ-Pentachlorocyclohexane	Mold	148)
		Pseudomonas putida	151)
		Bacillus	150)
		Pseudomonas	
		Penicillium	
		Clostridium	152)
	α-1,2,3,4-Tetrachlorocyclohexane	*Penicillium*	150)
		Pseudomonas putida	151)
	β-1,2,3,4-Tetrachlorocyclohexane	*Bacillus*	150)
		Pseudomonas	
		Penicillium	
	γ-1,2,3,4-Tetrachlorocyclohexane	*Penicillium*	150)
		Clostridium sphenoides	153)
	Pentachlorobenzene	*Bacillus*	150)
		Pseudomonas	
		Penicillium	
	Pentachlorophenol	Mold	77)
	Tetrachlorobenzene (3 isomeres)		
	Tetrachlorophenol (2 isomeres)		
	Trichlorobenzene (3 isomeres)		
	Trichlorophenol (4 isomeres)		

Dieldrin 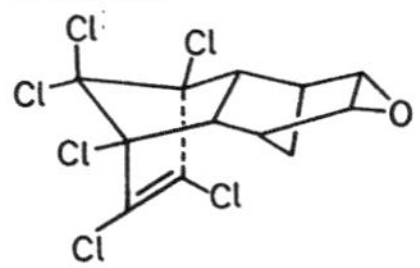	6,7-trans-Dihydroxy-6,7-dihydroaldrin 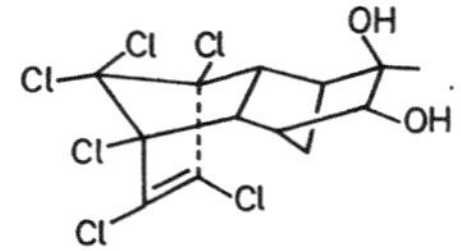	*Pseudomonas* *Bacillus* *Trichoderma viride*	77)
	Other metabolites 6,7-trans-Dihydroxy-6,7-dihydroaldrin Metabolite I 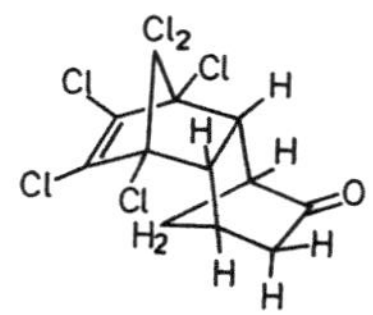 Metabolite II 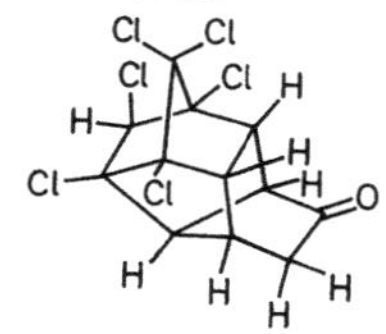	*Aerobacter aerogenes* *Pseudomonas*	154) 77)
	Other metabolites with suggested structures „Photodieldrin" 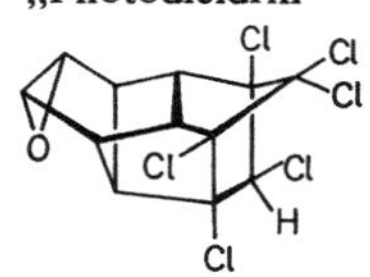	Microorganisms isolated from water, soil and rumen *Pseudomonas* *Neurospora* sp.	84) 77)
Aldrin 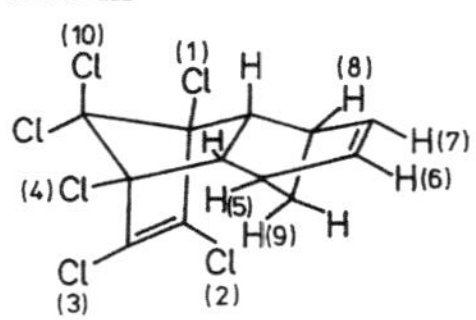	Dieldrin Other metabolites	*Penicillium* *Aspergillus*	77)
	Dieldrin	*Trichoderma viride* *Fusarium* *Penicillium*	155)

Table 3. (continued)

Pesticide	Metabolite (s)	Microorganism or Source	Refs.
	exo-Dieldrin	*Bacillus* *Mycobacterium* *Nocardia* *Streptomyces* *Pseudomonas* *Micromospora* *Thermoactinomyces*	[156]
Endrin 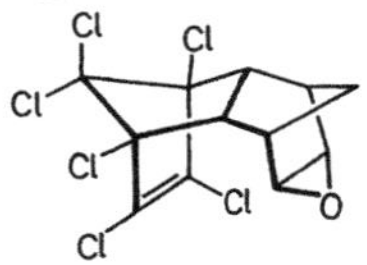	„Ketoendrin" 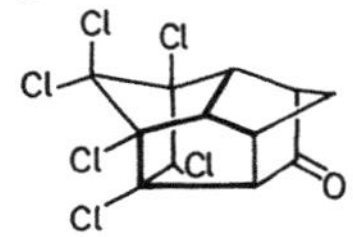	*Trichoderma viride* *Pseudomonas* *Arthrobacter* *Bacillus*	[157] [77]
	„Ketoendrin" Other metabolites	*Pseudomonas*	[83]
Heptachlor 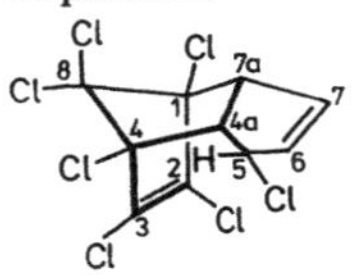	Heptachlor epoxide 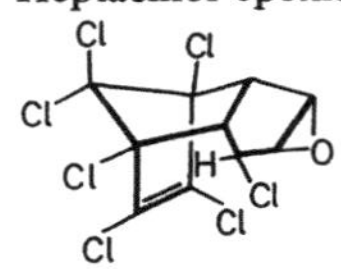 Chlordene 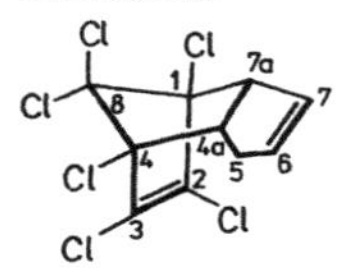 Chlordene epoxide 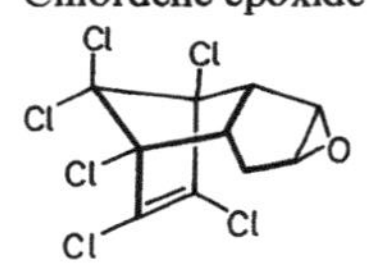	Fungi isolated from soil	[158]

	5-Hydroxy-6,7-epoxichlordene 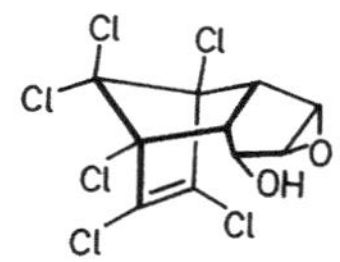		
cis-Chlordane *trans*-Chlordane 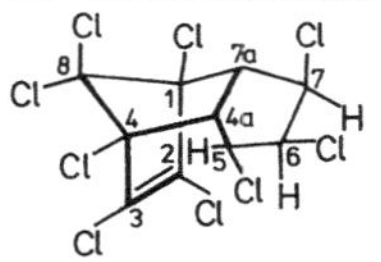 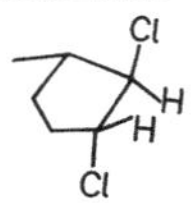	6,7-Dichlorochlordene Oxychlordane Heptachlor Heptachlor-endo-epoxide Chlordene chlorohydrin 5-Hydroxy-*trans*-chlordane	*Norcardiopsis* sp.	[159]
Endosulfan 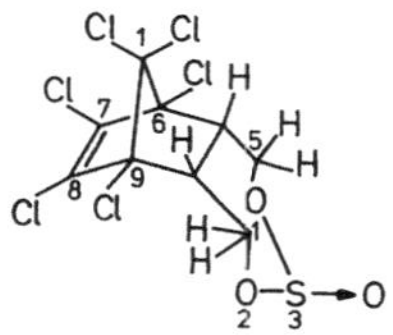	5,6-Bis-[hydroxymethyl]-1,2,3,4,7,7-hexachloro-bicyclo-[2.1.1]-heptene-(2) 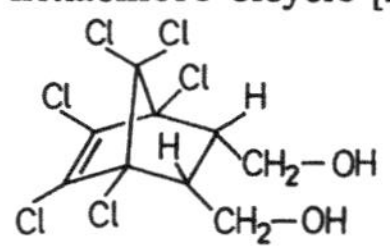	*Aspergillus niger* *Pseudomonas* sp.	[160] [161]
	„Endosulfan ether"	*Pseudomonas* sp.	[161]
Mirex 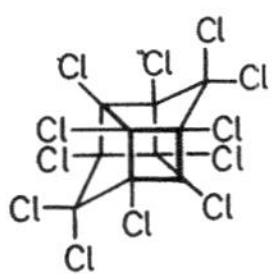	„10-Monohydroderivative"	Sewage sludge	[162]

Table 4. Aliphatic compounds

Pesticide	Metabolite (s)	Microorganism or Source	Refs.
Dalapon (H) $CH_3-CCl_2-COONa$	Dehalogenation	*Nocardia* *Pseudomonas*	163, 164)
	Pyruvic acid $CH_3COCOOH$	*Arthobacter* partly purified enzyme	163, 164)
	2-Chloropropionic acid $CH_3-CHCl-COOH$	*Alternaria*	163, 164)
TCA (H) $Cl-CCl_2-COONa$	Dehalogenation	*Pseudomonas*	163, 165)
cis- and *trans*-1,3-Dichloropropene (soil biocides) $Cl-CH=CH-CH_2-Cl$	cis- and trans-3-Chloroallylalcohol $Cl-CH=CH-CH_2OH$	Soil *Pseudomonas*	166, 167)
	3-Chloroacrylic acid, $CHO-CH_2-COOH$ CO_2	*Pseudomonas*	168)
Glyphosate (H) $HOOC-H_2C-HN-H_2C-P(=O)(OH)_2$	Aminomethylphosphonic acid $H_2N-H_2C-P(=O)(OH)_2$ CO_2	*Flavobacterium* sp. (*Flavobacterium* ssp.)	169)

Table 5. Anilides

Pesticide	Metabolite (s)	Microorganism or Source	Refs.
Propanil (H) 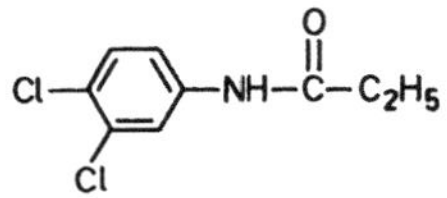	3,4-Dichloroaniline 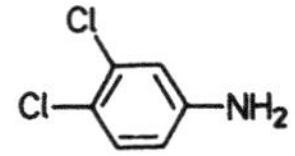	Soil *Fusarium solani* *Pseudomonas striata*, cell-free extract	170, 171)
		Bacillus sphaericus	172)
		Paecilomyces HUT 4017	173)
Karsil (H) 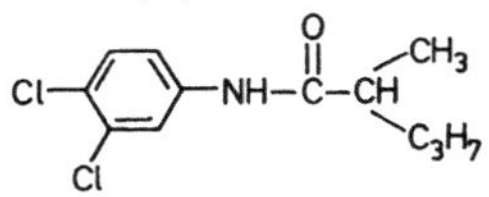	3,4-Dichloroaniline	*Penicillium* *Pullularia* *Penicillium*, cell-free extract	173)
		Bacillus sphaericus	172)
	3-Hydroxy-2-methyl-3′,4′-dichloropentananilide 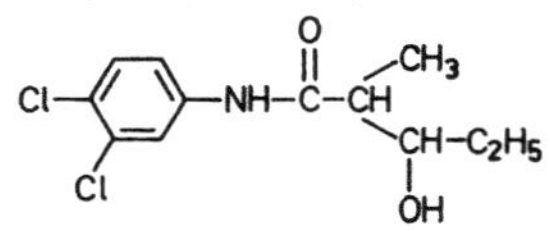	*Rhizopus japonicus*	33)
Dicryl (H) 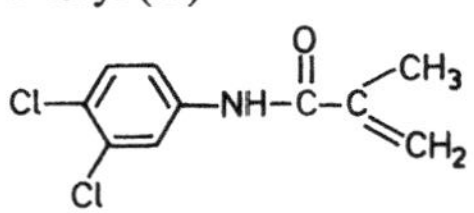	3,4-Dichloroaniline	*Bacillus sphaericus*	173)
	2,3-Dihydroxy-2-methyl-3′,4′-dichloropropananilide 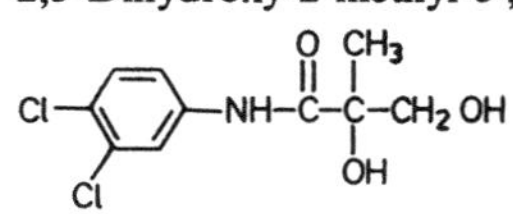	*Rhizopus japonicus*	34)

Table 5. (continued)

Pesticide	Metabolite (s)	Microorganism or Source	Refs.
Solan (H) 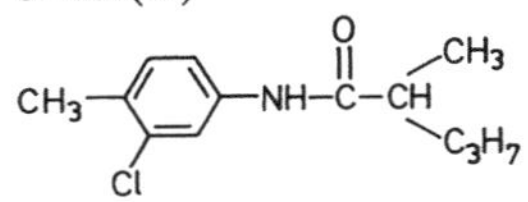	3-Chloro-4-methylaniline 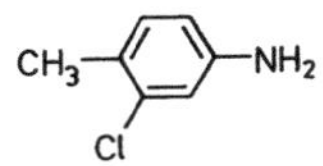	*Bacillus sphaericus*	172)
	3-Chloro-4-methylacetanilide 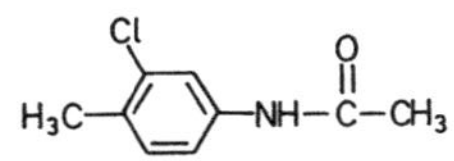	*Aspergillus niger*	174)
Alachlor (H) 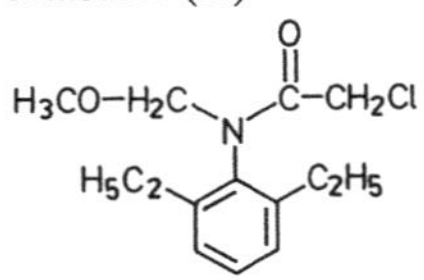	2,6-Diethyl-chloroacetanilide 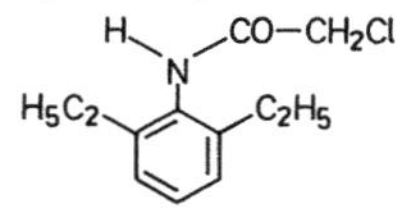	*Chaetomium globosum*	175)
	2,6-Diethylaniline 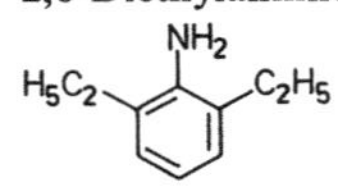		
	N-(Methoxymethyl)-2,6-diethylaniline 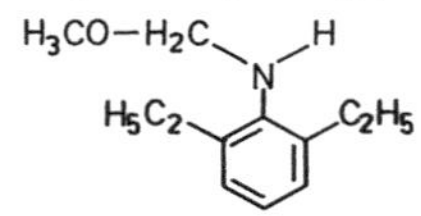		
	1-Chloroacetyl-2,3-dihydro-7-ethylindole 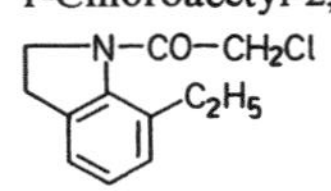		

Metolachlor (H) 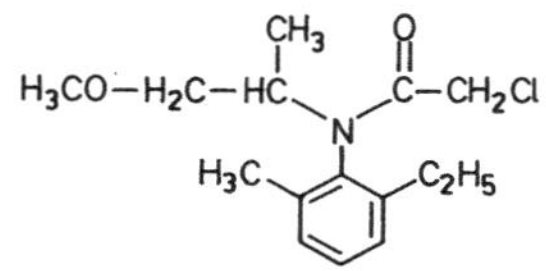	N-(2'-Hydroxy-1'-methylethyl)-2-ethyl-6-methylchloroacetanilide 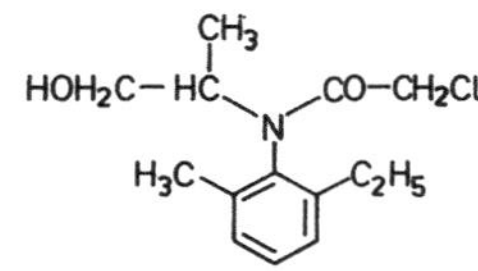	*Chaetomium globosum*	175)
	Hydroxylation: 2-Chloro-N-(2-ethyl-6-hydroxymethylphenyl)-N-(2-methoxy-1-methylethylacetamide (R_1=OH; R_2=H; R_3=CH_3) 2-Chloro-N-[2-(1-hydroxyethyl)-6-methylphenyl]-N-(2-methoxy-1-methylethyl)acetamide (R_1=H; R_2=OH; R_3=CH_3) 2-Chloro-N-[2-(1-hydroxyethyl)-6-hydroxy-methylphenyl]-N-(2-methoxy-methylethyl)acetamide (R_1, R_2=OH, R_3=CH_3) *Hydroxylation and Demethylation:* Compound A: R_1=OH; R_2, R_3=H Compound B: R_1, R_3=H; R_2=OH Compound C: R_1, R_2=OH; R_3=H 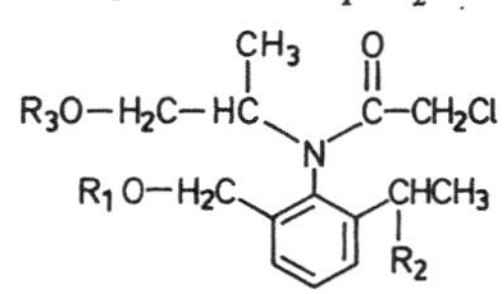	*Actinomyces* sp.	35)
Antor (H) 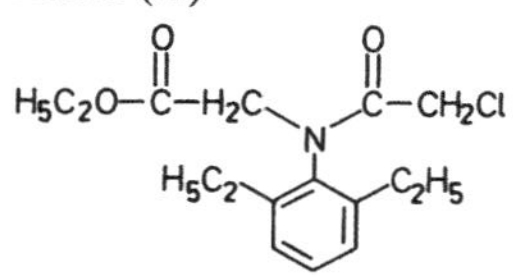	Ethyl-N-(2,6-diethylphenyl)iminoacetate 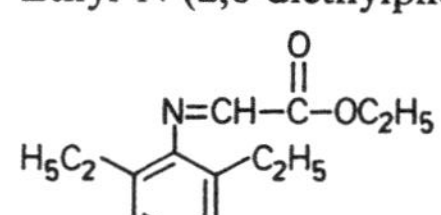 2,6-Diethyl-chloroacetanilide 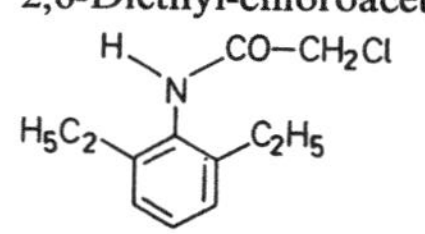	*Chaetomium globosum*	175)

Table 5. (continued)

Pesticide	Metabolite (s)	Microorganism or Source	Refs.
	Ethyl-N-(hydroxyacetyl)-N-(2-ethyl-6-vinylphenyl)glycinate 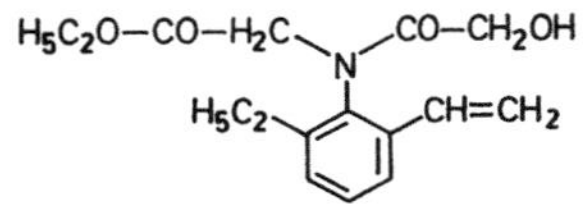		
	Ethyl-N-(chloroacetyl)-N-(2-ethyl-6-vinylphenyl)glycinate 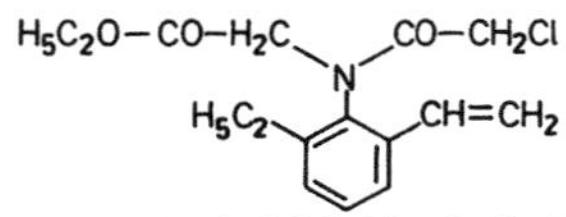		
	Ethyl-(7-ethyl-2,3-dihydroindol-1-yl)acetate 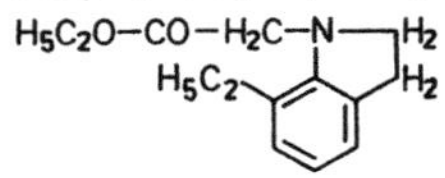		
Furacarbanil (F) 2,5-Dimethyl-3-furan-carboxanilide 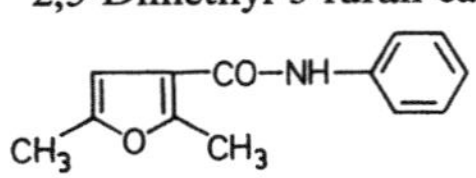	Aniline	*Bacillus sphaericus*	172)
	2,5-Dimethyl-3-furancarboxylic acid 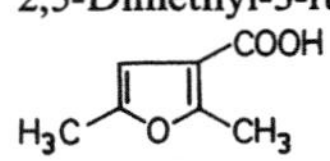		
	Acetanilide	*Aspergillus niger*	174)
	2-Hydroxymethyl-5-methyl-3-furancarboxanilide 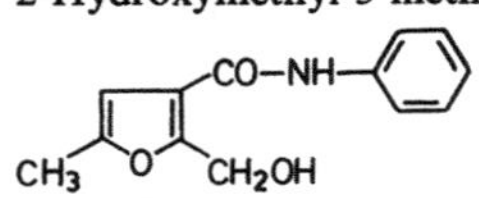	*Rhizopus japonicus* *Rh. peka* *Rh. nigricans*	176)
	5-Hydroxymethyl-2-methyl-3-furancarboxanilide 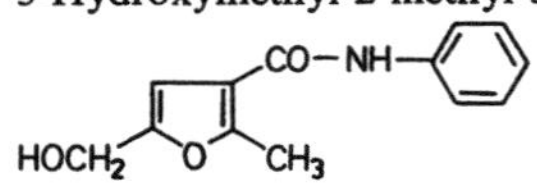		

2-Methylbenzanilide BAS-305 (F) 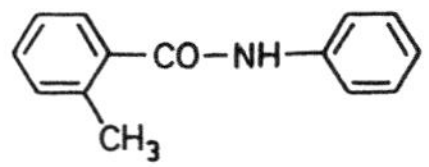	Aniline 2-Methylbenzoic acid 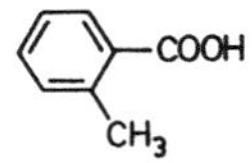	*Bacillus sphaericus*	172)
	4′-Hydroxy-2-methylbenzanilide 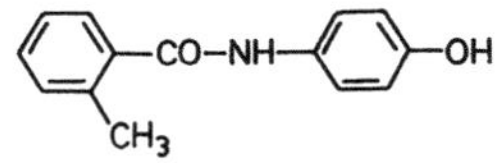	*Rhizopus japonicus*	177)
2-Chlorobenzanilide BAS-307 (F) 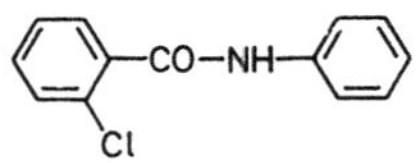	Aniline	*Bacillus sphaericus*	172)
	2-Chlorobenzoic acid 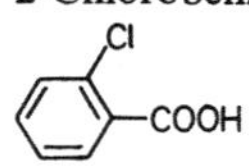		
	2-Chloro-4′-hydroxybenzanilide 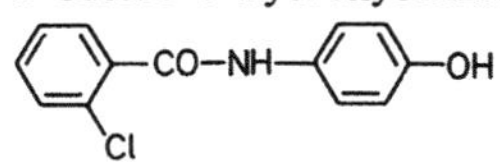	*Rhizopus japonicus*	177)
Carboxin Vitavax (F) 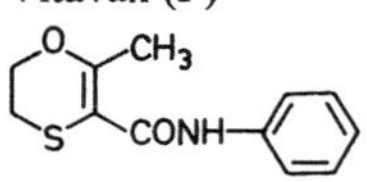	Aniline	*Bacillus sphaericus*	172)
	"Carboxin sulfoxide" 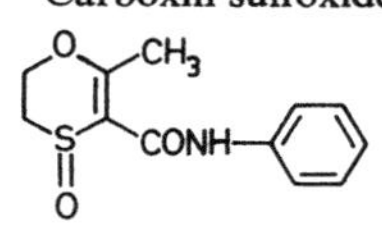 "Carboxin sulfon" 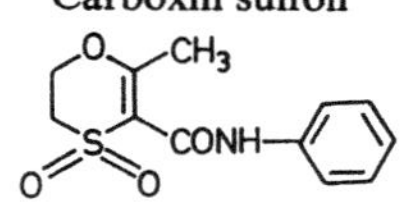	*Rhizopus japonicus*	178)

Table 6. Anilines derived from pesticides

Pesticide	Metabolite (s)	Microorganism or Source	Refs.
R_1 = H, R_2 = Cl: 3-Chloroaniline	3,3′-Dichloroazobenzene	Soil	[170]
R_1 = Cl, R_2 = H: 4-Chloroaniline	4,4′-Dichloroazobenzene		
R_1 = Cl, R_2 = Cl: 3,4-Dichloroaniline	3,3′,4,4′-Tetrachloroazobenzene		
R_1 = CH_3, R_2 = Cl: 3-Chloro-4-methylaniline	3,3′-Dichloro-4,4′-dimethylazobenzene	*Fusarium oxysporum* Schlecht	[171]
R_1 = Br, R_2 = H: 4-Bromoaniline	4,4′-Dibromoazobenzene	*Geotrichum candidum*	[141]
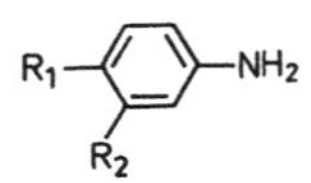	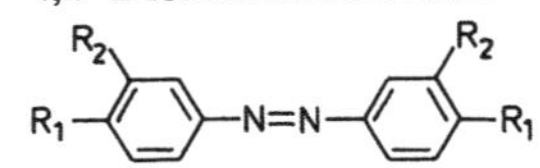		
4-Chloroaniline	Mixed azobenzenes	Soil	[170]
	4,4′-Dichloroazoxybenzene 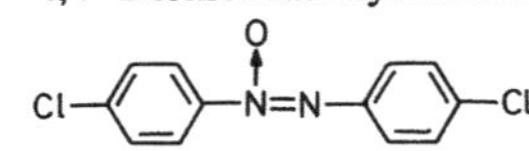	*Fusarium oxysporum* Schlecht	[171]
	4-Chlorophenylhydroxylamine 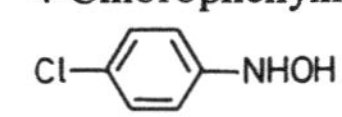		
	4-Chloronitrosobenzene 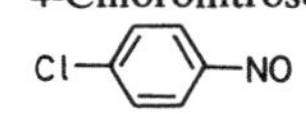		
	4-Chloroacetanilide 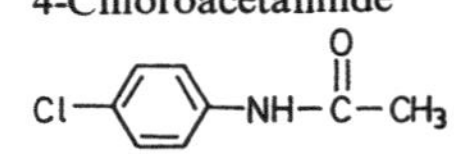	*Fusarium oxysporum* Schlecht *Micrococcus* sp. *Bacillus firmus*	[171] [179] [129]
	4-Chloropropionanilide 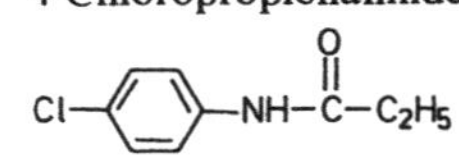	*Bacillus firmus*	[129]
	7-Chloro-2-amino-3H-phenoxazine-3-one 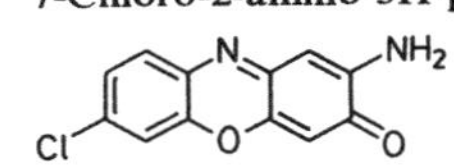	*Bacillus firmus* *Alcaligenes faecalis*	[129] [144]

	4,4′-Dichloro-diazoaminobenzene 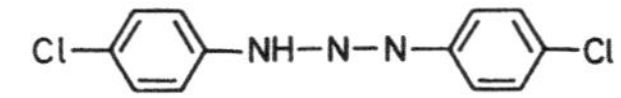	*Paracoccus* sp.	136)
	CO_2	*Pseudomonas* sp.	180)
	4-Chlorocatechol 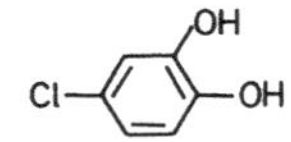	*Moraxella* sp. strain G.	181)
		Pseudomonas sp. OCA-1	59)
3,4-Dichloroaniline 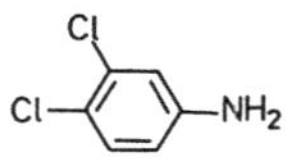	3,3′,4′-Trichloro-4-(3,4-dichloroanilino)-azobenzene 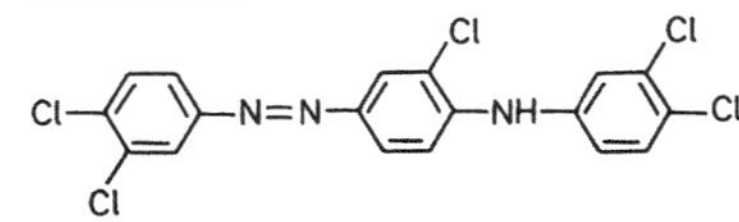	Soil	182)
	1,3-Bis(3,4-dichlorophenyl)triazene 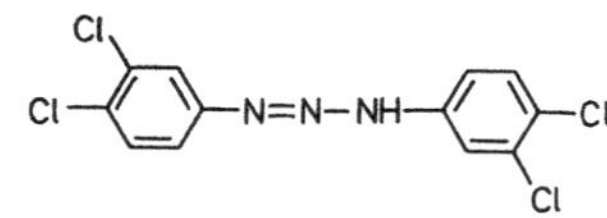	Soil	183)
	3,4-Dichloromuconic acid 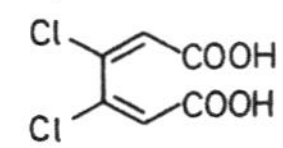	*Pseudomonas putida*	61)
	3-Chloro-4-(carboxymethylene)but-2-enolide 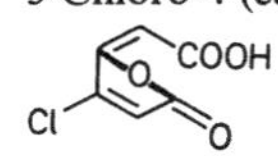		
4-Bromoaniline 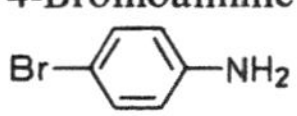	4-Bromoacetanilide 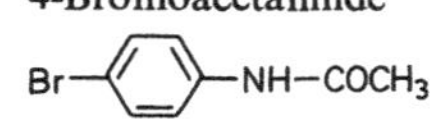	*Talaromyces wortmanni* *Fusarium oxysporum* *Chlorella vulgaris*	126)

Table 7. Aromatic herbicides

Pesticide	Metabolite (s)	Microorganism or Source	Refs.
Dichlobenil 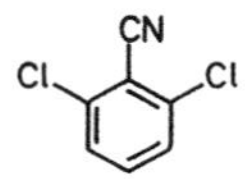	2,6-Dichlorobenzoic acid 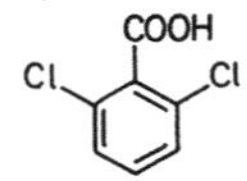	*Fusarium* *Geotrichum* *Penicillium* *Trichoderma*	184)
	2,6-Dichlorobenzamide 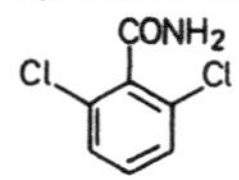	Soil	185)
Chlorthiamid 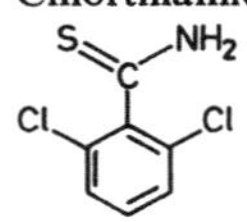	2,6-Dichlorobenzonitrile 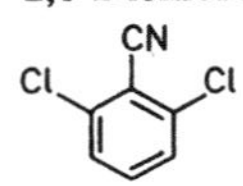	Soil	184)
Ioxynil 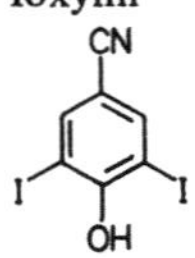	4-Hydroxy-3,5-diiodobenzamide 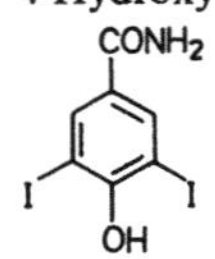	Soil *Fusarium solani* *Flexibacterium* BR 4	186) 187) 188)
	4-Hydroxy-3,5-diiodobenzoic acid 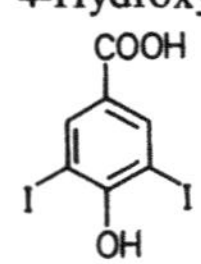	*Fusarium solani*	187)
Bromoxynil 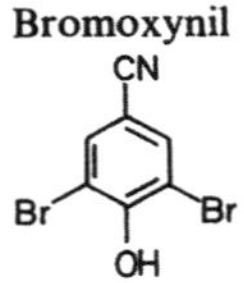	4-Hydroxy-3,5-dibromobenzoic acid 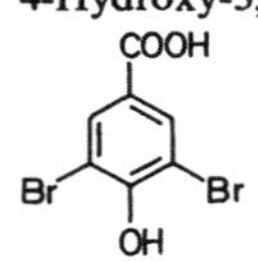	*Fusarium solani* *Klebsiella pneumoniae* ssp. *ozaenae*	187) 48)

	4-Hydroxy-3,5-dibromobenzamide 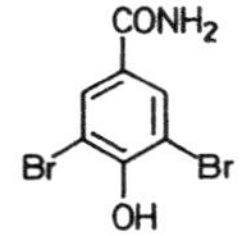	*Klebsiella pneumoniae* ssp. *ozeanae*	48)
Dinoseb 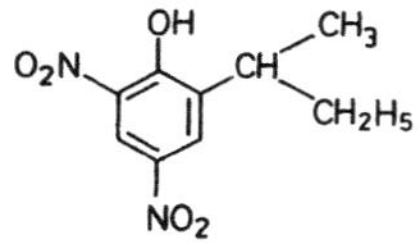	2-Amino-6-(1-methylpropyl)-4-nitrophenol 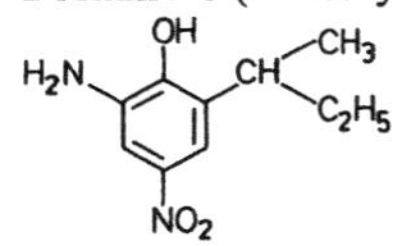	Mixed microflora of bovine rumen	189)
	2,4-Diamino-6-(1-methylpropyl)-phenol 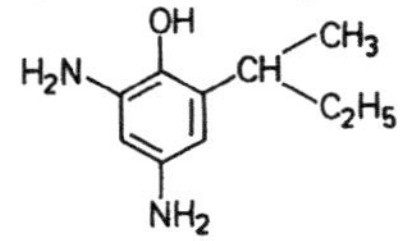		
	2-Hydroxy-3-(1-methylpropyl)-5-nitro-acetanilide 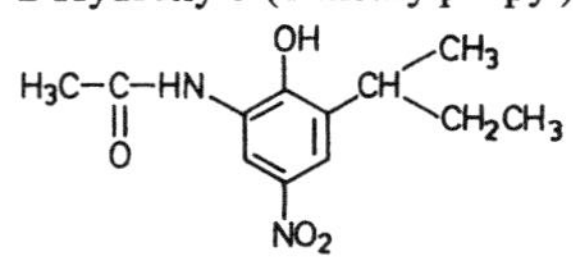	*Azotobacter* sp.	190)
DNOC 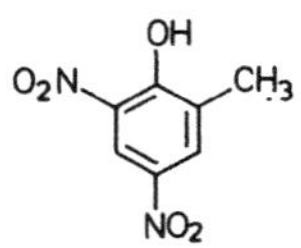	2-Amino-6-methyl-4-nitrophenol 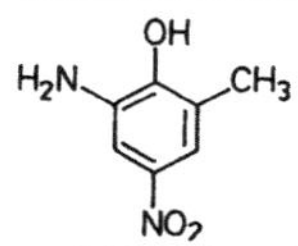	Mixed microflora of bovine rumen	189)
	2,4-Diamino-6-methylphenol 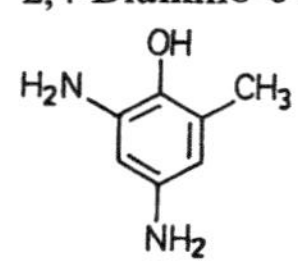		

Table 7. (continued)

Pesticide	Metabolite (s)	Microorganism or Source	Refs.
	2,3,5-Trihydroxytoluene (via intermediate reduction and hydroxylation products) 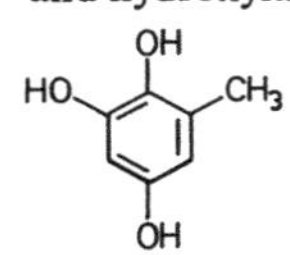	*Pseudomonas* *Arthrobacter simplex*	[25]
	2-Hydroxy-3-methyl-5-nitro-acetanilide 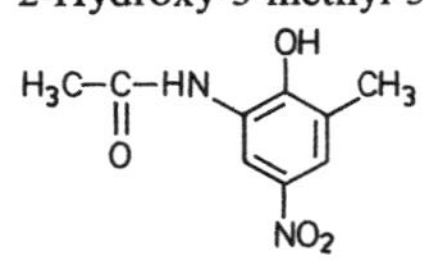	*Azotobacter* sp.	[190]
Trifluralin 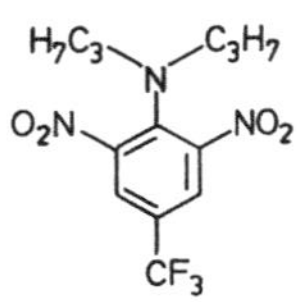	4-Trifluoromethyl-2,6-dinitro-N-propylaniline 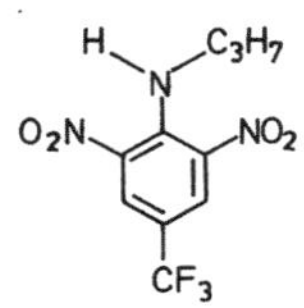	*Aspergillus niger* *A. carneus* *Fusarium oxysporum* *Trichoderma viride*	[192] [193]
	5-Trifluoromethyl-1,2,3-triaminobenzene and intermediate reduction and dealkylation products 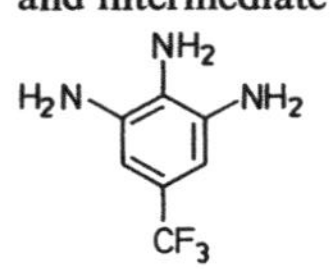	Soil (aerobic and anaerobic conditions)	[194, 192]

	4-Trifluoromethyl-2-amino-6-nitro-N,N-dipropyl-aniline 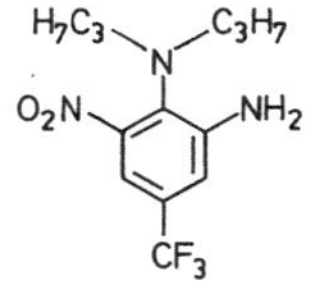	4-Trifluoromethyl-2,6-diamino-N,N-dipropyl-aniline 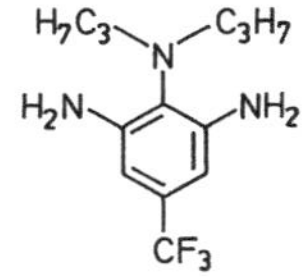	Rumen fluid	[195]
	2,6-Dinitro-4-trifluoromethylphenol 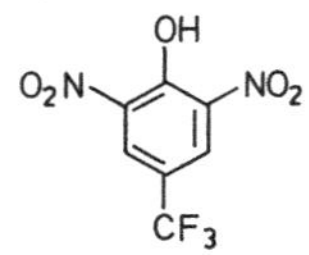 2,6-Dinitro-4-trifluoromethyl-aniline 2-Amino-6-nitro-4-trifluoromethyl-*p*-toluidine		*Aspergillus carneus* *Fusarium oxysporum* *Trichoderma viride*	[193]
Trifluralin ($X = CF_3$) Oryzalin ($X = SO_2NH_2$) Isopropalin ($X = CH(CH_3)_2$) 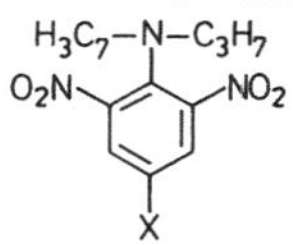	N,N-Dipropyl-2-amino-4-(X)-6-nitroaniline 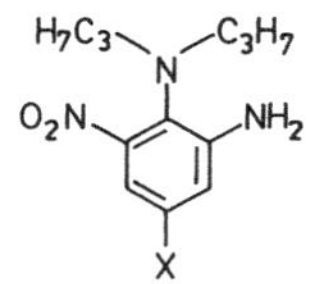	4-(X)-2,6-dinitroaniline [X = trifluoromethyl, = aminosulfonyl, = isopropyl] 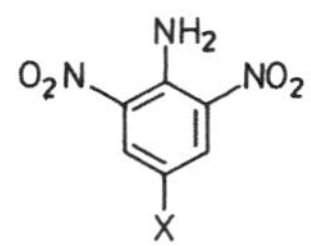	Soil	[196]
Trifluralin Oryzalin Isopropalin	N-Propyl-4-(X)-2,6-dinitroaniline 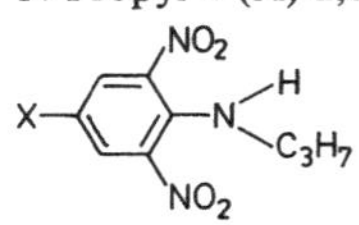		Soil	[196]

Table 7. (continued)

Pesticide	Metabolite (s)		Microorganism or Source	Refs.
	1-Propyl-2-ethyl-5-(X)-7-nitrobenzimidazole 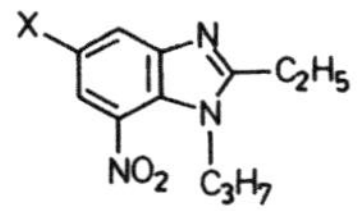	2-Ethyl-5-(X)-7-nitrobenzimidazole 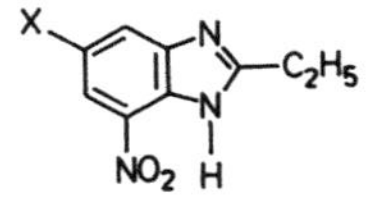		
Dinitramine 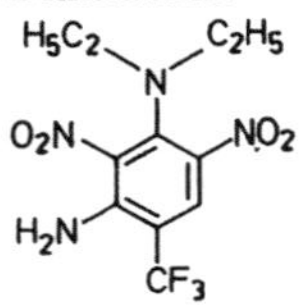	4-Trifluoromethyl-1,3-diamino-2,6-dinitrobenzene 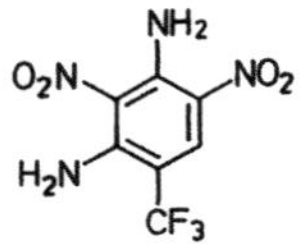	N-Ethyl-3-amino-4-trifluoromethyl-2,6-dinitroaniline 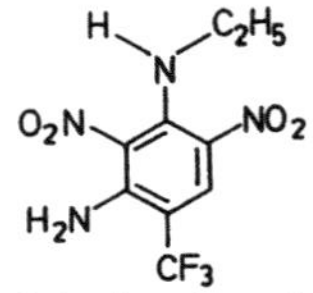	*Aspergillus fumigatus* *Fusarium oxysporum* *Paecilomyces* sp.	[197]
	6-Amino-1-ethyl-2-methyl-7-nitro-5-trifluoromethylbenzimidazole 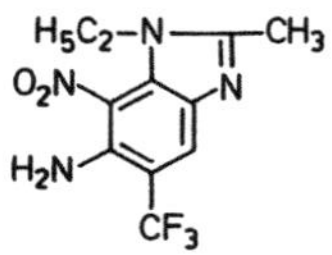	6-Amino-2-methyl-7-nitro-5-trifluoromethylbenzimidazole 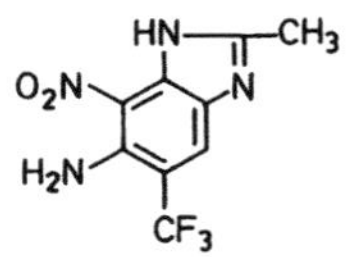		
Fluochloralin 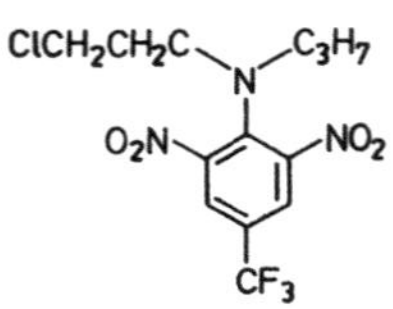	N-(2-chloroethyl)-2,6-dinitro-4-trifluoromethylaniline 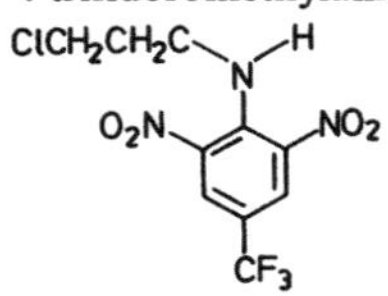	2,6-Dinitro-4-trifluoromethylaniline 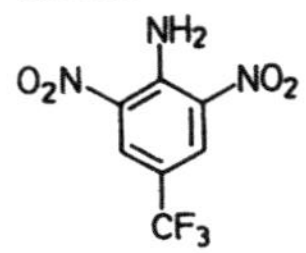	Soil	[198]

Nitralin 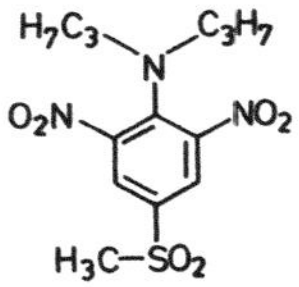	N-Propyl-4-(methylsulfonyl)-2,6-dinitroaniline 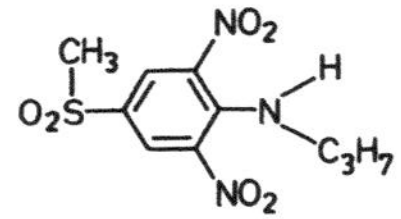	*Rhizoctonia solani*	199)
Fluorodiphen 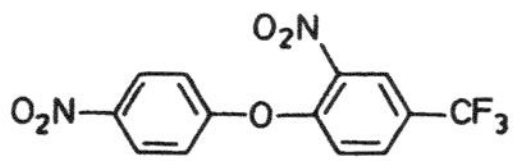	4-Nitrophenol 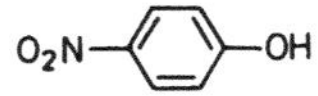	Soil bacteria	191)
Diclofop-methyl 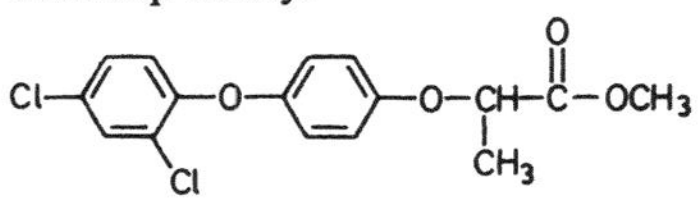	"Diclofop-acid" 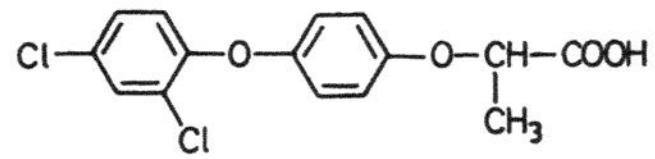 4-(2′,4′-Dichlorophenoxy)-phenetole 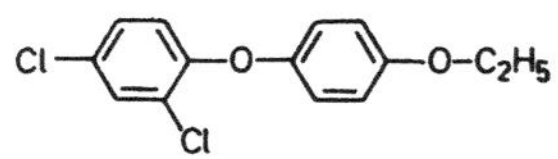 4-(2′,4′-Dichlorophenoxy)-phenol 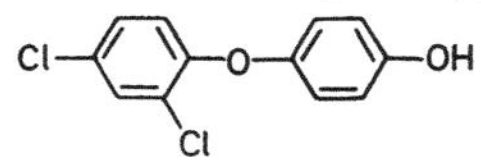	Soil	200)
Chlorphenpropmethyl 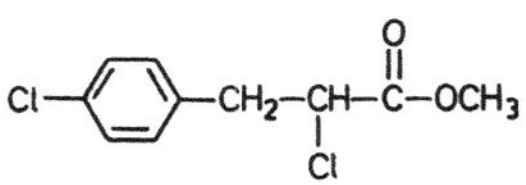	3-(4-Chlorophenyl)-2-chloro-propionic acid 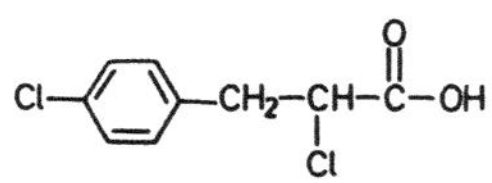 4-Chlorobenzoic acid 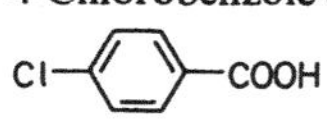	*Flavobacterium* *Brevibacterium*	201)

Table 7. (continued)

Pesticide	Metabolite (s)	Microorganism or Source	Refs.
2,3,6-TBA 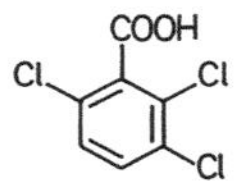	3,5-Dichlorocatechol (via some postulated intermediates) 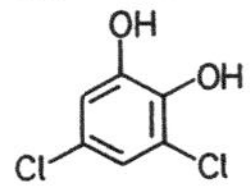	*Brevibacterium*	[5]
Pentachlorophenol 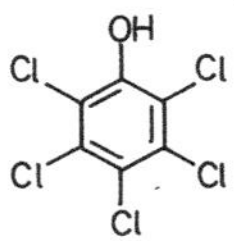	Pentachloroanisole 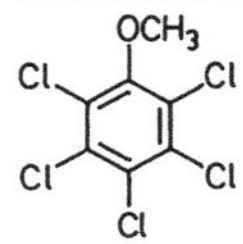	*Trichoderma virgatum*	[202]
	Various tetrachlorophenols	*Pseudomonas* sp. *Arthrobacter* sp. *Trichoderma* sp.	[79]
	Reductive dehalogenation CO_2, CH_4	Sewage sludge	[80]
N-(3-Methyl-but-1-in-3-yl)-3,5-dichlorobenzamide RH-315 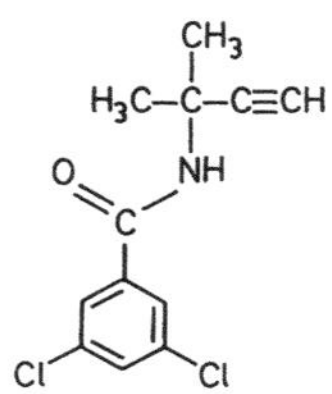	2-(3,5-Dichlorophenyl)-4,4-dimethyl-5-methyleneoxazoline 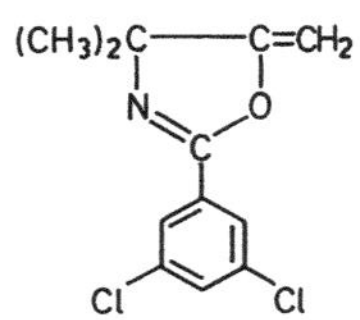	Soil	[203]

N-(2-Oxo-3-methyl-but-3-yl)-3,5-dichlorobenzamide

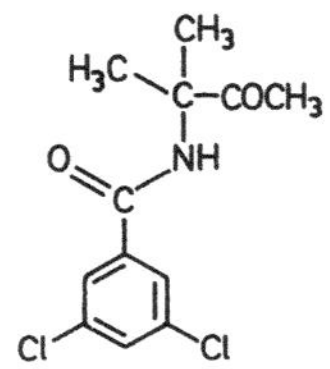

Table 8. Diphenylmethane (DDT-Type) compounds

Pesticide	Metabolite (s)	Microorganism or Source	Refs.
p,p'-DDT Ar–CH–Ar | CCl_3 Ar: 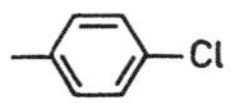	DDD (TDE) Ar–CH–Ar | $CHCl_2$	*Nocardia* *Streptomyces*, several ssp. *Escherichia coli* Yeast Rumen fluid Coliform bacteria from rat feces *Aerobacter aerogenes*	[204]
p,p'-DDT	DDD, DDE Ar–C–Ar ‖ CCl_2	Several strains isolated from rat gastrointestinal tract Several strains isolated from fly excreta Soil microorganisms (not defined)	[204] [205]
	DDD, DDE, DDNS Ar–CH–Ar | CH_3	Many microbial isolates from water	[204]
p,p'-DDT	DDD, DDNS, DDMS Ar–CH–Ar | CH_2Cl DDMU Ar–C–Ar ‖ CHCl	*Proteus vulgaris* (from mouse intestine)	[204]
	OH | Ar–C–Ar | CCl_3	Several soil microorganisms	[204]

	DDA Ar–CH(COOH)–Ar	*Trichoderma viride*	77)
p,p'-DDT	DDD, DDE, DDMU, DDMS, DDA, DDNU: Ar–C(=CH_2)–Ar DBP: Ar–C(=O)–Ar DDM, Ar–CH_2–Ar DBH: Ar–CH(OH)–Ar	*Aerobacter aerogenes* and cell-free extract	204)
p,p'-DDT	DDD, DDMU, DDMS, DDNU, DDA, DBP, DDB	*Bacillus cereus* *B. coagulans* *B. subtilis* *Escherichia coli* *Enterobacter aerogenes*	204)
p,p'-DDT	DDD, DDMU, DDA, DBP, Ar–CH(CH_2OH)–Ar	*Fusarium oxysporum* and partially purified enzymes	204)
	DDD, DDE, DDA 4,4'-Dichlorobenzophenon Benzophenon Phenylacetic acid Salicylic acid	*Pseudomonas aeruginosa*	206)
o,p'-DDT 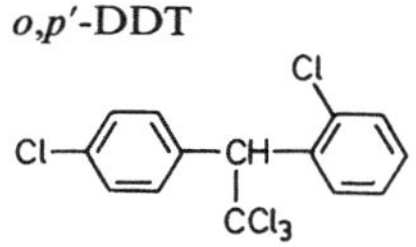	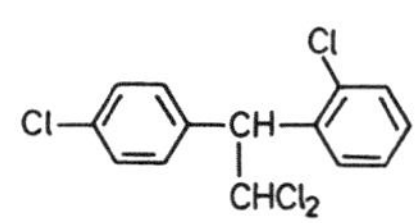	*Aerobacter aerogenes*	204)

Table 8. (continued)

Pesticide	Metabolite (s)	Microorganism or Source	Refs.
Methoxychlor 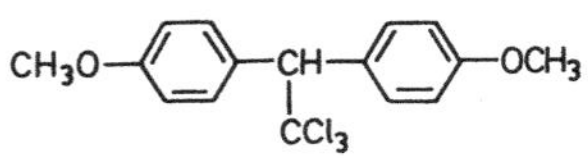	1,1-Bis-(4-methoxyphenyl)-2,2-dichloroethene 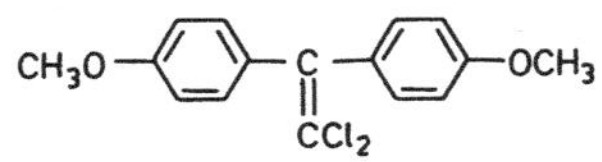	*Aerobacter aerogenes*	207)
	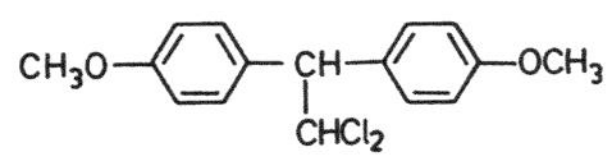	*Aerobacter aerogenes* *Klebsiella pneumoniae*	207) 78)
DPM and similar compounds related to DDT 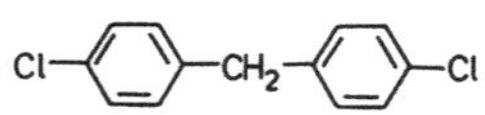	Phenylacetic acids and similar compounds	*Hydrogenomonas*	208, 209, 210)
Chlorobenzilate (I, A) 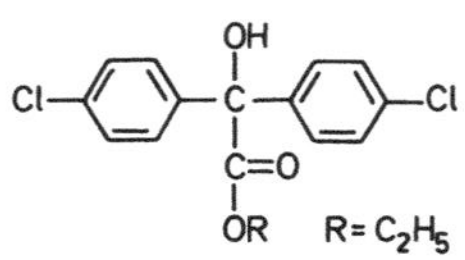	4,4′-Dichlorobenzilic acid 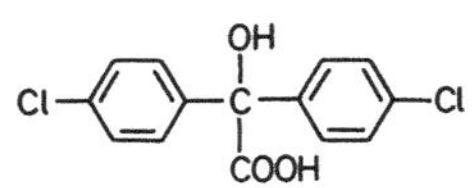 4,4′-Dichlorobenzophenone 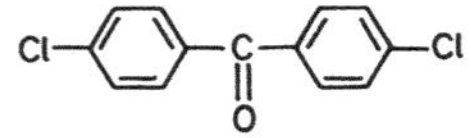	*Rhodotorula gracilis*	211, 212)
Chloropropylate (I, A) 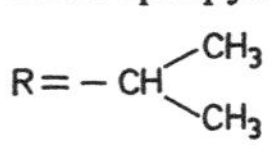	4,4′-Dichlorobenzilic acid 4,4′-Dichlorobenzophenone	*Rhodotorula gracilis*	211, 212)

Diphenamid (H) 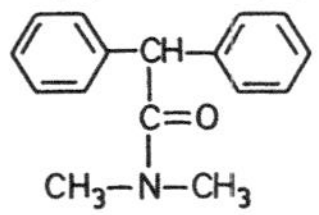	N-Methyl-2,2-diphenylacetamide (R = CH_3) 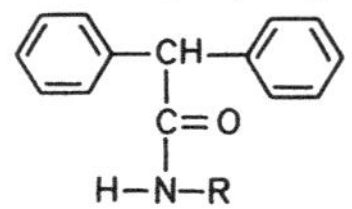 N-Methyl-2,2-diphenylacetamide (R = CH_3) 2,2-Diphenylacetamide (R = H)	*Trichoderma viride* *Aspergillus candidus*	213)

Table 9. Heterocyclic herbicides

Pesticide	Metabolite (s)	Microorganism or Source	Refs.
Paraquat	1-Methyl-4-carboxy pyridinium ion	Soil bacteria	[214]
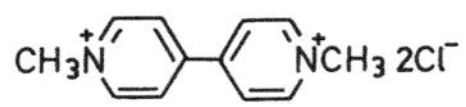	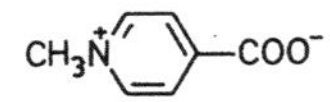	*Pseudomonas fluorescens*	[215]
		Aerobacter aerogenes	
		Streptomyces	[216]
		Nocardia	
	2,6-Dihydroxy-4-pyridinecarboxylic acid	*Achromobacter*	[116]
	Malic acid monoamide		
	1-Methyl-4,4′-bipyrilinium ion	Soil bacteria	[214, 215]
	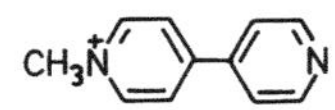		
	CO_2	*Lipomyces starkeyi*	[217]
Amitrole	CO_2	*Bacillus*	[218]
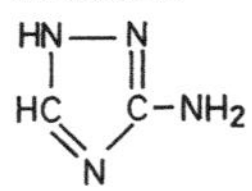		*Corynebacterium*	
		Pseudomonas	
	3-(3-Amino-1,2,4-triazole-1-yl)-alanine	*Escherichia coli*	[125]
	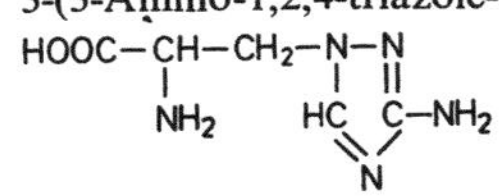		
Pyrazon	5-Amino-4-chloro-3(2H)-pyridazinone	Gram-negative coccus	[219]
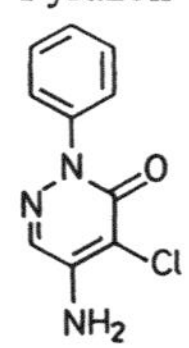	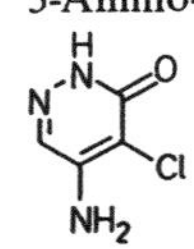	*Achromobacter*	[220]
		Pseudomonas sp.	[120]
	2-(2′,3′-Dihydroxyphenyl)-5-amino-4-chloro-3(2H)-pyridazinone	*Pseudomonas* sp.	[120]
	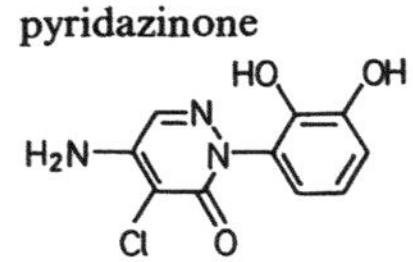		

	α-[5-Amino-4-chloro-3(2H)-pyridazinone-2yl]-*cis*, *cis*-muconic acid 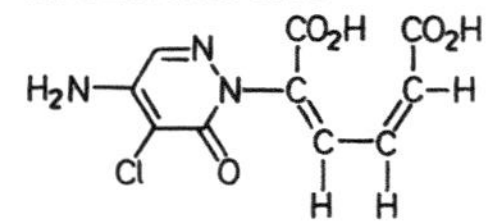 6-Carboxy-2-pyrone 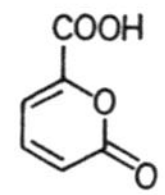		
Metamitron 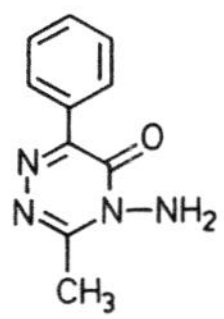	3-Methyl-6-phenyl-5(4H)-1,2,4-triazinone 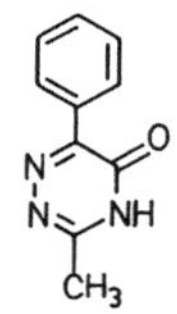	*Arthrobacter* *Pseudomonas* *Rhizopus japonicus* *Cunninghamella echinulata* Thaxter	221)
	3-Methyl-6-phenyl-2,3-dihydro-1,2,4,5-tetrazine-2-carboxylic acid 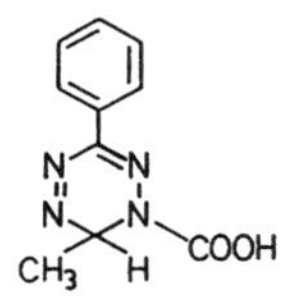 Benzoylformic acid acetylhydrazone 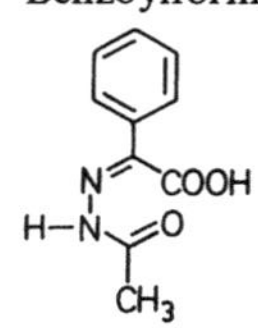	*Arthrobacter* sp. DSM 20389	122)

Table 9. (continued)

Pesticide	Metabolite (s)	Microorganism or Source	Refs.
	Benzoylformic acid 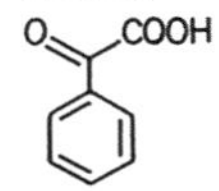		
Metribuzin 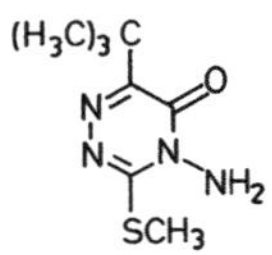	"Desaminometribuzin" 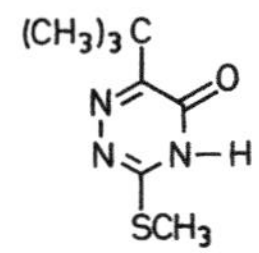 "Desaminodiketo metribuzin" 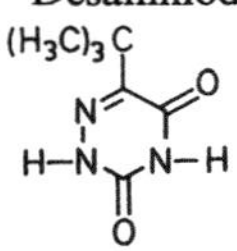	*Cunninghamella echinulata* Thaxter	64)
Picloram 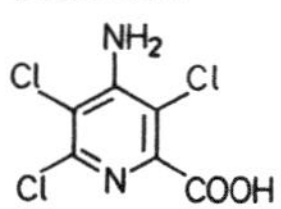	4-Hydroxy-3,5,6-trichloropyridine-2-carboxylic acid 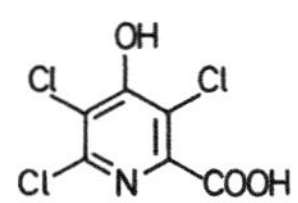	Soil	222, 207)
	4-Amino-2,3,5-trichloropyridine 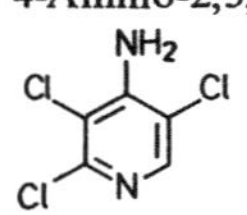	*Trichoderma*	222)
Atrazine 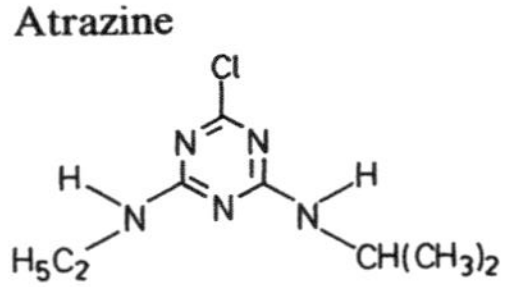	2-Hydroxy-4-ethylamino-6-isopropyl-amino-s-triazine 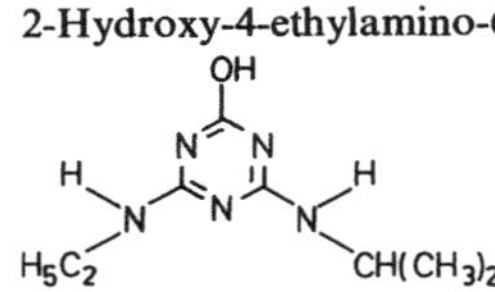	*Fusarium roseum*	223, 224, 30)

	2-Chloro-4-amino-6-isopropylamino-s-triazine	*Aspergillus fumigatus*	223,224)
	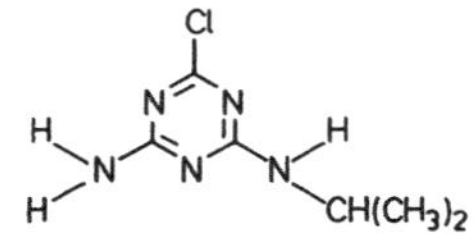	*Rhizopus stolonifer*	30,225)
		Other soil fungi	
		Pseudomonas sp.	31)
	2-Chloro-4-amino-6-ethylamino-s-triazine		
	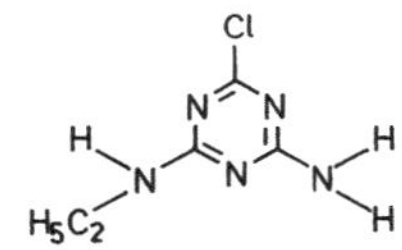		
	2-Chloro-4-amino-6-isopropylamino-s-triazine	*Nocardia* sp.	32)
	2-Chloro-4,6-diamino-s-triazine		
	4-Amino-2-chloro-s-triazine		
	4-Amino-1,2-dihydro-s-triazine-2-(1H)-one		
	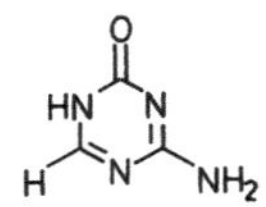		
	2-Hydroxy-4-amino-6-isopropylamino-s-triazine	*Pseudomonas* sp.	31)
	2-Hydroxy-4-amino-6-ethylamino-s-triazine		
Simazine	2-Chloro-4-amino-6-ethylamino-s-triazine	*Aspergillus fumigatus*	223,224,225)
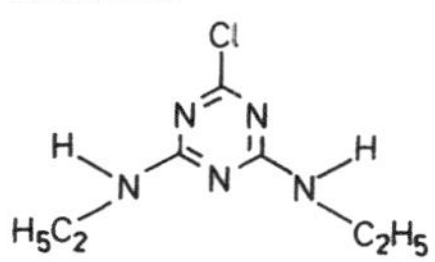	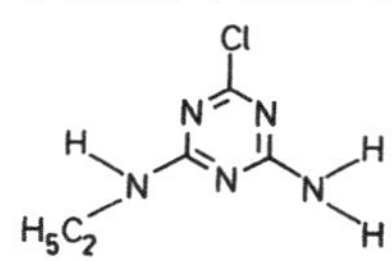		
	2-Hydroxy-4-amino-6-ethylamino-s-triazine	*Rhodococcus corallinus*	74)
	2,4-Dihydroxy-6-ethylamino-s-triazine	*Pseudomonas* sp. strain NRRL-B 12228	226)
	CO_2, NH_3		

Table 9. (continued)

Pesticide	Metabolite (s)	Microorganism or Source	Refs.
Prometryn 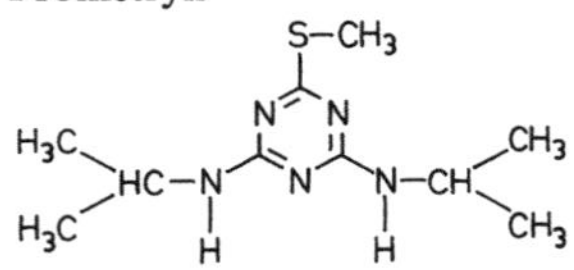	4-Ethylamino-6-isopropylamino-2-methylthio-s-triazine 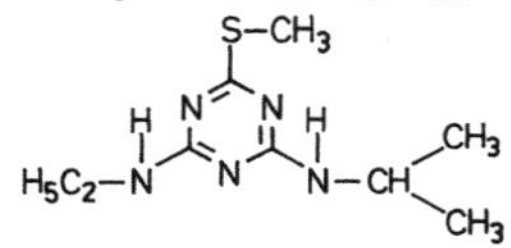	*Agrobacter radiobacter*	227)
	4-Amino-6-isopropylamino-2-methylthio-s-triazine 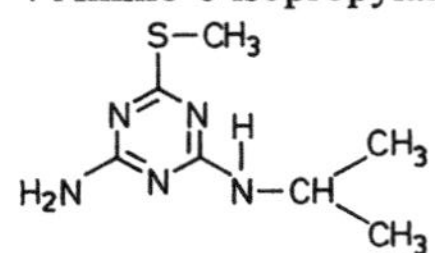		

Table 10. Methyl and phenyl carbamates

Pesticide	Metabolite (s)	Microorganism or Source	Refs.
CIPC (H) 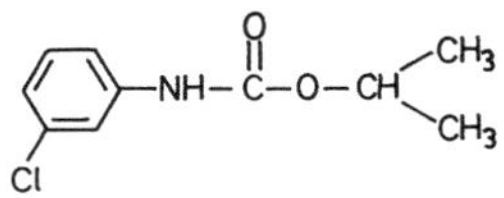	3-Chloroaniline 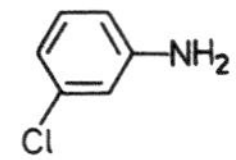	*Pseudomonas striata* cell-free-extract Soil organisms	228, 171) 204, 229)
CEPC (H) 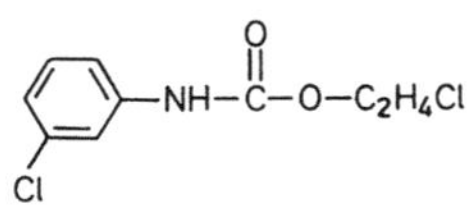	(Hydroxylated 3-chloroaniline) 3-Chloroaniline	*Bacillus sphaericus* *Pseudomonas striata*, cell-free-extract *Flavobacterium* *Agrobacterium* *Achromobacter*	172) 228, 171) 230)
IPC (H) 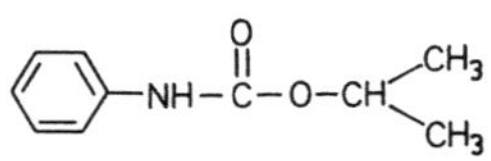	Aniline	*Pseudomonas*, cell-free-extract *Arthrobacter* *Achromobacter*	228) 171) 230)
	2-Propyl-N-(4-hydroxyphenyl)-carbamate	*Beauveria sulfurescens*	231)
Swep (H) 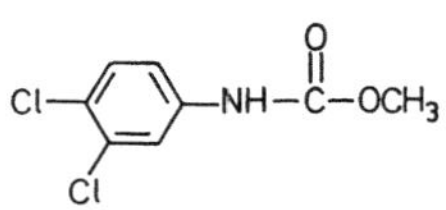	3,4-Dichloroaniline 3,3′,4,4′-Tetrachloroazobenzene 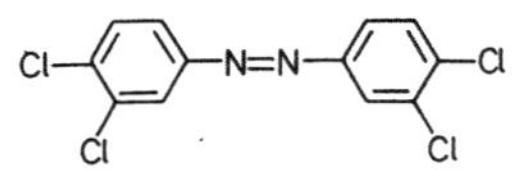	Soil	170)
Mobam (I) 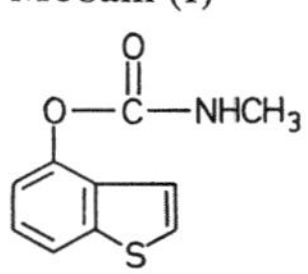	4-Hydroxy-benzo[b]thiophene 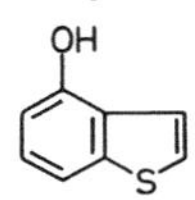	Rumen bacteria	232)

Table 10. (continued)

Pesticide	Metabolite (s)	Microorganism or Source	Refs.
Carbofuran (I) 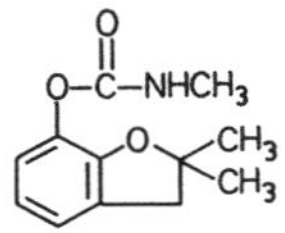	"3-Hydroxycarbofuran" 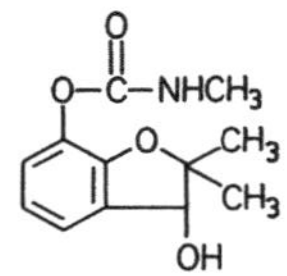	*Penicillium* sp.	233)
		Bacillus sp.	234)
		Pseudomonas cepacia	235)
		Nocardia sp.	
	"Carbofuran phenol" 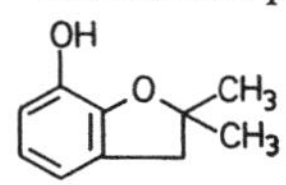	*Bacillus* sp.	234)
		Azospirillum lipoferum	
		Streptomyces ssp.	236)
		Nocardia sp.	235)
		Pseudomonas cepacia	
		Achromobacter sp.	237)
Carbaryl (I) 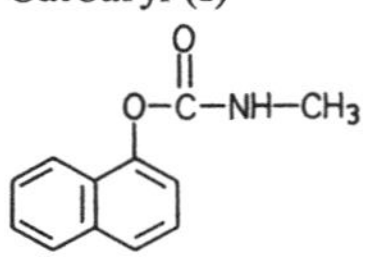	1-Naphthol 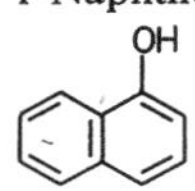	Gram-negative coccus	238)
		Gram-positive bacterium	
		Fusarium solani	
		Aspergillus terreus	
	1-Naphthol	*Rhodococcus* sp.	91)
	Methylamine, CO_2	*Pseudomonas* sp.	
	1-Naphthyl N-hydroxymethyl carbamate 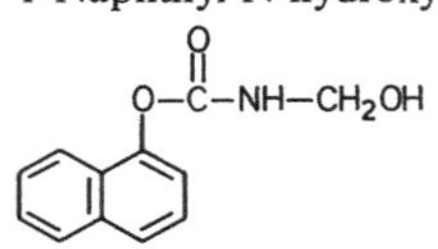	*Gliocladium roseum*	239)
		Aspergillus terreus	240)
	4-Hydroxy-1-naphthyl methylcarbamate 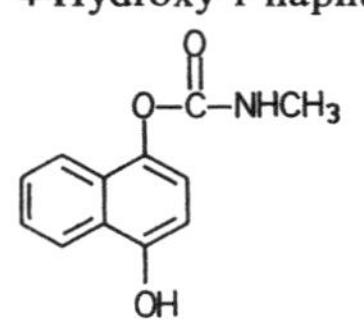	*Gliocladium roseum*	239)

	5-Hydroxy-1-naphthyl methylcarbamate 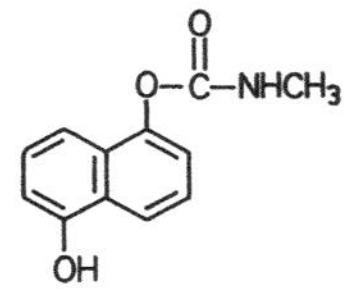		
	1-Naphthylcarbamate 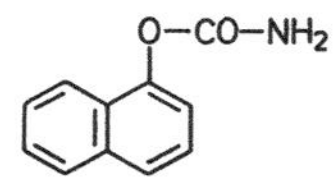	*Aspergillus terreus*	240)
Baygon (I) 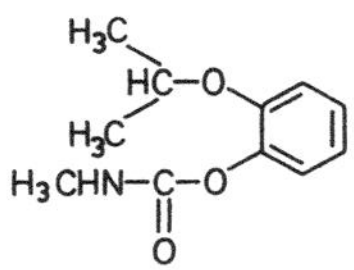	2-(Isopropoxy)phenol 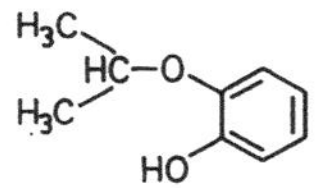	*Pseudomonas* sp.	118)
Mexacarbate (I) 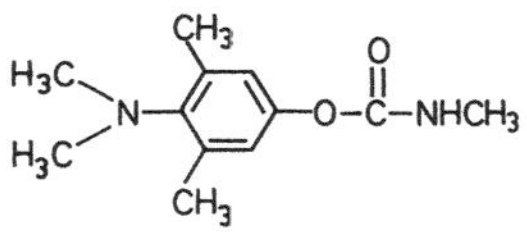	4-Dimethylamino-3,5-dimethylphenol 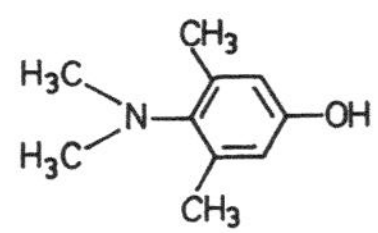	*Trichoderma viride* Bacterium H3	241)
	Methyl-(4-methylamino-3,5-dimethylphenyl)carbamate 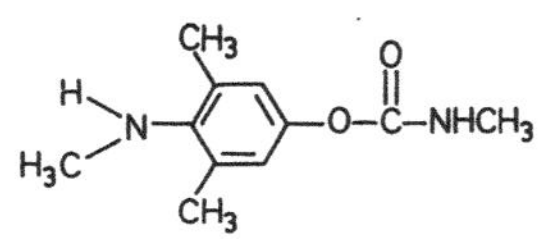		

Table 10. (continued)

Pesticide	Metabolite (s)	Microorganism or Source	Refs.
Aldicarb (I) 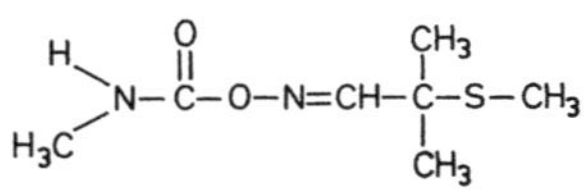	"Aldicarb sulfoxide" 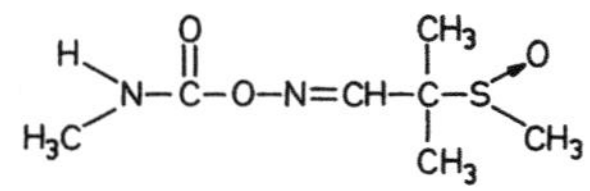 "Aldicarb sulfone" 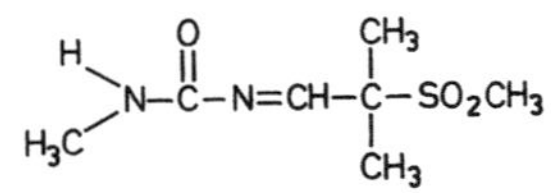	Soil	242)
	"Aldicarb sulfoxide" 2-Methyl-2-methylsulfinyl-propionitrile 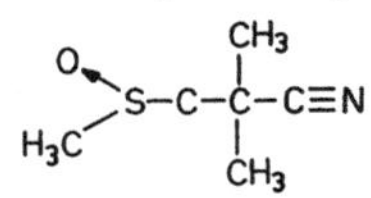 2-Methyl-2-methylsulfinyl-propanal oxime 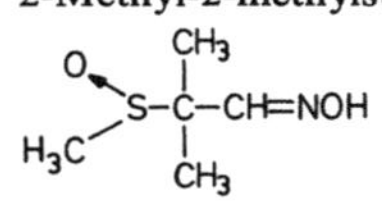	*Gliocladium catenulatum* *Cunninghamella elegans* *Trichoderma harzianum*	67)

Table 11. Miscellaneous fungicides and other pesticides

Pesticide	Metabolite (s)	Microorganism or Source	Refs.
Chloroneb (F) 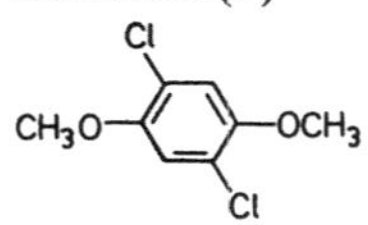	2,5-Dichloro-4-methoxyphenol 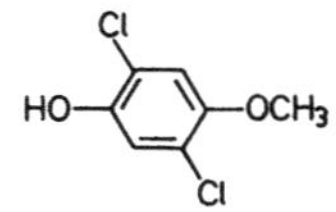	*Rhizoctonia solani* *Fusarium* sp. *Trichoderma viride* *Mucor ramannianus*	243) 244
	2,5-Dichlorohydroquinone 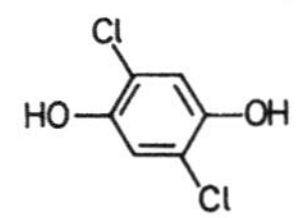	*Cephalosporium gramineum* *Rhizoctonia solani* *Mucor ramannianus* *Fusarium*	244
PCNB (F) 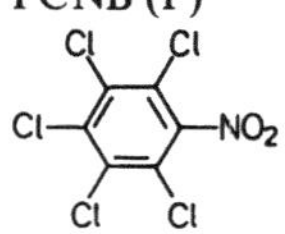	Pentachloroaniline 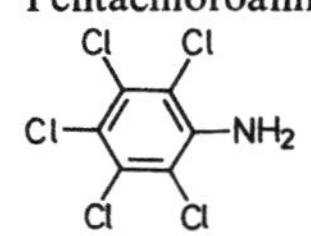	Soil *Streptomyces aureofaciens* Several fungi	245) 246)
	Pentachlorobenzene 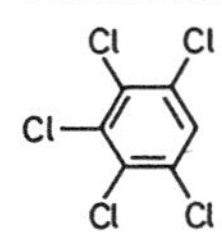	*Fusarium oxysporum* *Fusarium oxysporum* f. *lycopersici* Soil *Fusarium oxysporum* f. *lycopersici*	134) 247) 248) 249)
	Methylthio-pentachlorobenzene 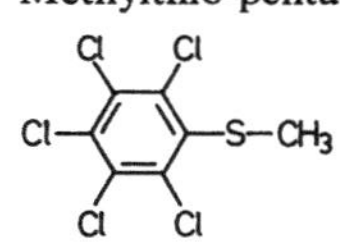	*Trichoderma lignorum* *Fusarium oxysporum* f. *lycopersici*	247)

Table 11. (continued)

Pesticide	Metabolite (s)	Microorganism or Source	Refs.
Botran (F) 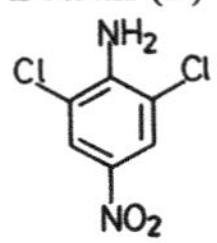	4-Amino-2,6-dichloroacetanilide 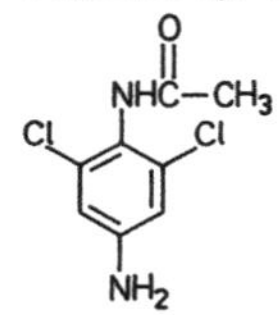	Soil	250)
Chlorphenamidine (A) 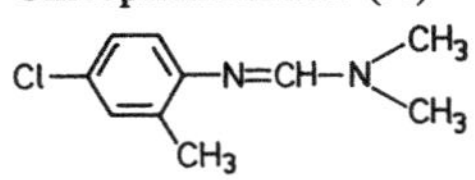	N-Formyl-4-chloro-2-methylaniline 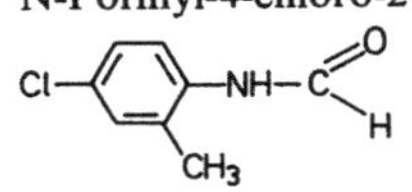	*Aerobacter aerogenes* *Serratia marcescens* *Fusarium moniliforme* *Rhizopus nigricans*	251)
	4-Chloro-2-methylaniline 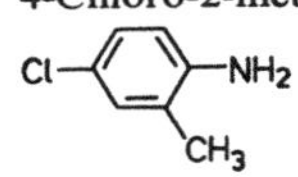	*Streptomyces griseus*	251)
Dexon (F) 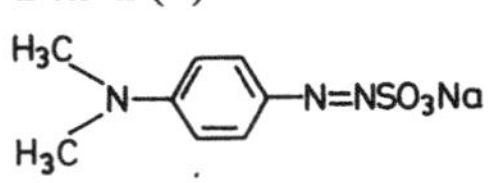	N,N-Dimethyl-1,4-diaminobenzene 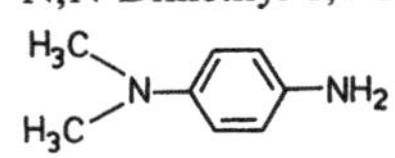	*Pseudomonas fragi* BK9	252)
Carbendazim (F) 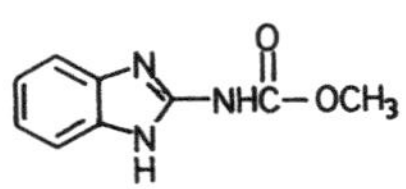	Methyl-(5-hydroxybenzimidazol-2-yl)-carbamate 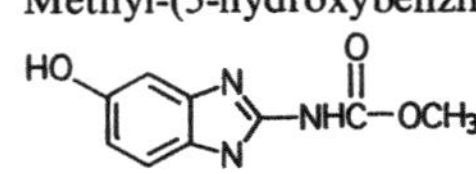	*Aspergillus nidulans*	252)
l hiophanate-methyl (F) 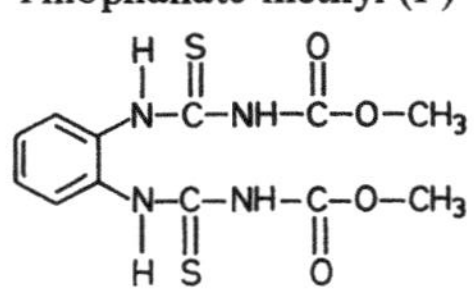	Methyl-(5-hydroxybenzimidazol-2-yl)-carbamate	*Alternaria mali*	253)

Tetrachloroisophthalonitrile TCIN (F) 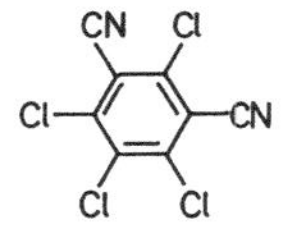	Pentachlorothioanisole 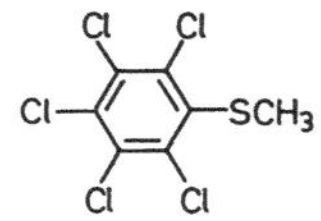	*Saccharomyces pastorianus*	254)
Captan (F) 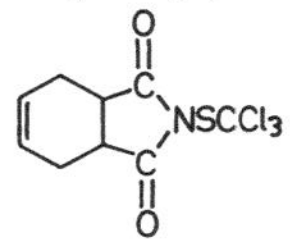	Thiol compounds Carbonyl sulfide O=C=S	*Neurospora crassa*, spores	255)
Folpet (F) 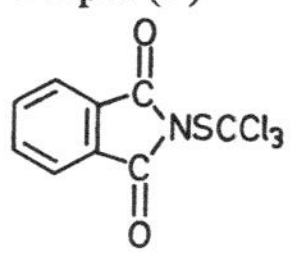	Phthalimide 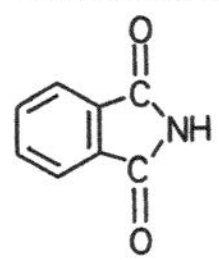	*Saccharomyces pastorianus*	256)
Triadimenol (F) 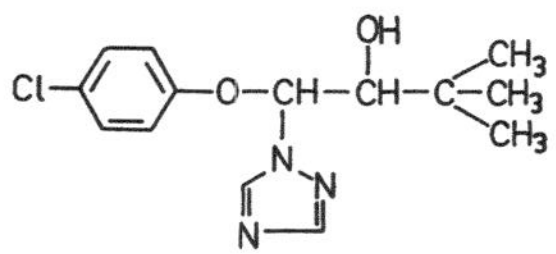	4-(4-Chlorophenoxy)-2,2-dimethyl-4-(1H-1,2,4-triazol-1-yl)-1,3 butandiol 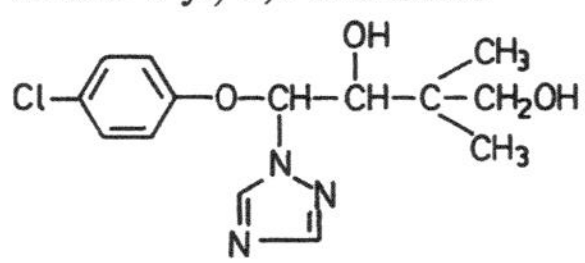	*Aspergillus niger* ATCC 24794	257)
O,O-Diphenyl dichloromethyl phosphonate (F) 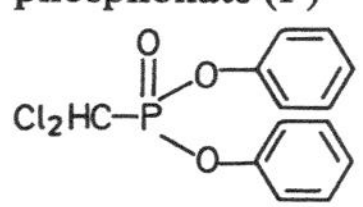	O-Phenyl-dichloromethyl phosphonic acid 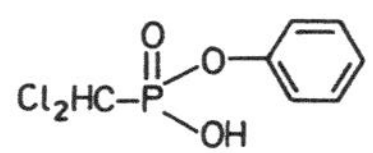	*Pyricularia oryzae*	258)

Table 11. (continued)

Pesticide	Metabolite (s)	Microorganism or Source	Refs.
	Dichloromethyl phosphonic acid 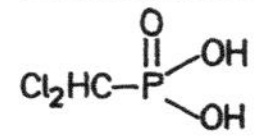		
Nabam (F) 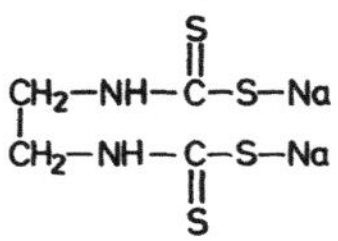	5,6-Dihydro-3H-imidazo[2,1]-1,2,4-dithiazole-3-thione 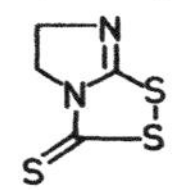 2-Thiono-1,3-diazolidine (ETU) 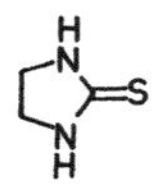	*Pseudomonas fluorescens*	259)
Dimethyldithiocarbamate (Na, Fe, Zn salt) DDC (F) 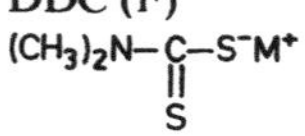	4-Dimethyldithiocarbamoyl-2-aminobuyric acid 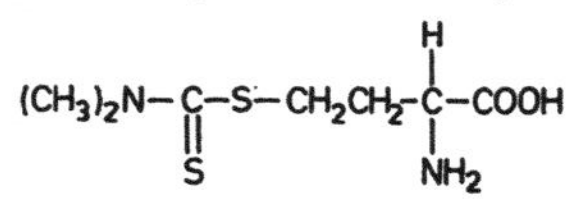 4-Dimethyldithiocarbamoyl-2-oxobutyric acid 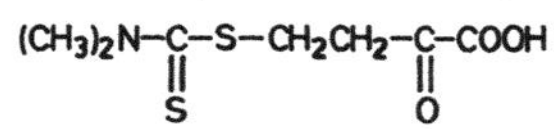	*Bacterium coli* *Saccharomyces cerevisiae* *Hansenula anomala* *Glomerella cingulata* *Aspergillus niger* *Cladosporium cucumerinum*	259)
TMTD (F) 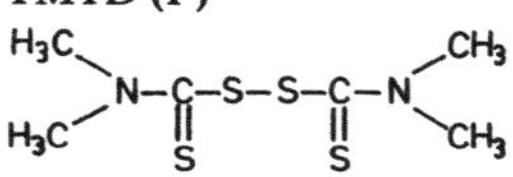	Carbondisulfide S=C=S 1,3-Dimethylthiourea 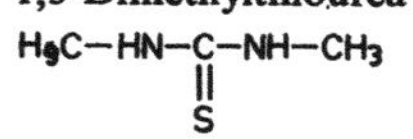	Soil	216)

	Dimethylamine Bis-(dimethylthiocarbamoyl) sulfide 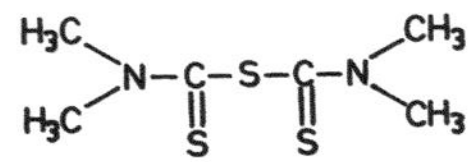		
Molinate (H) 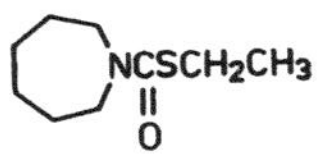	Molinate-sulfoxide Molinate ethanol Molinate acetic acid 4-Hydroxymolinate 3-Hydroxymolinate 4-Oxomolinate 3-Oxomolinate 2-Oxomolinate CO_2	*Mycobacterium* sp. *Flavobacterium* sp. *Streptomyces* sp. *Fusarium* sp.	260)
Permethrin (I) 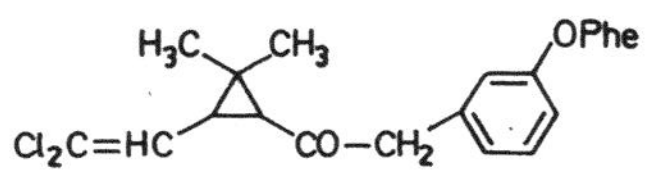	3-Phenoxybenzyl alcohol 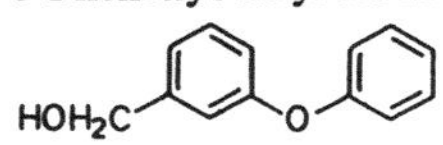	Soil	261)
Cypermethrin (I) 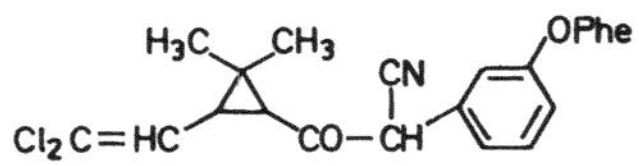	3-Phenoxy benzoic acid 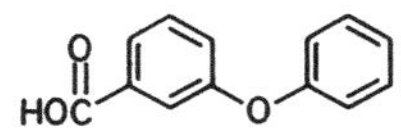 CO_2	Soil	262)

Table 12. Organophosphorus compounds

Pesticide	Metabolite (s)	Microorganism or Source	Refs.
Parathion (I) and some related organophosphorus compounds (e.g., methyl parathion, chlorthion) 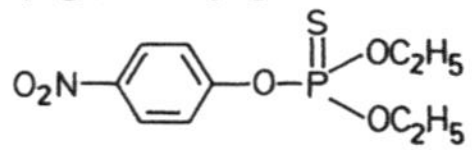	"Aminoparathion" 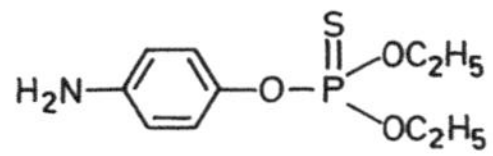	Rumen fluid	263, 264)
		Bacteria from water (Bacillus subtilis)	265, 266)
		Rhizobium japonicum melioti	267)
		Mixed bacterial culture	43)
		Penicillium waksmanni	268)
	4-Nitrophenol 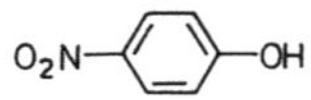	Rumen fluid	266)
		2 *Rhizobium* species	267)
		Flavobacterium sp.	43)
		Mixed bacterial culture	
		Pseudomonas sp. ATCC 29353	269)
		Flavobacterium sp. ATCC 27551	
	O,O-Diethyl-O-(4-nitrophenyl)-phosphate 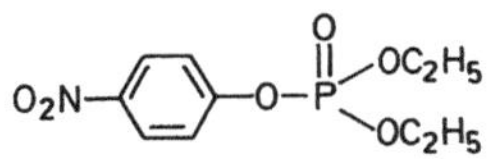	Mixed bacterial culture (*Pseudomonas, Brevibacterium, Azotomonas* sp., *Xanthomonas*)	43)
	4-Aminophenol 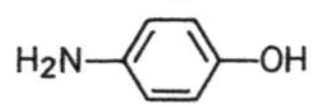		
	4-Nitrophenol		
	Hydroquinone 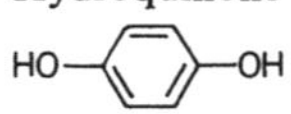		
Fenitrothion (I) 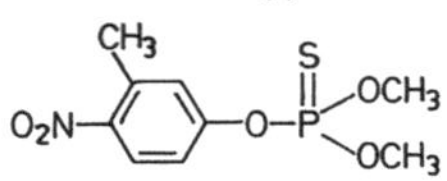	"Aminofenitrothion" 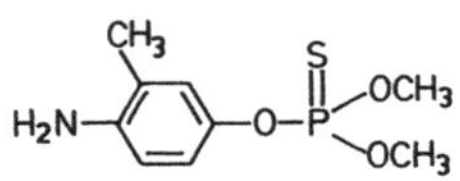	*Bacillus subtilis*	270)
	3-Methyl-4-nitrophenol 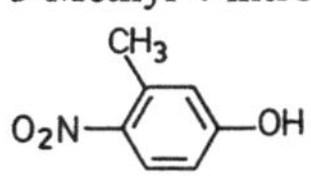	*Flavobacterium* sp. ATCC 27551	269)
		Trichoderma viride	271)

Fensulfothion (I) 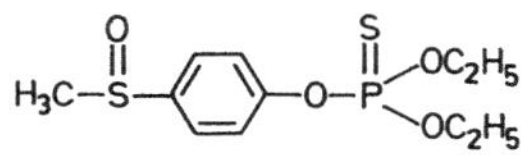	"Fensulfothion sulfide" 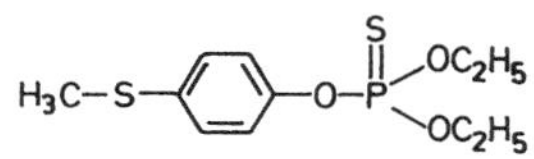	*Klebsiella pneumoniae* *E. coli* *Pseudomonas fluorescens* *Nocardia opaca*	272)
		Hafnia sp.	60)
	4-Methylsulfinylphenol 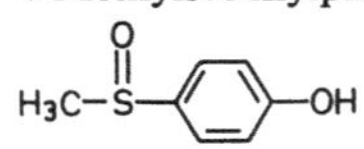	*Pseudomonas alcaligenes* *Alcaligenes* sp.	273)
	4-(Methylthio)-phenol 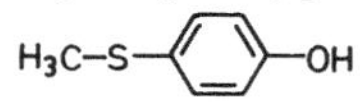	*Hafnia* sp.	65)
Fenthion (I) 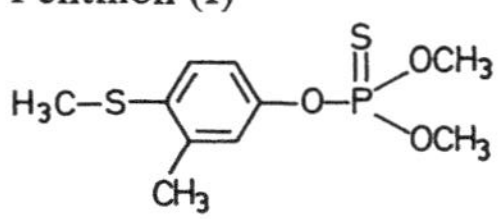	"Fenthion sulfoxide" 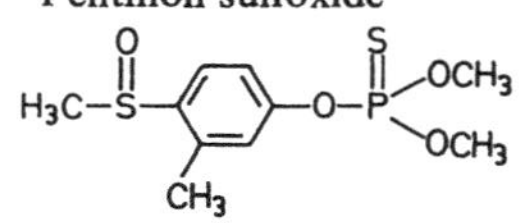 "Fenoxon sulfoxide" 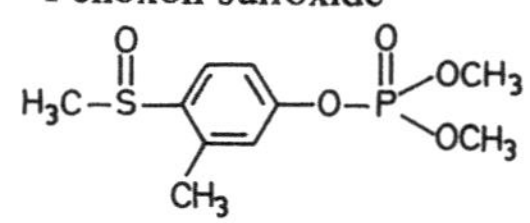 3-Methyl-4-(methylsulfinyl)phenol 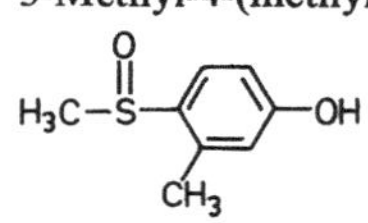	*Rhizopus japonicus*	274)
Nemacur (I) 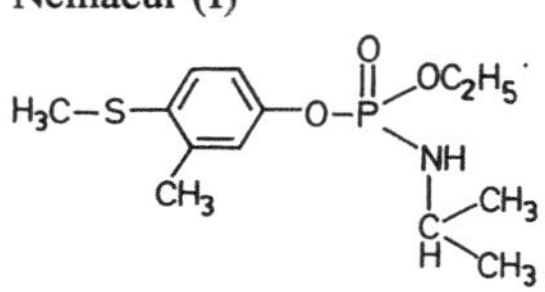	3-Methyl-4-(methylthio)-phenol 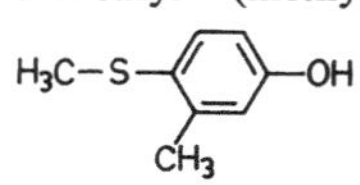 3-Methyl-4-(methylsulfinyl)phenol	Soil	275)

Table 12. (continued)

Pesticide	Metabolite (s)	Microorganism or Source	Refs.
	3-Methyl-4-(methylsulfonyl)-phenol 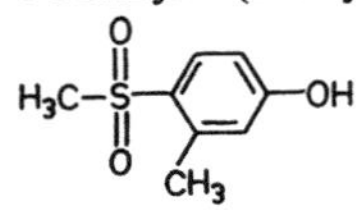		
Fonofos (I) 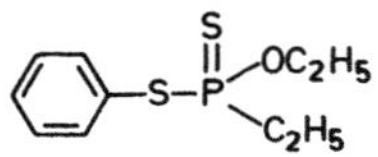	O-Ethyl-S-phenyl ethanephosphonothioate 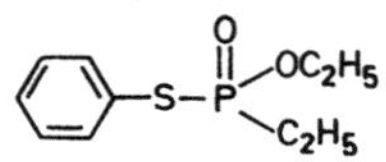 Thiophenol 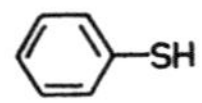 Diphenyldisulfide 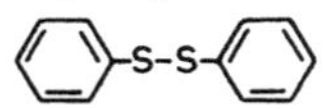	*Rhizopus arrhizus* *Mucor plumbeus* *Rhizopus japonicus*	44) 276)
Fenchlorphos (I) 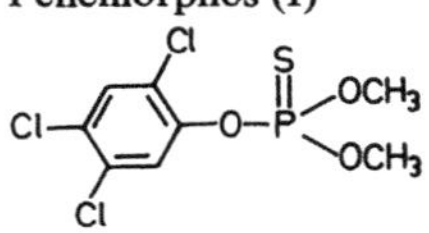	"Desmethyl-Fenchlorphos" 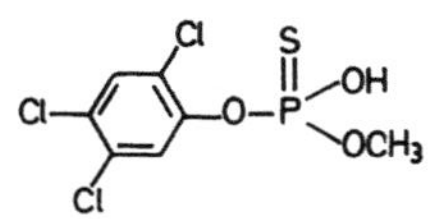	Rumen fluid	218)
Chlorfenvinphos (I) 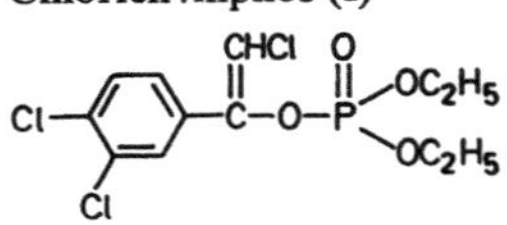	"Desethylchlorfenvinphos" 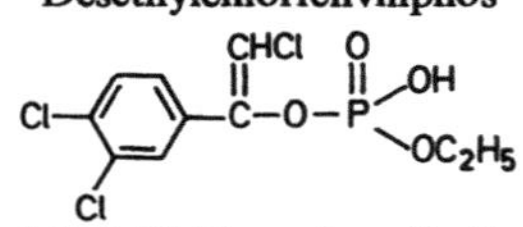 1-(3,4-Dichlorophenyl)-ethanol 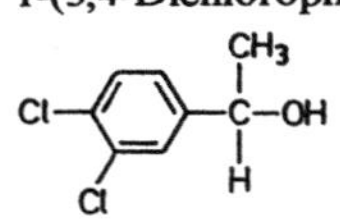	Soil	264)

	3,4-Dichloracetophenone; some related minor metabolites 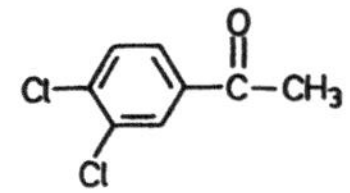		
Diazinon(I) 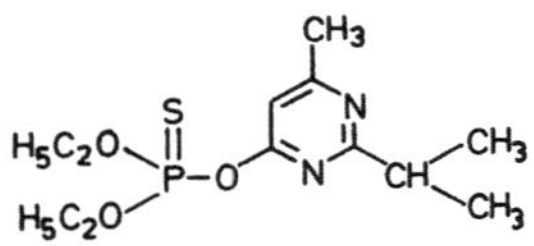	4-Hydroxy-2-isopropyl-6-methylpyrimidine 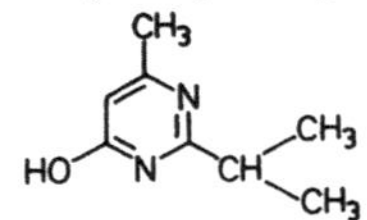	*Flavobacterium* sp. *Flavobacterium* sp. ATCC 27551 *Pseudomonas* sp. ATCC 29353	277) 269)
Azinphosmethyl (I) 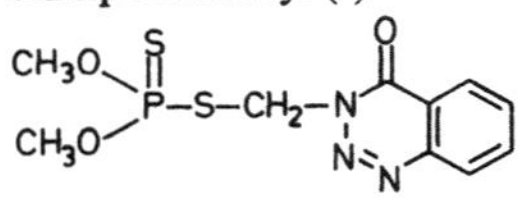	bis-(Benzazimidyl-methyl)-disulfide 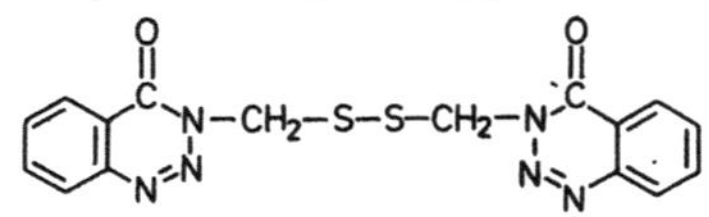	*Pseudomonas fluorescens* DSM 1976	121)
	Benzazimide 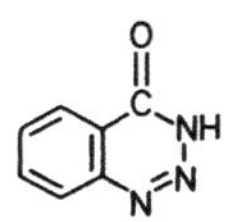		
	Anthranilic acid 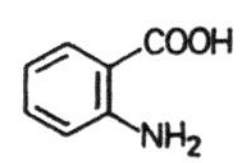		
Edifenphos 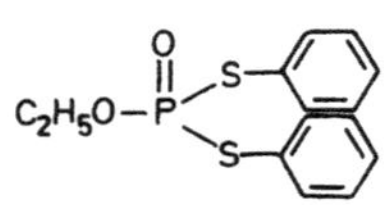	O-Ethyl-S-(4-hydroxyphenyl)-S-phenyl-dithiophosphate 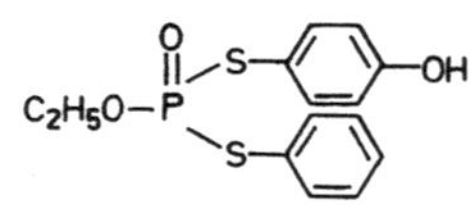	*Pyriculuria oryzae*, mycelial cells	278)

Table 12. (continued)

Pesticide	Metabolite (s)	Microorganism or Source	Refs.
	O-Ethyl-S-phenylthio-phosphate 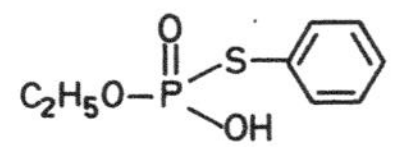		
	S,S-Diphenyl-dithiophosphate 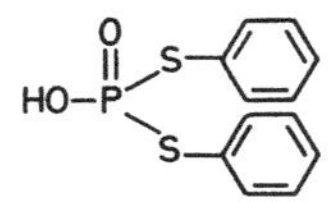		
Metasystox R (I) 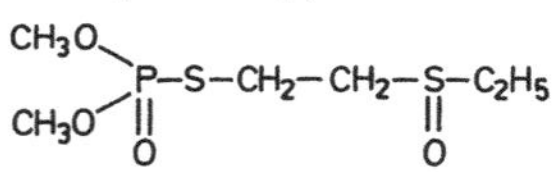	2-(Ethylsulfinyl)ethanethiol 2-(Ethylsulfonyl)ethanethiol Bis[2-(ethylsulfinyl)-ethyl]disulfide Bis[2-(ethylsulfonyl)-ethyl]disulfide Bis[2-(ethylsulfinyl)-ethyl]sulfide Bis[2-(ethylsulfonyl)-ethyl]sulfide	*Pseudomonas putida* 1453 *Nocardia* sp. DSM 43252	279)
Malathion (I) 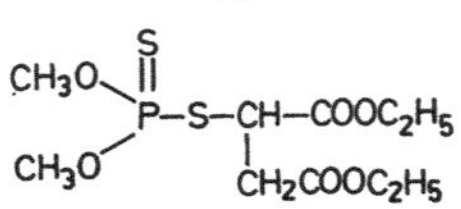	"Desmethylmalathion" Ester hydrolysis product of malathion	*Pseudomonas* *Trichoderma viride*	280)
	"Malathiondicarboxylic acid" 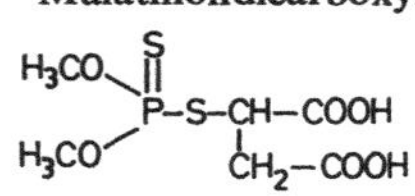	*Flavobacterium meningosepticum* *Xanthomonas* sp. *Comomonas terrigeri* *Pseudomonas striata*	281)
	Diethyl maleate 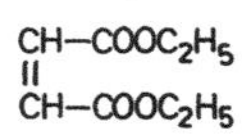	*Pseudomonas striata*	281)

	"Malathion-β-monocarboxylic acid" 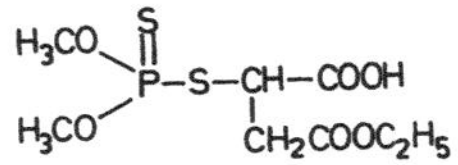		*Pseudomonas striata* *Aspergillus orycae* *Arthrobacter* sp. Salt marsh org.	281) 282) 283) 284)
Trichlorphon (I) 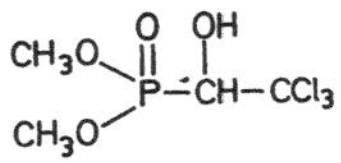	O-Methyl-(2,2,2-trichloro-1-hydroxyethyl) phosphonate 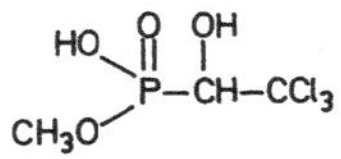	(2,2,2-Trichloro-1-hydroxy-ethyl) phosphonic acid 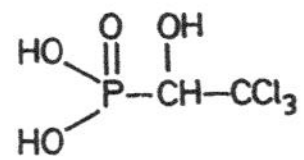	*Aspergillus niger* *Penicillium notatum* *Fusarium*	285)
Dichlorvos (I) 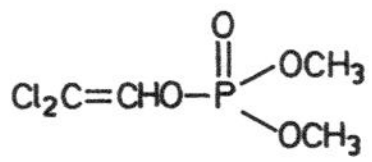	Dichloroethanol Dichloroacetic acid Ethyldichloro acetate		Sewage sludge	286)

Table 13. Phenols derived from pesticides

Pesticide	Metabolite (s)	Microorganism or Source	Refs.
4-Nitrophenol 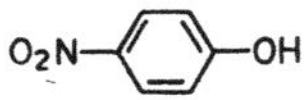	4-Nitroacatechol Hydroquinone 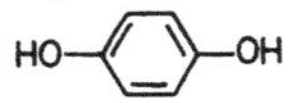	*Flavobacterium* sp. *Pseudomonas* sp.	287) 43)
4-(Methylthio)phenol (X = $-SCH_3$) 4-(Methylsulfinyl)phenol (X = $-\overset{\|\|}{S}CH_3$ with =O) 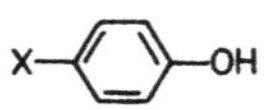	Mineralization 4-(X)-catechol X = Methylthio, Methylsulfinyl 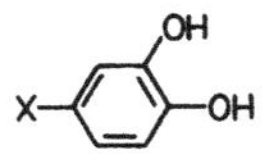 6-Oxo-2-hydroxy-5-(X)-cis, cis-2,4-hexadienoic acid X = methylthio, methylsulfinyl 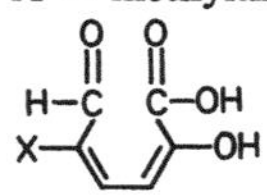	*Bacillus* sp. *Nocardia* DSM 43251	288) 27)
3-Methyl-4-(methylthio)phenol 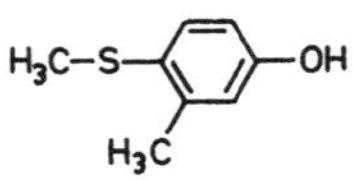	3-Methyl-4-(methylsulfinyl)phenol 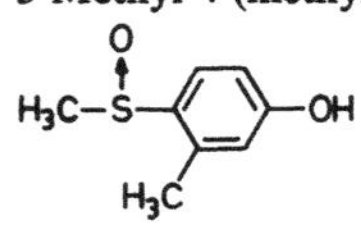 3-Methyl-4-(methylthio)catechol 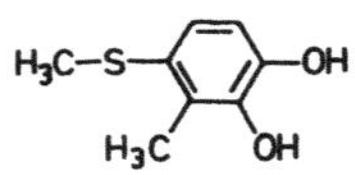 4-Methyl-5-(methylthio)-catechol 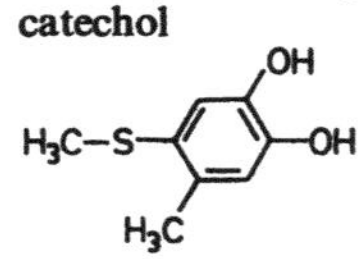 2,5-Dihydro-2-methyl-3-(methylthio)-5-oxofuran-2-acetic acid 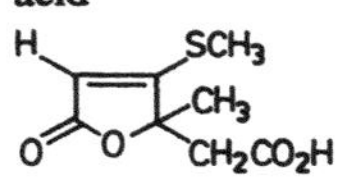	*Nocardia* DSM 43251	28)

	2,5-Dihydro-3-(methylthio)-4-methyl-5-oxo-furan-2-acetic acid 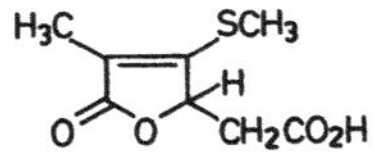		
2-Methyl-4-chlorophenol 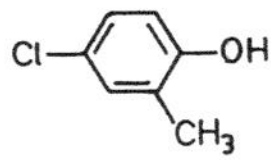	3-Methyl-5-chlorocatechol 2,5-Dihydro 3-methyl-2-oxo-furan-2-acetic acid 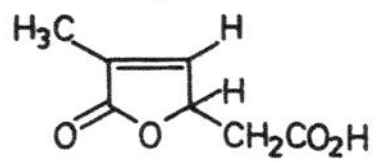	*Nocardia* sp. DSM 43252	289)
2,4-Dichlorophenol 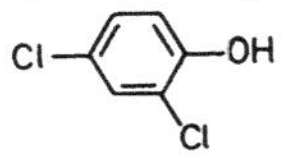	3,5-Dichlorocatechol *cis,cis*-2,4-Dichloromuconic acid 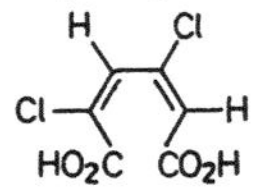	*Nocardia* sp. DSM 43252 *Arthrobacter* sp.	289)
	2,5-Dihydro-3-chloro-2-oxo-furan-5-acetic acid 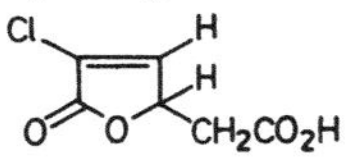	*Arthrobacter* sp.	
2,4-5-Trichlorophenol 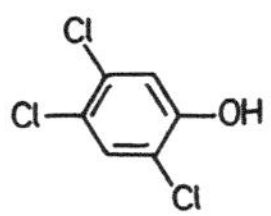	3,4,6-Trichlorocatechol	*Pseudomonas* sp. *Achromobacter* *Nocardia restricta* DSM 43199	289)

Table 14. Phenoxyherbicides

Pesticide	Metabolite (s)	Microorganism or Source	Refs.
2,4-D 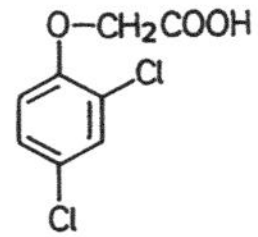	6-Hydroxy-2,4-dichlorophenoxyacetic acid 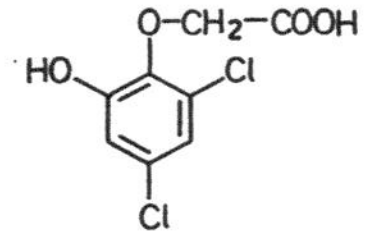	*Arthrobacter*	[26]
		Pseudomonas *Achromobacter*	[290]
	2,4-Dichlorophenol 3,5-Dichlorocatechol	*Nocardia*	[291]
		Arthrobacter, cell-free-extract	
	cis,cis-2,4-Dichloromuconic acid 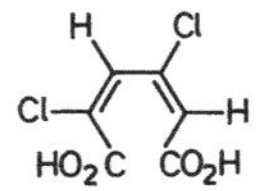 (4-Chloro-5-oxo-2-furanylidene)-acetic acid 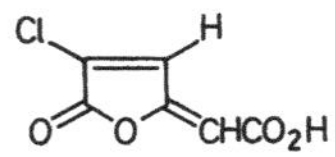 "Chloromaleylacetic acid" 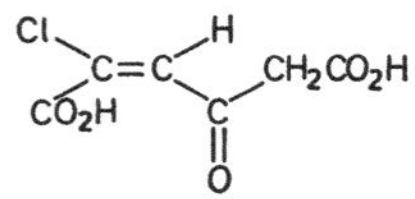	*Arthrobacter*, cell-free-extract	[292]
	Reductive dehalogenation CO_2, CH_4	Sewage sludge	[81]
MCPA 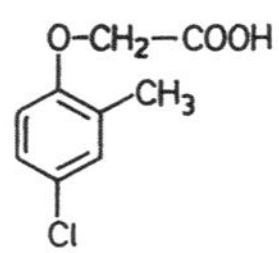	6-Hydroxy MCPA 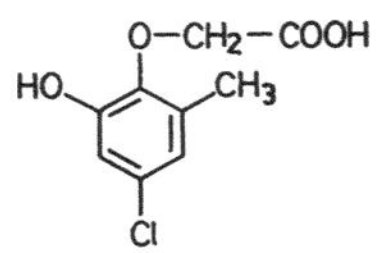	*Arthrobacter*	[26]
		Pseudomonas	[290]
		Flavobacterium pelegrinum	[293, 294]
		Arthrobacter, cell-free-extract	
		Aspergillus niger	

	4-Chloro-2-methylphenol 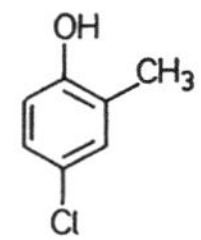	5-Chloro-3-methylcatechol 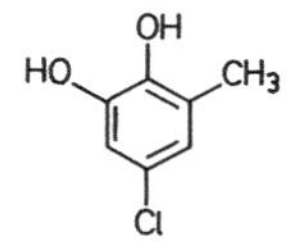		
4-Chlorophenoxyacetic acid 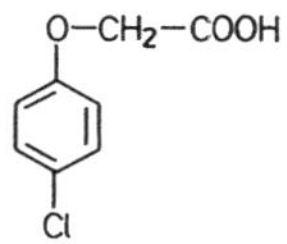	2-Hydroxy-4-chlorophenoxyacetic acid 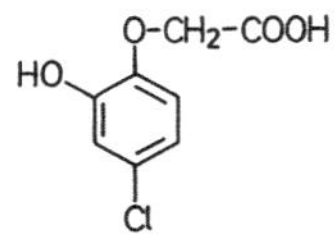		*Arthrobacter* *Pseudomonas*	295, 26) 290)
	4-Chlorophenol 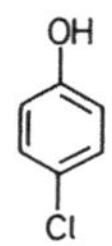	4-Chlorocatechol 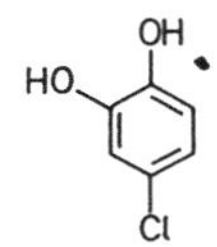		
Sesone 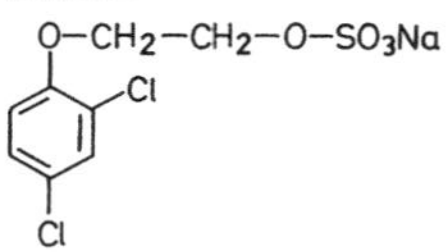	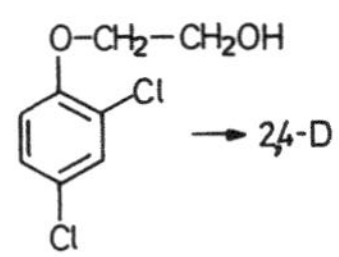		Soil *Bacillus cereus*	290, 26)
2,4,5-T 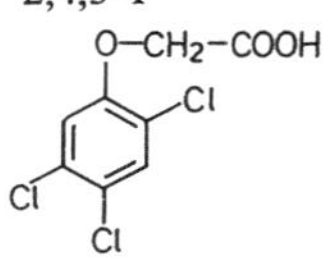	3,5-Dichlorocatechol 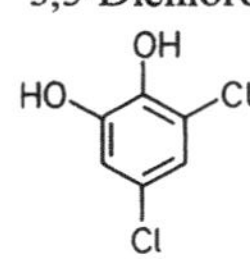		*Brevibacterium*	6)

Table 14. (continued)

Pesticide	Metabolite (s)		Microorganism or Source	Refs.
	2,4,5-Trichlorophenol 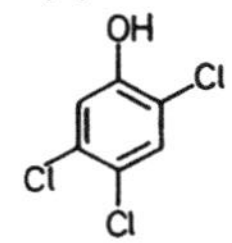		*Pseudomonas* sp.	296)
			Achromobacter sp.	
			Pseudomonas cepacia AS 1100	297)
	Dechlorination:		Anaerobic methanogenic Bacteria	298)
	2,5-Dichlorophenoxyacetic acid		*Pseudomonas cepacia* AC 1100	299)
	Complete dechlorination and mineralization			
	Reductive dehalogenation		Sewage sludge	81)
	CO_2, CH_4			
Dichlorophenoxyalkanoic acids 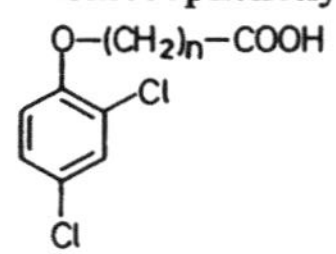	2,4-Dichlorophenol 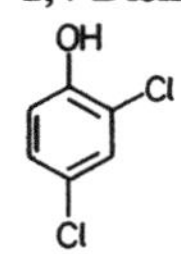		*Flavobacterium*	290, 26)
	Even acids (n = 3,5,7) β-oxidation	2,4-Dichlorophenoxyacetic acid	*Nocardia coeliaca*	290, 26)
	Odd acids (n = 2,4,6) β-oxidation	2,4-Dichlorophenol	Mixed soil microflora	

Table 15. Substituted urea herbicides

Pesticide	Metabolite (s)	Microorganism or Source	Refs.
Chloroxuron 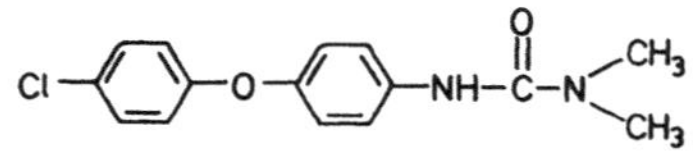	3-[4-(4′-Chlorophenoxy)-phenyl]-1-methylurea 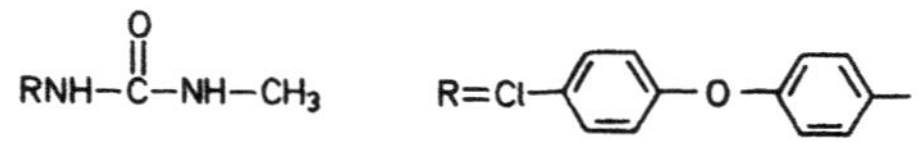 3-[4-(4′-Chlorophenoxy)-phenyl]urea 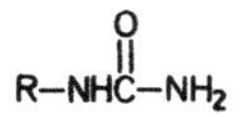 4-(4′-Chlorophenoxy)-aniline 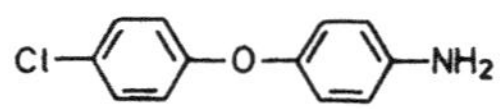	Enriched bacterial and fungal suspension from soil	300, 29)
Monuron 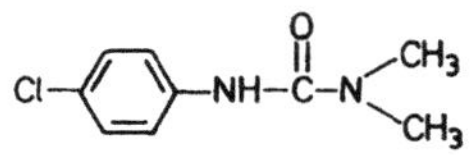	3-(4-Chlorophenyl)-1-methylurea 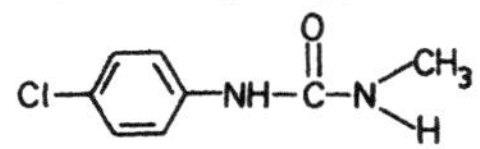	*Rhizopus japonicus*	301
Buturon 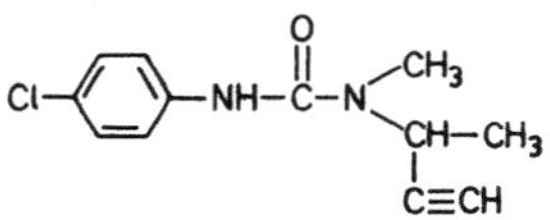	3-(4-Chlorophenyl)-1-methylurea	*Rhizopus japonicus*	301)
	3-(4-Chlorophenyl)-1-(3-butin-3-yl)urea 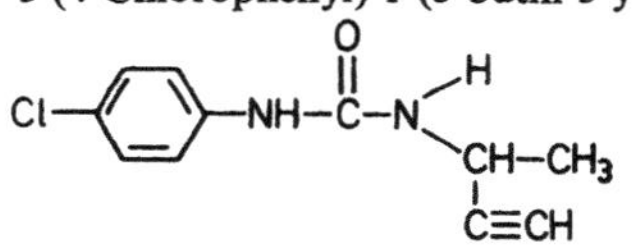 3-(4-Chlorophenyl)-1-methyl-1-(1-buten-3-yl)urea 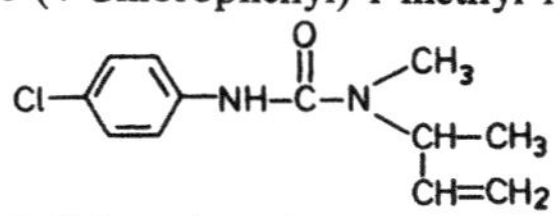 4-Chlorophenyl urea 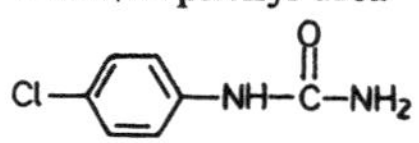	*Cunninghamella echinulata* Thaxter	302)

Table 15. (continued)

Pesticide	Metabolite (s)	Microorganism or Source	Refs.
Diuron 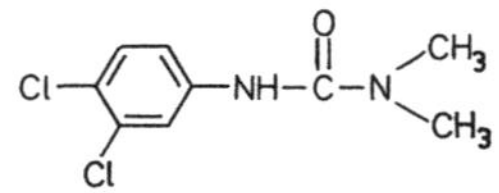	3,4-Dichloroaniline 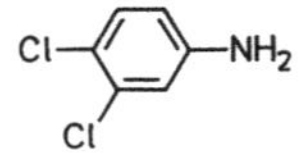	Soil	300, 29)
Siduron 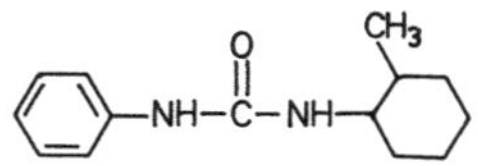	3-Phenyl-1-(4-hydroxy-2-methylcyclohexyl) urea 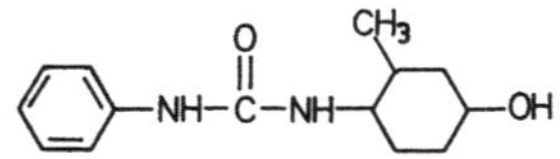	*Pseudomonas* Fungus imperfectus	29)
	3-(4-Hydroxyphenyl)-1-(2-methylcyclohexyl)urea 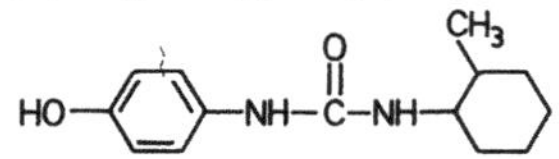		
	3-(4-Hydroxyphenyl)-1-(4-hydroxy-2-methylcyclohexyl)urea 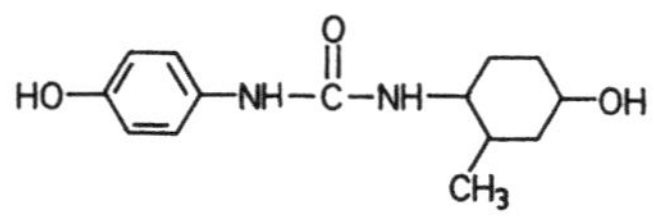		
Fluometuron 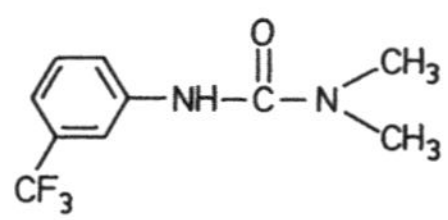	3-(3-Trifluoromethylphenyl)-1-methylurea 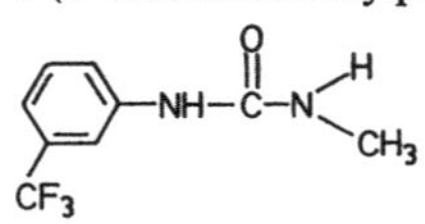	*Rhizopus japonicus*	301)
Monolinuron 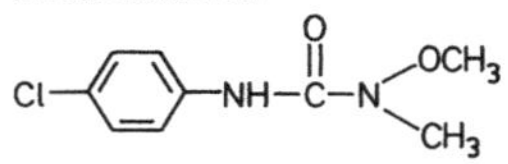	3-(4-Chlorophenyl)-1-methoxyurea 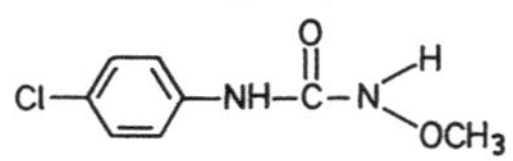	*Rhizopus japonicus*	301)

	4-Chloroaniline 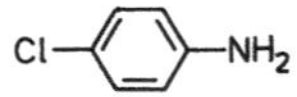		Soil *Bacillus sphaericus*	300, 29)
Linuron 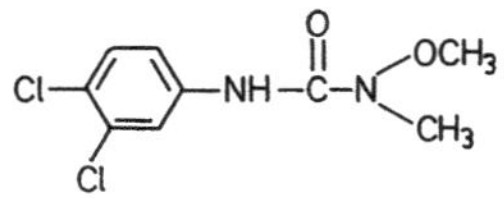	3-(3,4-Dichlorophenyl)-1-methoxyurea 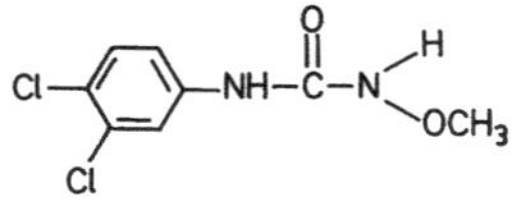		*Cunninghamella echinulata* Thaxter	302)
	3-(3,4-Dichlorophenyl)-1-methylurea 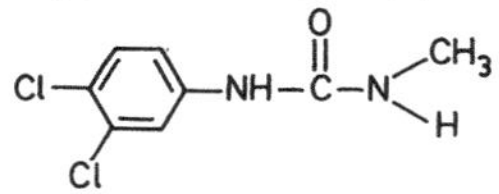			
	3-(3,4-Dichlorophenyl)-1-hydroxymethylurea 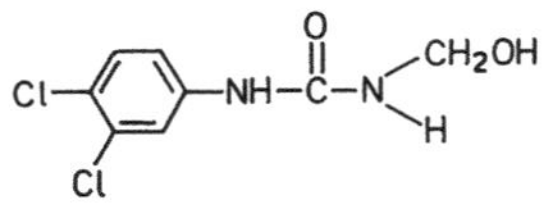			
	3,4-Dichloroaniline	N,O-Dimethylhydroxylamine	*Bacillus sphaericus*, cell-free extract	137)
Metobromuron 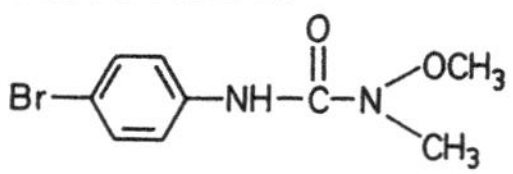	4-Bromoaniline 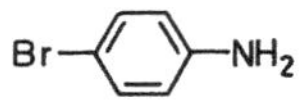		*Bacillus sphaericus*	29)
	3-(4-Bromophenyl)-1-methoxyurea 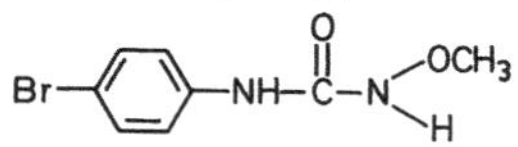		*Talaromyces wortmanni*	137)
	3-(4-Bromophenyl)-1-methylurea 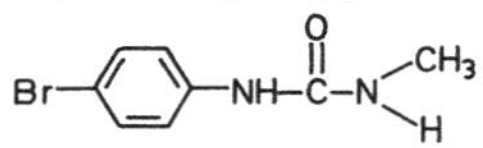			
	4-Bromophenylurea 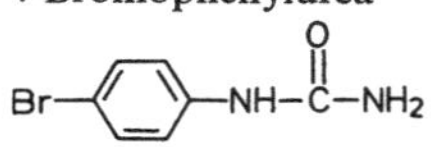			

Table 15. (continued)

Pesticide	Metabolite (s)	Microorganism or Source	Refs.
	4-Bromoacetanilide 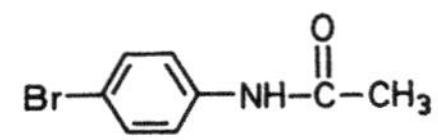		
Chlorbromuron 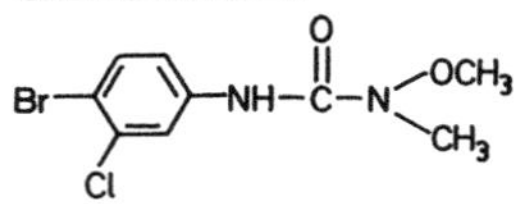	3-Chloro-4-bromoaniline 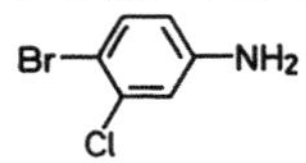	*Bacillus sphaericus*	29)
Methabenzthiazuron 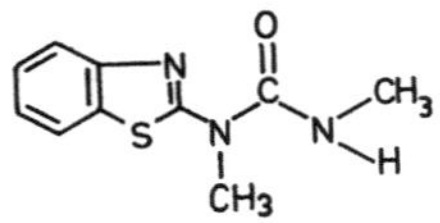	3-(2-Benzothiazolyl)-1-methylurea 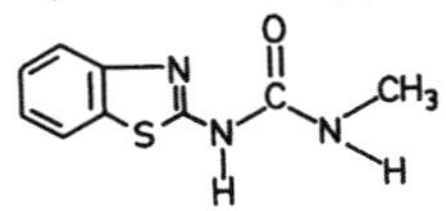	*Cunninghamella echinulata* Thaxter *Hypocrea* Cf-*pilulifera*	303) 304)
	3-(6-Hydroxy-2-benzothiazolyl) 1,3-dimethylurea 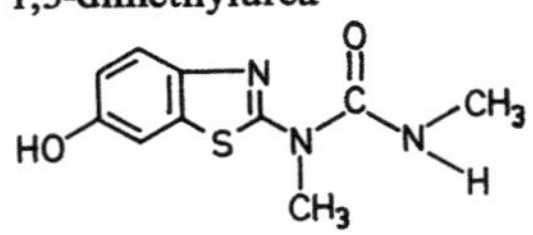		
	3-(2-Benzothiazolyl)-1-(hydroxymethyl)-3-methylurea 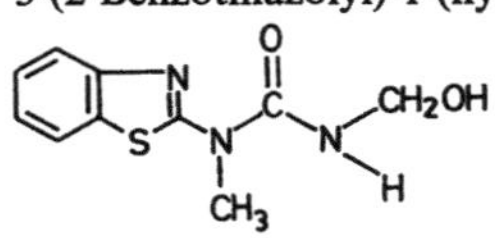	*Hypocrea* Cf. *pilulifera*	304)
	1-(2-Benzothiazyl)-1-methylurea 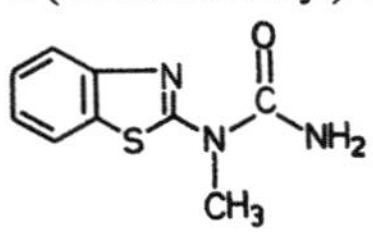		

Isouron 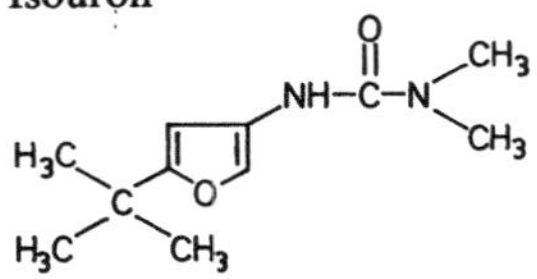	4-Amino-6-(1,1-dimethylethyl)-2(H) oxazin-2-one 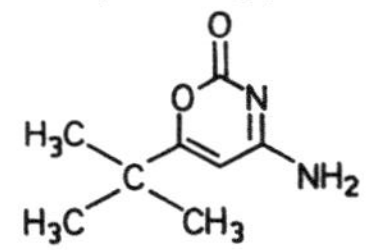	*Hansenula saturnus*	305)
	1-(1-Amino-4,4-dimethyl-3-oxo-1-pentenyl)-3,3-dimethylurea 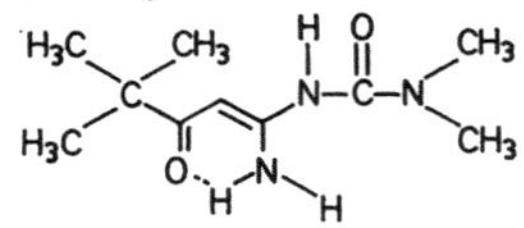	*Hansenula saturnus* *Pseudomonas putida*	305) 306)
	1-(4,4-Dimethyl-1,3-dioxopentyl)-3,3-dimethylurea 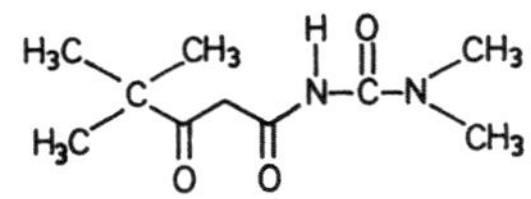		

The generally used common name and structure or formula are given for each compound. Whenever possible, the name suggested in Frear's Pesticide Index (4th ed., College Science Publishers. State College, Pennsylvania, 1969) is used. If available, recent review references are given, rather than individual papers.

Metabolic transformations by microorganisms are given only when the structures of the metabolites have been established with reasonable certainty. Care has to be taken, particularly in soil studies, that physical and chemical changes of pesticides (photochemical breakdown, hydrolysis) are not mistaken for microbial metabolism. Metabolites isolated from soil are listed in the table only if some evidence is presented for the biological nature of the change.

6 Significance of Pure Culture Studies with Microorganisms for the Ecotoxicological Assessment of Pesticides

6.1 Use of Microorganisms as Source for Pesticide Transformation Products

The use of selected microorganisms for laboratory culture studies of pesticide metabolism, as an aid to the interpretation of field data, is well established. They may also be used, however, as a tool for the production of transformation products of potential value in chromatographic identification of unknown products in extracts of soil. The solvents necessary for the efficient extraction of pesticides and their products from soil and aquatic sediments also extract a large amount of extraneous non-pesticidal materials. Very often these natural soil constituents prevent the isolation and purification of sufficient quantities of the pesticidal products for identification by gas-, high pressure liquid chromatography (HPLC), mass- and nuclear magnetic resonance (NMR) spectroscopy or render it at least more difficult. Preliminary identification by co-chromatography may then be necessary. Should speculative synthesis fail to produce reference compounds it may be necessary to find alternative sources of reference compounds. Such sources should enable the isolation and purification of sufficient quantities of any pesticide transformation product formed. They include aqueous or organic solutions subjected to irradiation, culture media incubated with pure or mixed microbial cultures, and eluates from perfusion columns or from simulated soil systems.

6.2 Use of Microorganisms and Microbial Enzymes for the Degradation of Pesticides

Unwanted discharge of xenobiotics, e.g. pesticides containing aromatic or heteroaromatic structures, into the environment can occur numerous sources. Primarily, xenobiotic compounds applied, e.g., to agricultural areas which through drift, water run off, or soil erosion do not remain at the target site undesirably enter the environment and may cause pollution.

Another major source of undesired chemical discharge occurs at the pesticide production and formulation plants as well as in any other plant producing commercial chemicals or drugs.

With regard to controlling discharges from production plants, physical procedures such as solvent extraction or distillation coupled to conventional biological treatment processes are generally not able to meet zero-discharge requirements. To meet this level or low-ppb discharge new technologies are required. Biological systems involving cell-free enzymatic degradation of undesirable chemicals may represent a promising new technology for disposal of dicharges.

The ability of particular bacteria and fungi to change a pesticide molecule into less complex metabolites which may then be further metabolized by the same organism or by a wider range of secondary microorganisms existing in the soil environment, plays an important role in the degradation of xenobiotic chemicals. Generally, once the initial enzymatic degradative step has occured, the substance loses its special toxicity and the metabolites formed can be degraded by a wider range of microorganisms. In research involving development of cell-free enzyme systems for the detoxification of pesticides, the initial enzymatic reaction with the parent molecule is of great importance. If enzymes can be found which can greatly decrease the toxicity of the starting compound by a simple hydrolysis or other enzymatic reactions, then these enzymes could possibly be used as potent tools for the detoxification of pesticides. The possibility of obtaining cell-free enzymes capable of detoxifying various pesticides containing different classes of aromatic and heterocyclic moieties has been described[45]. Many enzymes responsible for aromatic metabolism are encoded on plasmids. Research can thus make use of genetic engineering techniques: Multiple gene copies, multiple expression, and movement of enzymes to a more suitable industrial strain may lead to a potential application of aromatic and heteroaromatic degrading enzymes, e.g., in soil-spill clean-up.

6.3 Manipulation of Degradative Genes of Microorganisms

Since microorganisms are the major degraders and recyclers of organic molecules in nature, their failure to degrade novel compounds like pesticides contributes significantly to the pollution problem. To overcome the inherent limitations in microorganisms, particularly bacteria, in vivo and in vitro genetic engineering techniques are suggested to be used to construct bacteria with novel degradative abilities. The biological degradation of pesticides is a result of catabolism by individual microbial members of bacterial communities or the concerted catabolism by consortia of different microbes. The net result can be the total conversion of a compound to CO_2 and biomass at one extreme or its partial metabolism to less toxic materials at the other. However, there are many compounds which are too recalcitrant to be attacked by natural microorganisms and it is difficult, if not impossible, to enrich cultures which have any effect upon them. The theoretical possibility exists to degrade these compounds by constructing strains through the introduction of a number of enzyme activities from different bacteria into a certain chosen bacterium which under natural conditions might have little or no chance to exchange genetic information. Such hybrid metabolic pathways could be constructed in a number of ways, but the two most promising methods are: a) the cloning of catabolic enzymes onto plasmid or bacteriophage vectors by the techniques of recombinant DNA technology, and b) the use of naturally cloned catabolic DNA located on degradative plasmids[307].

The use of degradative plasmids will probably prove most effective, at least in the near future. This statement is based upon the requirement for a fairly complex sequence of reactions for the complete conversion of the kind of molecules which cause environmental problems, many of which are heterocyclic aromatic compounds. Recombinant DNA technology is still somewhat limited by the size of the DNA fragments which can be inserted into a vector, and this obviously limits the number of genes which can be interdependently cloned. Some plasmids have been found to contain genes for integrated and regulated degradative pathways in a form which can be transferred or mobilized between bacterial hosts. As microbiologists are becoming aware of the existence of such extrachromosomal catabolic genes, more are being discovered. Most degradative plasmid genes found in saprophytic cultures are of esoteric biochemical nature and are likely to be used to degrade the more recalcitrant chemicals.

In order to optimize the construction of strains by these methods a complete understanding of the biochemistry of these pathways, of the regulation and chemical nature of the inducers, the number of operons, and the location of all the relevant genes on the vector genome, is necessary [72].

7 Conclusion

Soil microorganisms are responsible for the degradation and ultimate dissipation of most pesticides from soil. Microorganisms have the potential for adapting to the metabolism of various kinds of synthetic organic chemicals. Many pesticides are degraded by direct attack of the adapted microorganisms. Also enzymatic degradation of pesticides mostly belonging to the class of chlorinated hydrocarbons may occur by cometabolic processes involving coincidental reactions with constitutive enzymes. Some pesticides tend to preceed toward ecologically acceptable products, i.e. natural products which serve as cellular building material for growth of the microbial population involved.

Much of the information on metabolic pesticide degradation existing today has been obtained with isolated cultures of soil microorganisms. Analogies have been made to metabolism in soil when similar products or metabolites are isolated from field soil sample. With few exceptions, the pathways established with isolated cultures can also be recognized in soil [66]. It should be noted that in such a complex system like the soil the numerous and complicated chemical-physical interactions interfering with the soil microbiota will also have a significant effect on the ultimate dissipation of a pesticide and its residues. It should be recognized too that transformation or mineralization of a pesticide may occur in isolated culture systems with much more ease, whereas the incorporation of aromatic or fragment moieties into the soil organic matter as "bound residues" may decrease more extensive degradation of the pesticide molecule.

8 References

1. Bollag J-M (1974) Adv. Appl. Microbiol. *18*: 75
2. Jensen HL (1963) Acta Agr. Scand. *13*: 404
3. Alexander M (1967) in: Agriculture and the Quality of our Environment Publ. No. 85, pp 331–342. Amer. Ass. Advanc. Sci. Washington D.C.

4. Horvath RS (1972) Bacteriol. Rev. *36*: 146
5. Horvath RS (1971) Bull. Environ. Contam. Toxicol. *6*: 537
6. Horvath RS (1971) J. Agric. Food Chem. *19*: 291
7. Horvath RS (1970) Biochem. J. *119*: 871
8. Anderson JR (1978) in: Pesticide Microbiology pp 313. Hill IR, Wright SJL (eds) London: Academic Press
9. Smith WK (1962) J. Appl. Bacteriol. *25*: 83
10. Morill LG, Dawson JE (1964) Proc. Soil Sci. Soc. Am. *28*: 710
11. Sharp RF, Taylor BF (1969) Soil Biochem. *1*: 191
12. Audus LJ (1964) in: The Physiology and Biochemistry of Herbicides, chapter 5. Audus LJ (ed) New York: Academic Press
13. Audus LJ (1950) Nature (London) *166*: 356
14. Gray TRG, Williams ST (1971) Soil Microorganisms. Edinburgh: Oliver & Boyd
15. Hill IR, Arnold DJ (1978) in: Pesticide Microbiology. pp 203. Hill IR, Wright SJL (eds) London: Academic Press
16. Johnson JB, Cain RD (1973) Abs. Commun. Meet. FEBS Abs. No. *137*
17. Cox CS (1968) Nature (London) *220*: 1139
18. Cunningham JL (1973) Cryobiology *10*: 361
19. Davis RJ (1963) J. Bacteriol. *85*: 486
20. Daily WA, Higgins CE (1973) Cryobiology *10*: 364
21. Mazur P (1962) in: Culture Collections: Perspectives and Problems. pp 59 Martin SM (ed) Toronto: Univ. Toronto Press
22. Ashwood-Smith MJ (1965) Cryobiology *2*: 39
23. Fewson CA (1981) in: Microbial Degradation of Xenobiotics and Recalcitrant Compounds. pp 141 Leisinger T, Cook AM, Hütter R, Nüesch J (eds) London: Academic Press
24. Hill IR (1978) in: Pesticide Microbiology. pp 137 Hill IR, Wright SJL (eds) London: Academic Press
25. Wallnöfer PR, Hutzinger O, Engelhardt G, Ziegler W. (1984[2]) in: Handbook of Microbiology, Vol VI b. pp 491 Laskin AJ, Lechevalier HA (eds) Cleveland: CRC Press
26. Loos MA (1975[2]) in: Degradation of Herbicides, vol 1, pp 1. Kearney PC, Kaufman DD (eds) New York: Marcel Dekker
27. Engelhardt G, Rast HG, Wallnöfer PR (1977) Arch. Microbiol. *114*: 25
28. Rast HG, Engelhardt G, Wallnöfer PR, Oehlmann L, Wagner K (1979) J. Agric. Food Chem. *27*: 699
29. Geissbühler H, Martin H, Voss G (1975[2]) in: Degradation of Herbicides, vol. 1, pp 210 Kearney PC, Kaufman DD (eds) New York: Marcel Dekker
30. Esser HO, Dupuis G, Ebert E, Marco G, Vogel C (1975[2]) in: Degradation of Herbicides, vol. 1 pp 130 Kearney PC, Kaufman DD (eds) New York: Marcel Dekker
31. Behki RM, Khan SU (1986) J. Agric. Food Chem. *34*: 746
32. Giardi MT, Giardina MC, Filacchioni G (1985) Agric. Biol. Chem. *49*: 1551
33. Wallnöfer PR, Safe S, Hutzinger O (1972) Chemosphere *1*: 155
34. Wallnöfer PR, Safe S, Hutzinger O (1973) J. Agric. Food Chem. *21*: 502
35. Krause A, Hancock WG, Minard RD, Freyer AJ, Honeycutt RC, LeBaron HM, Paulson DL, Liu S-Y, Bollag J-M (1985) J. Agric. Food Chem. *33*: 384
36. Chapman PJ (1972) in: Degradation of Synthetic Organic Molecules in the Biosphere. pp 17 Proc. Conference, Natl. Acad. Sci. Washington, DC
37. De Frank JJ, Ribbons DW (1977) J. Bacteriol. *129*: 1365
38. Gibson DT (1972) in: Degradation of Synthetic Organic Molecules in the Biosphere. pp 116 Proc. Conference Natl. Acad. Sci. Washington, DC
39. Engelhardt G, Wallnöfer PR, Rast HG, Fiedler F (1976) Arch. Microbiol. *109*: 109
40. Gibson DT (1971) CRC Crit. Rev. Microbiol. *1*: 199
41. Gibson DT, Mahadevan V, Jerina DM, Yagi H, Yeh HJC (1965) Science *189*: 295
42. Matsumura F, Benezet HJ (1978) in: Pesticide Microbiology. pp 623 Hill IR, Wright SJL (eds) New York: Academic Press
43. Munnecke DM, Hsieh DPH (1976) Appl. Environ. Microbiol. *31*: 63
44. Flashinski SJ, Lichtenstein EP (1974) Can. J. Microbiol. *20*: 399

45. Munnecke DM (1981) in: Microbial Degradation of Xenobiotic and Recalcitrant Compounds. pp 251 Leisinger T, Cook AM, Hütter R, Nüesch J (eds) London: Academic Press
46. Engelhardt G (1972) Dissertation, Technische Universität München
47. Serdar CM, Gibson DT, Munnecke DW, Lancaster JH (1982) Appl. Environ. Microbiol. *44*: 246
48. McBride KE, Kenny JW, Stalker DM (1986) Appl. Environ. Microbiol. *52*: 325
49. Kaufman DD, Kearney PC (1976) in: Herbicides. Physiol. Biochem. Ecology. vol 2 pp 29 Audus LJ (ed). London: Academic Press
50. Spain J, Wyss O, Gibson DT (1979) Biochem. Biophys. Res. Commun. *88*: 634
51. Sudhakar B, Siddaramappa R, Sethunathan N (1976) Antonie van Leeuwenhoek J. Microbiol. Serol. *42*: 461
52. Wymann JF, Guard HE, Won WD, Quay JH (1979) Appl. Environ. Microbiol. *37*: 222
53. Esaak EG, Matsumura F (1984) Int. Encycl. Pharmacol. Ther. *113*: 265
54. Sudhakar B, Siddaramappa R, Wahid PA, Sethunathan N (1978) Antonie van Leeuwenhoek J. Microbiol. Serol. *44*: 171
55. Walker N, Harris D (1969) J. Appl. Bacteriol. *32*: 457
56. Bachofer R, Lingens F, Schäfer W (1975) FEBS Lett. *50*: 288
57. Surovtseva EG, Vasil'Eva GK, Baskunov BP, Vol'Nova AI (1981) Mikrobiologiya *50*: 740
58. Reber H, Helm V, Karanth NGK (1979) Eur. J. Appl. Microbiol. Biotechnol. *7*: 181
59. Latorre J (1982) Dissertation, Universität Göttingen
60. Surovtseva EG, Vasil'Eva GK, Vol'Nova AI, Baskunov BP (1980) Dokl. Akad. Nauk. SSSR *254*: 226
61. You I-S, Bartha R (1982) J. Agric. Food Chem., *30*: 247
62. Latorre J, Reinecke W, Knackmuss H-J (1984) Arch. Microbiol. *140*: 159
63. Que L, Jr (1978) Biochem. Biophys. Res. Commun. *84*: 123
64. Schilling R, Engelhardt G, Wallnöfer PR (1985) Chemosphere *14*: 267
65. MacRae IG, Cameron AJ (1985) Appl. Environ. Microbiol. *49*: 236
66. Engelhardt G, Oehlmann L, Wagner K, Wallnöfer PR, Wiedemann M (1984) J. Agric. Food Chem. 32: 102
67. Stevens-Jones A (1976) J. Agric. Food Chem. *24*: 115
68. Coppedge JR, Lindquist DA, Bull DL, Dorough HW (1967) J. Agric. Food Chem. *15*: 902
69. Bull DL (1968) J. Econ. Entomol. *61*: 1598
70. Janke D, Fritsche W (1978) Z. Allg. Mikrobiol. *18*: 365
71. 5. Dechema Fachgespräch Umweltschutz, Apr. 29–30, Frankfurt 1987
72. Knackmuss H-J (1981) in: Microbial Degradation of Xenobiotic and Recalcitrant Compounds. pp 189 Leisinger T, Cook AM, Hütter R, Nüesch J (eds) London: Academic Press
73. Schreiber A, Hellwig M, Dorn E, Reinecke W, Knackmuss H-J (1980) Appl. Environ. Microbiol. *39*: 58
74. Cook AM, Hütter R (1986) FEMS Microbiol. Lett. *34*: 335
75. Kearney PC, Kaufman DD, Beall ML (1964) Biochem. Biophys. Res. Comm. *14*: 29
76. Bollag JM, Alexander M (1971) Soil Biol. Biochem. *3*: 91
77. Lal R, Saxena DM (1982) Microbiol. Rev. *46*: 95
78. Baarschers WH, Bharath AI, Elvish J, Davies M (1981) Can. J. Microbiol. *28*: 176
79. Engelhardt G, Wallnöfer PR, Mücke W, Renner G (1986) Toxicol. Environ. Chem. *11*: 233
80. Mikesell MD, Boyd SA (1986) Appl. Environ. Microbiol. *52*: 861
81. Gibson SA, Suflita JM (1986) Appl. Environ. Microbiol. *52*: 681
82. Benezet HJ, Matsumura F (1973) Nature (London) *241*: 480
83. Matsumura F, Khanvilkar VG, Patil KC, Boush GM (1971) J. Agric. Food Chem. *19*: 27
84. Matsumura F, Patil K.C, Boush GM (1970) Science *170*: 1206
85. Higgins SJ, Mandelstam J (1972) J. Biochem. *126*: 917
86. Cook AM, Fewson CA (1972) Biochem. Biophys. Acta *290*: 384
87. Thayer JR, Wheelis ML (1976) Arch. Microbiol. *110*: 37
88. Dagley S (1978) in: The Bacteria. Vol IV pp 305. Ornston LJ, Sokatch JR (eds) London: Academic Press
89. Dagley S (1978) Naturwissenschaften *65*: 85
90. Dagley S (1978) Q. Rev. Biophys. *11*: 577
91. Larkin MJ, Day MJ (1986) J. Appl. Bacteriol. *60*: 233

92. Kieslich K (ed) (1976) in: Microbial Transformations of Non-Steroid Cyclic Compounds. Stuttgart: Georg-Thieme-Verlag
93. Rogers JE, Gibson DT (1977) J. Bacteriol. *130*: 1117
94. Hayaishi O (ed) (1974) in: Molecular Mechanism of Oxygen Activation. New York: Academic Press
95. Boyer PD (ed) (1975[3]) in: The Enzymes, Vol. XII, New York: Academic Press
96. Hou CT, Patel R, Lillard MO (1977) Appl. Environ. Microbiol. *33*: 725
97. Engelhardt G, Rast HG, Wallnöfer PR (1979) FEMS Microbiol. Lett. *5*: 377
98. Stanier RY, Palleroni NJ, Doudoroff M (1966) J. Gen. Microbiol. *43*: 159
99. Keyser P, Pujar BG, Eaton RW, Ribbons DW (1976) Environ. Health Perspectives *18*: 159
100. Tarvin D, Buswell AM (1934) J. Am. Chem. Soc. *56*: 1751
101. Evans WC (1977) Nature *270*: 17
102 Dutton PL, Evans WC (1970) Arch. Biochem. Biophys. *136*: 288
103. Guyer M, Hegeman G (1969) J. Bacteriol. *99*: 906
104. Oshima T (1973) Z. Allg. Mikrobiol. *5*: 386
105. Taylor BF, Campbell WL, Chinoy I (1970) J. Bacteriol. *102*: 430
106. Bakker G (1977) FEMS Microbiol. Lett. *1*: 103
107. Williams RJ, Evans WC (1975) Biochem. J. *148*: 1
108. Zeyer J, Kuhn EP, Schwarzenbach RP (1986) Appl. Environ. Microbiol. *52*: 844
109. Aftring RP, Chalker BE, Taylor BF (1981) Appl. Environ. Microbiol. *41*: 1177
110. Keith CL (1972) Diss. Abstr. Intern. B *33*: 3214
111. Balba MT, Evans WC (1977) Biochem. Soc. Trans. *5*: 302
112. Fina LR, Bridges RL, Coblentz TH, Roberts FF (1978) Arch. Microbiol. *118*: 169
113. Keith CL Bridges RL, Fina LR, Iverson KL, Cloran JA (1978) Arch. Microbiol. *118*: 173
114. Healy JB, Jr, Young LY (1979) Appl. Environ. Microbiol. *38*: 84
115. Cook AM, Hütter R (1981) J. Agric. Food Chem. *29*: 1135
116. Callely AG (1978) in: Progress of Industrial Microbiology, pp 205. Bull MJ (ed) Amsterdam: Elsevier
117. Dagley S (1972) in: Degradation of Synthetic Organic Molecules in the Biosphere. pp. 1. Proc. Conference, Natl. Acad. Sci., Washington, DC
118. Gupta KG, Sud RK, Aggarwal PK, Aggarwal JC (1975) Plant Soil *42*: 317
119. Watson GK, Caine RB (1975) Biochem. J. *146*: 157
120. Eberspächer J, Mang S, Blobel F, Sauber K (1976) Zbl. Mikrobiol. Abt. 1 *162*: 145
121. Engelhardt G, Ziegler W, Wallnöfer PR, Oehlmann L, Wagner K (1981) FEMS Microbiol. Lett. *11*: 165
122. Engelhardt G, Ziegler W, Wallnöfer PR, Jarczyk HJ, Oehlmann L (1982) J. Agric. Food Chem. *30*: 278
123. Summerer J (1983) Diplomarbeit, Technische Universität München
124. Engelhardt G (1983) Personal Commun.
125. Williams AK, Cox ST, Eagon RE (1965) Biochem. Biophys. Res. Commun. *18*: 250
126. Tweedy BG, Loeppky C, Ross JA (1970) Science *168*: 482
127. Torstensson L (1977) Weeds and Weed Contr., 18th Swedish Weed Conf., pp 16 Uppsala
128. Lee JK, Kim KC (1978) J. Korean Agr. Chem. Soc. *21*: 197
129. Engelhardt G, Wallnöfer PR, Fuchsbichler G, Baumeister W (1977) Chemosphere *6*: 85
130. Ross JA, Tweedy BG (1973) Bull Environ. Contam. Toxicol. *10*: 234
131. Cerniglia CE, Freeman JP, Van Baalen Ch (1981) Arch. Microbiol. *130*: 272
132. Lyons CD, Katz SE, Bartha R (1985) Bull. Environ. Contam. Toxicol. *35*: 704
133. Loos MA, Roberts RN, Alexander M (1967) Can. J. Microbiol. *13*: 691
134. Nakanishi T, Oku H (1969) Phytopathology *59*: 1761
135. Corke CT, Nigel JB, Beaumont AL, Merrick RL (1979) J. Agric. Food Chem. *27*: 644
136. Minard RD, Russel S, Bollag J-M (1977) J. Agric. Food Chem. *25*: 841
137. Cripps RE, Roberts TR (1978) in: Pesticide Microbiology. pp 669 Hill IR, Wright SJL (eds) London: Academic Press
138. Kaufman DD, Plimmer JR, Iwan J, Klingebiel UI (1972) J. Agric. Food Chem. *20*: 916
139. Kaufman DD, Plimmer JR, Klingebiel UI (1973) J. Agric. Food Chem. *21*: 127
140. Bordeleau LM, Bartha R (1972) Can. J. Microbiol. *18*: 1857
141. Bordeleau LM, Bartha R (1972) Can. J. Microbiol. *18*: 1873

142. Bordeleau LM, Bartha R (1972) Can. J. Microbiol. *18*: 1865
143. Hoff T, Liu S-Y, Bollag J-M (1985) Appl. Environ. Microbiol. *49*: 1040
144. Briggs GG, Walker N (1973) Soil Biol. Biochem. *5*: 695
145. Bartha R, You JS, Saxena A (1982) in: Pesticide Chemistry: Human Welfare and the Envi-.ronment. vol. 3 pp 343. Matsumura F, Hutson DH, Murphy SD (eds) Oxford: Pergamon Press
146. Bollag J-M, Sjöblad RD, Minard RD (1977) Experientia *33*: 1564
147. Bollag J-M, Liu S-Y (1985) Pesticide Biochem. Physiol. *23*: 261
148. Macholz RM, Kujawa M (1985) Residue Rev. *94*: 119
149. Haider K, Jagnow G, Rohr K (1976) Landwirt. Forsch. Sonderh. *32*: 147
150. Tu CM (1975) Arch. Microbiol. *105*: 131
151. Matsumura F, Benezet HJ, Patil KC (1976) J. Pestic. Sci. *1*: 3
152. Sethunathan W, Bautista EM, Yoshida T (1969) Can. J. Microbiol. *15*: 1349
153. Heritage AD, MacRae IG (1977) Appl. Environ. Microbiol. *34*: 222
154. Wedemeyer G (1968) Appl. Microbiol. *16*: 661
155. Tu CM, Miles JRW, Harris CR (1968) Life Sci. *7*: 311
156. Ferguson JA, Korte F (1977) Appl. Environ. Microbiol. *34*: 7
157. Patil KC, Matsumura F, Boush GM (1971) Appl. Microbiol. *19*: 879
158. Miles JRW, Tu CM, Harris CR (1969) J. Econ. Entomol. *62*: 1331
159. Beeman RW, Matsumura F (1981) J. Agric. Food Chem. *29*: 84
160. El Zorgani GA, Omer MEH (1974) Bull. Environ. Contam. Toxicol. *12*: 182
161. Perscheid M, Schlüter H, Ballschmiter K (1973) Z. Naturforsch. *28*: 761
162. Andrade P, Jr, Wheeler WB, Carlson D (1975) Bull. Environ. Contam. Toxicol. *14*: 473
163. Foy CL (1975[2]) in: Degradation of Herbicides, vol. 1 pp 399 Kearney PC, Kaufman DD (eds) New York: Marcel Dekker
164. Kearney PC, Harris CI, Kaufman DD, Sheets T (1965) J. Adv. Pest. Control Res. *6*: 1
165. Kearney PC, Kaufman DD, Von Endt DW, Guardia FS (1969) J. Agric. Food Chem. *17*: 581
166. Belser NO, Castro CE (1971) J. Agric. Food Chem. *19*: 23
167. Castro GE, Belser NO (1966) J. Agric. Food Chem. *14*: 69
168. Yang RSH (1986) Residue Rev. *97*: 19
169. Balthazor TM, Hallas LE (1986) Appl. Environ. Microbiol. *51*: 432
170. Bartha R, Pramer D (1970) Adv. Appl. Microbiol. *13*: 317
171. Still GG, Merrett RA (1976[2]) in: Degradation of Herbicides, vol 2, pp 609 Kearney PC, Kaufman DD (eds) New York: Marcel Dekker
172. Engelhardt G, Wallnöfer PR, Plapp R (1973) Appl. Microbiol. *26*: 709
173. Akatsuka T, Uchiyama T (1973) Jbaraki Daigaku Nogakubu Gakujutsu Mokaku *21*: 49
174. Wallnöfer PR, Tillmanns G, Engelhardt G (1977) Pestic. Biochem. Physiol. *7*: 481
175. McGahen LL, Tiedje JM (1978) J. Agric. Food Chem. *26*: 414
176. Wallnöfer P, Königer M, Safe S, Hutzinger O (1972) J. Agric. Food Chem. *20*: 20
177. Wallnöfer P, Safe S, Hutzinger O (1971) Pestic. Biochem. Physiol. *1*: 458
178. Wallnöfer PR, Königer M, Safe S, Hutzinger O (1972) Intern. J. Environ. Anal. Chem. *2*: 37
179. Bollag J-M, Russel S (1976) Microb. Ec. *3*: 65
180. Zeyer J, Wasserfallen A, Timmis K (1985) Appl. Environ. Microbiol. *50*: 447
181. Zeyer J, Kearney PC (1982) Pestic. Biochem. Physiol. *17*: 215
182. Linke HAB (1970) Naturwissenschaften *57*: 307
183. Plimmer JR, Kearney PC, Chisaka H, Yount JB, Klingebiel UI (1970) J. Agric. Food Chem. *18*: 850
184. Frear DS (1976[2]) in: Degradation of Herbicides, vol 2, pp 542 Kearney PC, Kaufman DD (eds) New York: Marcel Dekker
185. Verloop A, Nimmo WB (1970) Weed Res. *10*: 65
186. Zaki MA, Taylor HF, Wain RL (1967) Ann. Appl. Biol. *59*: 481
187. Hsu JM (1976) Diss. Abstr. Int. B *36*: 3780
188. Smith AE, Cullimore DR (1974) Can J. Microbiol. *20*: 773
189. Fröslie A, Karlog O (1970) Acta Vet. Scand. *11*: 114
190. Wallnöfer PR, Ziegler W, Engelhardt G, Rothmeier H (1978) Chemosphere *7*: 967
191. Tewfik MS, Hamdi YA (1975) Soil Biol. Biochem. *7*: 79

192. Probst GW, Tepe JB (1969) in: Degradation of Herbicides. pp 255 Kearney PC, Kaufman DD (eds) New York: Marcel Dekker
193. Zayed SM, Mostafa IY, Farghaly MM, Attaby HSH, Adam YM, Mahdy FM (1983) J. Environ. Sci. Health, Part B *B 18*: 253
194. Golab T, Herberg RJ, Gramlich JV, Raul AP, Probst GW (1970) J. Agric. Food Chem. *18*: 838
195. Golab T, Herberg RJ, Day EW, Raun AP, Holzer FJ, Probst PW (1969) J. Agric. Food Chem. *17*: 576
196. Golab T, Amundson ME (1975) in: Pesticides, Suppl. vol. III of EQS, pp 258 Coulston F, Korte F (eds) Stuttgart: Georg Thieme-Verlag
197. Laanio TL, Kearny PC, Kaufman DD (1973) Pestic. Biochem. Physiol. *3*: 271
198. Otto S (1975) in: Pesticides, Suppl. vol. III of EQS, pp 277 Coulston F, Korte F (eds) Stuttgart: Georg Thieme-Verlag
199. Camper ND, Ellers KL (1978) Weed Res. *18*: 99
200. Smith AE (1977) J. Agric. Food Chem. *25*: 893
201. Köcher H, Lingens F, Koch W (1976) Weed Res. *16*: 93
202. Cserjesi AJ, Johnson EL (1972) Can. J. Microbiol. *18*: 45
203 Yih RY, Swithenbank C (1971) J. Agric. Food Chem. *19*: 314
204. Johnsen RE (1976) Residue Rev. *61*: 1
205. Kokke R (1970) Nature *226*: 977
206. Golovleva LA, Skryabin GK (1981) in: Microbial Degradation of Xenobiotic and Recalcitrant Compounds. pp 287 Leisinger T, Cook AM, Hütter R, Nüesch J (eds) London: Academic Press
207. Mendel JL, Klein AK, Chen JT, Walton MS (1967) J. Assoc. Anal. Chem. *50*: 897
208. Focht DD, Alexander M (1970) Appl. Microbiol. *20*: 608
209. Focht DD, Alexander M (1970) Science *170*: 91
210. Focht DD, Alexander M (1971) J. Agric. Food Chem. *19*: 20
211. Miyazaki S, Boush GM, Matsumura F (1969) Appl. Microbiol. *18*: 972
212. Miyazaki S, Boush GN, Matsumura F (1970) J. Agric. Food Chem. *18*: 87
213. Kesner CD, Ries SK (1967) Science *155*: 210
214. Calderbank A, Slade P (1976^2) in: Degradation of Herbicides, vol 2, pp 501 Kearney PC, Kaufman DD (eds) New York: Marcel Dekker
215. Funderburk HH, Jr, Bozarth GA (1967) J. Agric. Food Chem. *15*: 563
216. Nitsche I, Siemrova J, Ballschmiter K, Selenka F (1975) in: Pesticides, Suppl. vol. III of EQS. pp 29 Coulston F, Korte F (eds) Stuttgart: Georg Thieme-Verlag
217. Carr RJG, Bilton RF, Atkinson T (1985) Appl. Environ. Microbiol. *49*: 1290
218. Campacci EF, New PB, Tschan YT (1977) Nature *266*: 164
219. Engvild KC, Jensen HL (1969) Soil Biol. Biochem. *1*: 295
220. Fröhner C, Oltmann O, Lingens F (1970) Arch. Mikrobiol. *94*: 82
221. Engelhardt G, Wallnöfer PR (1978) Chemosphere *7*: 463
222. Goring CAI, Hamaket JW (1971) Down Earth *27*: 12
223. Harris CI (1967) J. Agric. Food Chem. *15*: 157
224. Kaufman DD, Kearney PC (1970) Residue Rev. *32*: 235
225. Shimabukuro RH, Lamourex GL, Frear DS, Bakke JE (1971) IUPAC Symposium on Pesticide Terminal Residues, Tel Aviv
226. Cook AM, Hütter R (1984) J. Agric. Food Chem. *32*: 581
227. Giardina MC, Buffone R (1977) Chemosphere *6*: 589
228. Barth A, Munch E (1969) Pharmazie *24*: 713
229. Clark CG, Wright SJL (1970) Soil Biol. Biochem. *2*: 217
230. Kaufman DD, Kearney PC (1965) Appl. Microbiol. *13*: 443
231. Vigne B, Archelas A, Fourneron JD, Furstoss R (1986) Tetrahedron *4*: 2451
232. Williams PP, Robbins JD (1969) Bacteriol. Proc. A *107*: 17
233. Williams IH, Pepin HS, Brown MJ (1976) Bull. Environ. Contam. Toxicol. *15*: 244
234. Rajagopal BS, Rao VR, Nagendrappa G, Sethunathan N (1984) Can. J. Microbiol. *30*: 1457
235. Venkateswarlu K, Sethunathan N (1985) Plant Soil *84*: 445
236. Venkateswarlu K, Sethunathan N (1984) Bull. Environ. Cont. Toxicol. *33*: 556
237. Karns JS, Mulbry WW, Nelson JO, Kearney PC (1986) Pestic. Biochem. Physiol. *25*: 211

238. Bollag J-M, Liu S-Y (1971) Soil Biol. Biochem. *3*: 337
239. Liu S-Y, Bollag J-M (1971) J. Agric. Food Chem. *19*: 487
240. Liu S-Y, Bollag J-M (1971) Pestic. Biochem. Physiol. *1*: 366
241. Benezet HJ, Matsumura F (1974) J. Agric. Food Chem. *22*: 427
242. Richey FA, Jr, Bartley WI, Sheets KP (1977) J. Agric. Food Chem. *25*: 47
243. Hock WK, Sisler HD (1969) J. Agric. Food Chem. *17*: 123
244. Wiese MV, Vargas IM, Jr (1973) Pestic. Biochem. Physiol. *3*: 214
245. Ko WH, Farley JD (1969) Phytopathology *54*: 64
246. Chacko CJ, Lockwood JL, Zabick M (1966) Science *154*: 893
247. Igarashi H, Uchiyama M, Sato RI (1975) Pestic. Sci. *3*: 88
248. Rhodes RC (1977) J. Agric. Food Chem. *25*: 528
249. Nakanishi T (1973) Ann. Phytopath. Soc. Jpn. *38*: 249
250. Van Alfen NK, Kosuge T (1976) J. Agric. Food Chem. *24*: 584
251. Knowles CO (1970) J. Agric. Food Chem. *18*: 1038
252. Davidse LC (1976) Pestic. Biochem. Physiol. *6*: 538
253. Yasuda Y, Hashimoto S, Soeda Y, Noguchi T (1973) Ann. Phytopath. Soc. Jpn. *39*: 49
254. Vincent PG, Sisler HD (1968) Physiol. Plant. Pathol. *21*: 1249
255. Somers E, Richmond DV, Pickard JA (1967) Nature *215*: 214
256. Siegel MR, Sisler HD (1968) Phytopathology *58*: 1123
257. Engelhardt G, Wallnöfer PR (1981) Bayer AG Report
258. Bedi S, Roy NK (1980) J. Environ. Sci. Health B, *15*: 259
259. Sijpesteijn AK, Vonk JW (1975) in: Pesticides, Suppl. vol. III of EQS. pp 57 Coulston F, Korte F (eds) Stuttgart: Georg Thieme-Verlag
260. Imai Y, Kuwatsuka S (1986) J. Pestic. Sci *11*: 245
261. Lord KA, McKinley M, Walker N (1982) Environ. Pollut. Ser. A *29*: 81
262. Sakata S, Mikami N, Tadashi M, Miyamoto J (1986) Nippon Noyaku Gakkaishi *11*: 71
263. Fest S, Schmidt K (1970) in: Chemie der Pflanzenschutz- und Schädlingsbekämpfungsmittel, vol 1. pp 246 Wegler R (ed) Berlin: Springer Verlag
264. Spencer EY (1971) JUPAC Symposium on Pesticide Terminal Residues, Tel Aviv
265. Yasuno M, Hirakoso S, Sasa M, Uchida M (1965) Jpn. J. Exp. Med. *35*: 545
266. Ahmed MK, Casida JE, Nichols RE (1958) J. Agric. Food Chem. *6*: 740
267. Dahm PA (1970) in: Biochemical Toxicology of Insecticides. pp 51 O'Brien RD, Yamamoto I (eds) New York: Academic Press
268. Rao AV, Sethunathan N (1974) Arch. Microbiol. *97*: 203
269. Adhya TK, Sudhakar B, Sethunathan N (1981) J. Appl. Bacteriol. *50*: 167
270. Miyamoto J, Kitabawa K, Sato Y (1966) Jpn. J. Exp. Med. *36*: 211
271. Baarschers WH, Heitland HS (1986) J. Agric. Food Chem. *34*: 707
272. Timms P, MacRae JC (1982) Aust. J. Biol. Sci. *35*: 661
273. Sheela S, Pai SB (1983) Appl. Environ. Microbiol. *46*: 475
274. Wallnöfer PR, Söhlemann F, Oehlmann L (1976) Pflanzensch.-Nachr. Bayer *29*: 236
275. Waggoner BT, Khasawinah AM (1974) Residue Rev. *53*: 79
276. Lichtenstein EP, Parlar H, Korte F, Süß A (1977) J. Agric. Food Chem. *25*: 445
277. Sethunathan N, Yoshida T (1973) Can. J. Microbiol. *19*: 873
278. Uesugi Y, Tomizawa C (1971) Agric. Biol. Chem. *35*: 941
279. Ziegler W, Engelhardt G, Wallnöfer PR, Oehlmann L, Wagner K (1980) J. Agric. Food Chem. *27*: 1102
280. Matsumura F, Boush GM (1966) Science *153*: 1278
281. Paris DF, Lewis DL (1975) in: Pesticides, Suppl. vol. III of EQS. pp 228 Coulston F, Korte F (eds) Stuttgart: Georg Thieme-Verlag
282. Lewis DL, Paris DF, Baughman GL (1975) Bull. Environ. Cont. Toxicol. *13*: 596
283. Walker WW, Stojanovic BI (1974) J. Environ. Qual. *3*: 4
284. Bourquin AW (1977) Appl. Environ. Microbiol. *33*: 356
285. Zayed SMAD, Mostafa IY, Hassan A (1965) Arch. Mikrobiol. *51*: 118
286. Lieberman MT, Alexander M (1983) J. Agric. Food Chem. *31*: 265
287. Raymond DGM, Alexander M (1971) Pestic. Biochem. Physiol. *1*: 123
288. Siddaramappa R, Rajavam KP, Sethunathan N (1973) Appl. Microbiol. *26*: 846

289. Wallnöfer PR, Engelhardt G (1984) in: Biotechnology vol 6a pp 279 Rehm H-J, Reed G (eds) Weinheim: Verlag Chemie
290. Loos MA (1971) IUPAC Symposium on Pesticide Terminal Residues, Tel Aviv
291. Evans WC, Smith BSW, Fernley HN, Davies JI (1971) Biochem. J. *122*: 543
292. Sharpee KW, Duxbury IM, Alexander M (1973) Appl. Mircobiol *26*: 445
293. Gamar Y, Gaunt J.K. (1971) Biochem. J. *122*: 527
294. Gaunt JK, Evans WC (1971) Biochem. J. *122*: 519
295. Evans WC, Smith BSW, Moss P, Fernley HM (1971) Biochem. J. *122*: 509
296. Ou LT, Sikka HC (1977) J. Agric. Food Chem. *25*: 1336
297. Karns JS, Duttagupta S, Chakrabarty AM (1983) Appl. Environ. Microbiol. *46*: 1182
298. Suflita JM, Stout J, Tiedje JM (1984) J. Agric. Food Chem. *32*: 218
299. Karns JS, Kilbane JJ, Chatterjee DK, Chakrabarty AM (1984) Basic Life Sci. *28*: 3
300. Geissbühler H, Voss G (1971) JUPAC Symposium on Pesticide Terminal Residues, Tel Aviv
301. Wallnöfer PR, Safe S, Hutzinger O (1973) Pestic. Biochem. Physiol. *3*: 253
302. Tillmanns G, Wallnöfer PR, Engelhardt G, Olie K, Hutzinger O (1978) Chemosphere *7*: 59
303. Wallnöfer PR, Tillmanns G, Thomas R, Wünsche C, Kurz J, Jarczyk HJ (1976) Chemosphere *5*: 377
304. Goettfert J, Parlar H, Korte F (1978) J. Agric. Food Chem. *26*: 628
305. Ozaki M, Hayase Y, Kobayashi S, Kuwatsuka S (1984) Nippon Noyaku Gakkaishi *9*: 285
306. Ozaki M, Kuwatsuka (1986) J. Pestic. Sci. *11*: 427
307. Franklin FGH (1981) in: Microbial Degradation of Xenobiotic and Recalcitrant Compounds. pp 109 Leisinger T, Cook AM, Hütter R, Nüesch J (eds) London: Academic Press

Enzymes Involved in the Metabolism of Organophosphorus, Carbamate and Pyrethroid Insecticides

U. Schoknecht and D. Otto
Institut für Pflanzenschutzforschung Kleinmachnow der Akademie der Landwirtschaftswissenschaften der DDR, 1532 Kleinmachnow, Stahnsdorfer Damm 81, DDR

Enzyme-catalyzed metabolism is an important factor influencing insecticide efficiency and causing differences between organisms. The main insecticide metabolizing enzymes — monooxygenases, esterases and glutathione S-transferases — are introduced and characterized with special emphasis on their occurrence and features in arthropods. Enzyme-catalyzed metabolism of three insecticide classes — organophosphorus compounds, carbamates and pyrethroids — by these enzymes is generally explained and illustrated for several compounds. Enzyme-related differences in insecticide sensitivity depending on developmental stage, species, strain and environmental conditions are regarded. Genetic basis of enzyme changes and possibilities to control insecticide metabolism are discussed.

Chemistry of Plant Protection 2

1 Introduction

The question: 'What happens to insecticide molecules, when they are applied in the field' is of interest from several aspects, namely toxicology, environmental safety, development of new effective and selective insecticide strategies to prevent resistance and elaboration of integrated pest management advice.

Insecticide molecules may be degraded not only by non-enzymatic, but also by enzyme-catalyzed reactions which occur in microorganisms as well as in plants and animals. There are three groups of enzymes that are especially important in the metabolism of insecticides: monooxygenases, esterases and glutathione S-transferases.

Physiological importance, ability to convert various xenobiotics including insecticides, reaction mechanisms, genetic basis and regulation of activity have been subjects of investigation on these enzymes for about 25 years.

Enzyme-catalyzed reactions are subdivided into two phases. Usually xenobiotics are at first modified by oxidation or hydrolysis, forming more polar products (phase I) that can be conjugated by phase II reactions [1]. However, this is not compulsory, compounds may be only converted by phase I reactions and immediately excreted on the one hand, or at once conjugated on the other hand.

Enzymes protecting organisms from harmful chemicals have to fulfil several suppositions:

— location in those organs particularly exposed to foreign compounds (e.g. liver, lung and intestinal tract of vertebrates, fatbody and gut of insects)
— catalysis of analogous reactions at substrates of different structures with uniform effect, namely production of less toxic compounds that can be more readily excreted
— regulation of enzyme activity due to changes of enzyme concentration during the life cycle on the one hand, and enzyme induction on the other hand, thus reflecting phases of especially high exposure to substrates.

Detoxication systems composed of only a few favoured enzymes of multiple function and a remarkable degree of nonspecificity possess important economical advantages by saving information storage capacity and protein synthesis. Indeed, low substrate specificity causes relatively low catalytic activity, however, this is compensated by high enzyme concentration in the tissues [2].

It must be considered that man-made chemicals could not be selection factors for the evolution of such complex systems since organisms have been exposed to them only for about the last century. However, xenobiotic-metabolizing enzymes occur in all the sub-kingdoms of living organisms, and therefore it is likely that there were other selection pressures. Natural chemicals wording off enemies and competitors could represent such factors [3]. An example is the coevolution of plant allelochemicals and enzymatic defense mechanisms of herbivorous insects [3–6].

This review deals with insecticide metabolism in mammals and insects. The authors intend to summarize general knowledge on the mentioned enzyme systems of mammals and insects, molecular sites of insecticides that are attacked by these enzymes, and relations between insecticide metabolism and insecticide susceptibility. While the characterization of the xenobiotic degrading enzymes is valid to pesticide metabolism in general, enzyme reactions in insecticides are concentrated on some main types, namely organophosphorus compounds, carbamates and pyrethroids.

2 Characterization of Insecticide-Metabolizing Enzymes

2.1 Monooxygenases

Monooxygenases (mixed function oxidases/mfo) belong to the most important xenobiotic-metabolizing enzymes because of their wide-spread occurrence, reaction diversity and ability to convert a wide range of substrates.

Mixed function oxidases are evident in bacteria, fungi, plants and animals. They are mostly membrane bound, predominantly in microsomal membranes, smooth endoplasmic reticulum having higher specific activity than rough. In some cases they were also found in mitochondrial membranes or in a soluble form. [6-9] Highest activity was determined in the microsomes of mammalian liver, so rat liver is the main object of mfo research.

In general reaction of mixed function oxidases is the introduction of one oxygen atom of molecular oxygen into the substrate molecule (RH). The second oxygen atom forms water (see Eq. (1)).

Eq. 1:

$$RH + NADPH + H^+ + O_2 \xrightarrow{mfo} ROH + NADP^+ + H_2O$$

In case of cytochrome P-450 dependent monooxygenases, which are of special interest with respect to xenobiotic oxidation, NADPH is the preferred electron donor. Electron transfer from NADH via cytochrome b_5 may also be involved [10].

The following types of reactions are catalyzed by several monooxygenases:

- deamination,
- N-, O-, S-dealkylation,
- desulfuration,
- dehydrogenation,
- aromatic hydroxylation,
- aliphatic hydroxylation
- heterocyclic ring hydroxylation,
- N-hydroxylation,
- ester cleavage,
- epoxidation,
- sulfoxidation,
- dioxole ring cleavage,
- oxidation of alcohols and aldehydes to acids.

Various endogenous and exogenous compounds, like steroids, bile acids, fatty acids, hydrocarbons, drugs, carcinogens and pesticides are known monooxygenase substrates.

This way mfo's are involved in biosynthesis, metabolism and regulation of various endogenous compounds and represent a defense line against a variety of environmental and dietary toxicants.

Usually lipophilic drugs and insecticides are converted into less toxic and more hydrophilic metabolites. The products are either directly or after secondary conjugative processes eliminated by the water-based excretory systems in organisms.

However, there are some exceptions, e.g. the oxidative activation of organophosphorothionates, cyclodienes, aflatoxins, carcinogens, pyrrolizidine alkaloids and N-dealkylation of nitrosamines. These reactions produce more toxic metabolites and could be considered as 'suicidal mistakes' [6,7,11-13].

Cytochrome P-450 Dependent Monooxygenases

Cytochrome P-450 dependent monooxygenases contain three constituents of functional importance, the hemoprotein cytochrome P-450, the NADPH-cytochrome P-450 reductase and the surrounding phospholipids of the endoplasmic reticulum. Molecular oxygen and electrons are needed for reaction.

Cytochrome P-450 was demonstrated as a carbon monoxide-binding pigment in 1958 by Klingenberg [14] and Garfinkel [15], its protoporphyrin structure was described by Omura and Sato [16] in 1962, and Cooper [17] proved its participation in liver microsomal drug oxidation in 1965. Its name is related to the typical peak of the reduced cytochrome P-450-CO-complex at 450 nm. Methods for determination of cytochrome P-450 content are based on this property. This hemoprotein with a molecular weight between 40 and 50 kDa [18,19] contains binding sites for oxygen and various substrate molecules and acts as terminal oxidase of this electron transport system.

The second component, the cytochrome P-450 reductase, is a flavin-containing protein with a molecular weight of about 80 kDa [20] that mediates the transport of reducing equivalents from NADPH via reduced flavin adenine dinucleotide (FAD) and flavin mononucleotide (FMN) to cytochrome P-450.

The surrounding phospholipids, mainly phosphatidylcholin, not only constitute a catalytically active arrangement but also modulate the electron transfer and stabilize and/or induce a functionally active conformation of cytochrome P-450. (see reviews [6,7,10,13,19,21])

There are four main phases of the overall reaction: substrate binding, reduction of the enzyme-substrate complex, oxygen activation process and oxygen transfer (see Fig. 1).

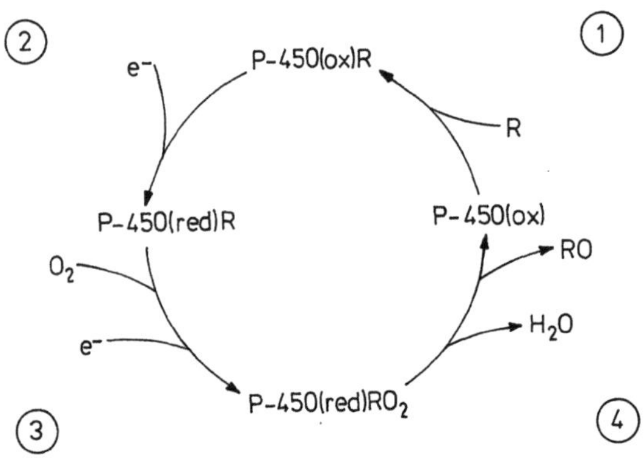

Fig. 1. Phases of monooxygenase reaction (modified according to Ruckpaul and Bernhardt [21])

With respect to the ability of mixed function oxidases to convert a wide variety of endogenous and exogenous compounds it is worth discussing substrate binding and specificity as well as occurrence of isoenzymes.

Because of the heme moiety substrate binding at cytochrome P-450 causes spectral changes in the soret region around 400 nm. Different types of substrate binding spectra are distinguished by means of absorption maxima and minima at characteristic wave lengths.

Type I spectra have a peak at about 385–390 nm and a trough at 420 nm. Binding is believed to take place at a lipophilic site at the protein moiety away from the heme iron center. Type I spectra are formed by most substrates with the oxidized form of cytochrome P-450. It is suggested that the substrate-induced type I spectral change represents a manifestation of an enzyme-substrate complex, and high oxidase activity depends on the degree of type I binding more than on cytochrome P-450 content.

Type II spectra have a peak at 425–435 nm and a trough at about 390–410 nm. In this case binding occurs at the O_2-binding site, i.e. the heme iron. The basis requirements for type II spectra are a sp^2 or sp^3 nitrogen atom in the substrate molecule, therefore these spectra are mainly obtained with compounds containing a primary amine group (e.g. aniline, n-octylamine). A lot of inhibitors give type II spectra.

In addition to these two main types of spectra there were found reversed type I spectra with a maximum at 420 nm and a trough at about 390 nm. They are formed, for instance, with several alcohols and some steroids.

Contrary to these types of binding spectra type III spectra are yielded with reduced cytochrome P-450. They have two peaks at about 430 and 455 nm that are in a pH-dependent equilibrium. These spectra, e.g. yielded with ethyl cyanide, may represent complexes with retarded dissociation of enzyme and reaction product (compare with Fig. 1).

Each factor influencing complex formation or spin state of the heme iron (e.g. membrane structure, temperature, exposure to inducers) effects monooxygenase reaction. (see reviews [7, 10, 13, 22, 23)])

Today it is generally accepted that the broad substrate spectrum is closely connected to multiplicity of cytochrome P-450s in the organisms, whereby each isoenzyme can possess a different substrate preference, and in this way the system as a whole can handle a wide range of compounds. Not only chromatographic, electrophoretic and immunological methods, but also induction studies with various organisms (e.g. mammals and insects) led to the suggestion that there really exists a number of monooxygenases with different but overlapping reaction and substrate specificity in the species examined. There is a lot of evidence for the existence of cytochrome P-450 patterns differing between species, strains, sexes, organs and tissues, depending on age, environmental and nutritional conditions [7, 10, 13, 23, 24)]. Synthesis of cytochrome P-450s can be induced by several hundred compounds of diverse chemical structure [25, 26)]. It was found that various inducers increase distinct isoenzymes and certain monooxygenase activities even in special organs, but can decrease others. Phenobarbital and 3-methylcholanthrene are well known and often used inducers causing typical isoenzyme patterns. Many substrates are described as inducers stimulating there own metabolism by means of this mechanism. Among these compounds there are insecticides with different structure, e.g. organohalogens, organophosphates, carbamates and pyrethroids as well as synergists which are shown in Table 1. Several publications [13, 24, 26 – 30, 34)].deal with this phenomenon. Some more aspects of induction of cytochrome P-450s are discussed in connection with insect monooxygenases.

Table 1. Insecticides and synergists that induce monooxygenases (in vertebrates/compounds marked with (i) also caused induction of insect monooxygenases)

Compounds	Refs.
Organohalogens	
22 compounds	27)
Chlordane	28)
Dieldrin (i)	26)
DDT (i)	26)
Phosphororganics	
Chlorpyriphos	28)
Diazinon	29)
Etrimphos	28)
Malathion	28, 29)
Sumithion	30)
Tetrachlorvinphos	31)
Trichlorfon	30)
Carbamates	
Carbaryl	29)
Pyrethroids	
Deltamethrin	32)
Pyrethrum	29)
Chlordimeform	29)
Synergists	
Piperonyl butoxide	26, 28)
Sesamex	28)

Having in view restriction of monooxygenase-catalyzed detoxication reactions, and thus increase insecticide toxicity, inhibitors of these enzymes are of special interest. Compounds of various structures were found to be inhibitors of this enzyme system [3, 22, 25, 35–40]. Among them are metal ions and organic sulfhydryl reagents — both probably inhibit the NADPH-dependent cytochrome P-450 reductase, competitive substrates, compounds that complex with the heme iron and traditional synergists like the methylene dioxyphenols, piperonyl butoxide and sesamex, the aryloxylamine SKF 525 A and plant secondary compounds like safrole.

The inhibition mechanism of the mentioned synergists seems to be founded on formation of stable complexes with cytochrome P-450. It is known that piperonyl butoxide and derivatives form type I spectra with oxidized cytochrome P-450 (corresponding to enzyme-substrate complex) and type III spectra with the reduced cytochrome [22,41,42]. This complex occurs during reaction cycle (see picture 1) and prevents further reaction because of its relative stability [43].
Moreover, 1-aminobenzotriazole was described as a suicide substrate that causes self-catalyzed destruction of cytochrome P-450 [44].

FAD-Dependent Monooxygenases

It must be mentioned, that mammals possess an additional oxidation system taking part in insecticide detoxication: flavin adenine dinucleotide (FAD) dependent monooxygenase (microsomal flavin-containing monooxygenase/MFMO; EC 1.14.13.8) which is also localized in the endoplasmic reticulum and requires NADPH and O_2 [45]. The molecular weight amounts to 64 kDa. Catalytically active enzymes are aggregating units of one kind of monomers. This enzyme system catalyzes oxidative attack on nitrogen- and sulfur-containing compounds [45] and at the phosphorus atom in phosphonates [46] to produce compounds that have increased water-solubility and serve as substrates for further reactions. Note that contrary to cytochrome P-450 dependent monooxygenases — forming S(—)-sulfoxides — sulfoxidation by FAD-dependent monooxygenases yield R(+)-sulfoxides. Products of oxidative desulfuration by these two enzymes also possess opposite chiral character.

FAD-dependent monooxygenases have not been reported in insects until now. For further information on this enzyme system several papers may be consulted [46-50].

Cytochrome P-450 Dependent Monooxygenases in Arthropods

Cytochrome P-450s and related monooxygenase activities were detected in a wide variety of insect species (see table 3 for examples), mainly in microsomes of fatbody and midgut, Malpighian tubules, carcass and other organs, but also in mitochondria [51]. The diptera *Musca domestica*, *Phormia regina*, *Sarcophaga bullata*, *Drosophila melanogaster* and the lepidoptera *Heliothis* spp. and *Spodoptera* spp. are well established species.

In contrast, NADPH-dependent reductases were examined in only a few insect species [20,52-59]. By means of immunological studies it was shown that insect reductases are very similar to one another, but differ from those of mammals [54,55].

However, in general insect cytochrome P-450 dependent monooxygenases resemble those of mammals. There are only a few exceptions: an apparent lack of an absolute phospholipid requirement [60], differences in ionic strength optimum, antigenic determinants, amino acid composition [55].

Multiplicity of cytochrome P-450s was suggested for several species [18,52,53,56,60-65], so one can assume that it is common in insects.

Occurrence and activity of monooxygenases depend on age, developmental stage and/or sex of insects [10,11,25,66-73], insect strain [56,62,74], organ [75] and nutritional factors. Moreover, endogenous inhibitors, for instance, produced by tyrosinase [66] may influence monooxygenase activity.

In the last decade insect cytochrome P-450s were intensively studied with respect to their function in biosynthesis and regulation of hormones and pheromones [6,51,76] as well as degradation of both harmful food constituents (e.g. plant allelochemicals) and pesticides.

Monooxygenase reactions are of special importance in herbivorous insects which are faced to various plant toxins and allelochemicals of different structure, like steroids, phenols, quinones, alkaloids, terpenes, indoles, flavones, aldehydes, glycosides [6,36,77-79]. Monooxygenases are qualified for detoxication of these compounds and in this way play an important role in biochemical insect-host plant relationships. Preferred subjects for investigation are lepidoptera. Their larvae live on plant leaves and their guts contain especially high amounts of cytochrome P-450 [6]. Studies with several lepidopterous species [65,77-88] showed that various plant constituents are able to induce insect mixed function oxidases. Plant species with mfo-inducing components are, for example peppermint [86], corn, cowpea, cabbage and soybean [81], tobacco and lucerne [65], peanut [79], tomato [84], carrot [89].

Comparative studies with 58 lepidopterous species illustrated that monooxygenase activity varies between mono-, oligo- and polyphagous larvae, but also depends on the consumed host [83].

Like the mammalian enzyme system, the insect monooxygenases are also induced by phenobarbital, β-naphthoflavone [8, 58, 62] and methylene dioxyphenyl compounds [90]. Although induction of enzyme proteins is a very complex mechanism — involving not only uptake and transfer of the inducer, but also interaction with gene-regulating systems, protein synthesis and incorporation of the new protein into the membrane [10, 25, 34] — it can occur in a few hours (about 12 hr). In one case in polyphagous lepidopterous larvae a first increase in enzyme activity was already found after 30 minutes [52].

Various qualitative and quantitative modifications were described in connection with induction:

- increase in cytochrome P-450 content [8, 53, 62, 67, 80, 82, 84–87, 89, 91, 93]
- increase in several monooxygenase activities [8, 52, 53, 77, 78, 81–83, 86, 87, 89, 91–93]
- increase in cytochrome P-450 reductase activity [8, 52, 53, 89]
- changes in cytochrome P-450 CO-difference spectra because of enhanced content of distinct cytochromes [8, 18, 62, 65, 84]
- increase in size and shift of maximum wave length of substrate binding spectra [84, 87]
- change of substrate specificity [53, 65].

It must be mentioned that overall cytochrome P-450 content not necessarily correlates with any monooxygenase activities, since activity depends on the content of distinct cytochrome P-450 isoenzymes.

Induction effects depend on inducer and investigated tissue [65] and may be reversible [52]. The effects described support the suggestion that genes for distinct cytochrome P-450s may be activated, the isoenzyme pattern synthesized varies (see the measured qualitative changes) and in this way the detoxication capacity due to mfo's increases. Although the effects depend on inducer, the increased cytochromes are not strongly specific for metabolism of the inducing substance [87]. Other processes, such as changes in the membrane environment of the cytochrome, particularly lipid composition and/or fluidity, could be involved [87].

These results show that insects are able to adapt to chemical stress through regulation of their monooxygenase activity. Moreover, several authors reported that host plant induction of insect mfo's affects insect susceptibility to certain insecticides. For example and references see Chapt. 4, Table 10. It is also known that some insecticides induce cytochrome P-450s (see Table 1), but one cannot assume that they induce their own metabolism in insects, since they cause lethal effects before an inducing dose can be reached [93].

Changes of the gene activity due to induction are not inherited, whereas the ability of the genetic material to respond to inducers might be [3, 93]. It seems likely that not only the originally available monooxygenases and their ability to convert insecticide molecules, but also their regulation determine tolerance to insecticides.

For more detailed information on insect monooxygenases reviews [3, 10, 12, 25, 93] can be recommended.

2.2 Esterases

Another group of enzymes involved in insecticide metabolism are hydrolases with phosphorotriesterases and carboxylesterases as the most important ones. Esterases were classified by Aldridge [95] as early as 1953.

Phosphotriesterases (EC 3.1.1.2) belong to the A-esterases/arylesterases; whereas carboxylesterases (EC 3.1.1.1; aliesterases, non-specific esterases) are B-esterases. Other hydrolases taking part in insecticide metabolism are carboxylamidases (EC 3.5.1.4) and epoxide hydrolases (EC 3.3.2.3). They all catalyze cleavage of different bonds with water acting as the acceptor in the transfer of specific groups. [2,13,97]

A- and B-esterases may be distinguished by means of substrate specificity and effect of inhibitors. Although both types possess broad substrate specificity A-esterases hydrolyze p-nitrophenyl phosphate faster than p-nitrophenyl butyrate and B-type esterases vice versa. A-esterases are inhibited by sulfhydryl reagents (e.g. p-chloromercuribenzoate), heavy metal ions and EDTA (chelator), and B-esterases are inhibited by several organophosphates due to irreversible phosphorylation of the active site serin. [2,96,98]

Phosphorotriesterases cleave phosphorus ester or anhydride bonds. Possible reactions are demonstrated in Eq. 2..

Eq. 2:
R: alkyl group
X: halide/alkoxy or aryloxy group/another substituted phosphorus group linked by an anhydride bond

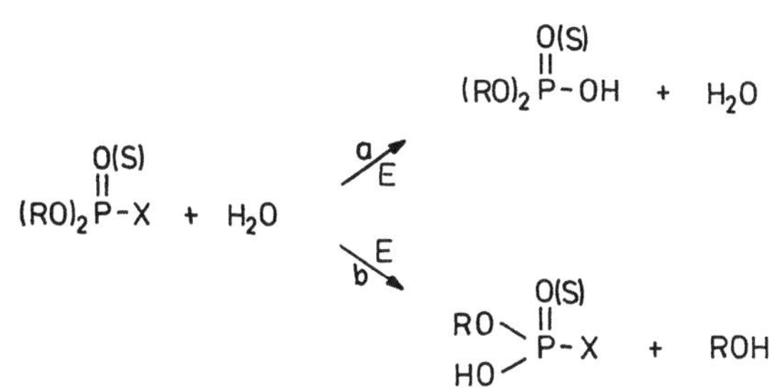

Phosphorotriesterase reactions involve two steps, after initial phosphorylation of the enzyme the substrate is cleaved. Substrate cleavage depends on pH, it is enhanced at high pH [13,99]. In general pH-optima of A-esterases range from pH 7.5 to pH 7.8 [100].

A-esterases were mainly found in mammalian plasma, but are also present in other mammalian tissues. Molecular weights range between 35 and 50 kDa. Phosphorotriesterases are stimulated by either Ca^{2+} or Mn^{2+} and Co^{2+}. Their biological significance consists in the formation of a phosphorus-containing metabolite in an anionic form at neutral pH. Since these metabolites are poor cholinesterase inhibitors, in this way parent compounds are detoxified [96].

Carboxylesterases catalyze cleavage of aliphatic and aromatic esters, according to the scheme in Eq. 3.

Eq. 3:
R^1: various possibilities
R^2: alkyl group rather than aryl group

$$R^1\text{-}C(=O)\text{-}O\text{-}R^2 + H_2O \xrightarrow{E} R^1\text{-}COOH + R^2\text{-}OH$$

Carboxylesterases are widely distributed in different mammalian tissues [96]. They attack non-ionic compounds and form more polar products. This increase in polarity decreases biological activity and promotes excretion of the substrates.

By means of electrophoretic separation and determination of enzyme activity with coloured reaction products — a suitable and often used method for investigation of esterases — existence of multiple forms of carboxylesterases was proved. Probably all species contain several isoenzymes with somewhat different substrate specificity [2].

In mammals carboxylesterases are bound to the cytosolic site of the endoplasmic reticulum and amount to 5–7% of the total microsomal protein. They may be glycosylated [98]. Carboxylesterases occur as monomers of about 60 kDa and as oligomers. Often they are trimers of identical subunits [98,101–103], each subunit containing one serin-bearing active site.

Their physiological importance might be the metabolism of a variety of endogenous as well as detoxication of foreign compounds [2,96,97].

The occurrence of carboxylesterases was found to be dependent on age [101], tissue, species and sex.

Induction by phenobarbital [101] and allelochemicals [104] was also described for esterases.

Organophosphates, carbamates, chelators and sulfhydryl reagents are inhibitors of esterases [105,106]. With the help of the indolalkaloid eserin the B-esterases carboxylesterase and cholinesterase may be distinguished, since only cholinesterases are inhibited by this compound [98]. Well known and often used esterase inhibitors are S,S,S-tributylphosphorotrithioate (DEF) [103,107–109], triphenylphosphate [103], DFP [105] and 0,0-diisopropyl-benzyl phosphorotrithiolate (IBP) [110,111]. Phenylsaligenin cyclic phosphonate was described as a selective esterase inhibitor [112]. Piperonyl butoxide was found to slightly inhibit esterases, although it is thought to be a specific inhibitor of monooxygenases [108].

Esterases in Arthropods

Compared with mammals, esterase titer is low in arthropods [97], esterases are mainly B-esterases, but A-esterases were also detected [96,97,99,113,114]. However, activity of phosphorotriesterases and carboxylesterases were proved in a lot of species, see Table 1. The physiological role of esterases is connected with metabolism of endogenous substrates [115,116] — e.g. juvenile hormone esterases regulate the content of this hormone [117] — but they take also part in the detoxication of xenobiotics with ester and anhydrid bonds [97].

Esterases of insects were found in cytosol, microsomes, mitochondria and nuclei, mainly in gut and thoracic muscles [118–122].

As in mammals, post-translational glycosylation of the enzyme protein occurs also in insects [123]. pH-Optima were described to be pH 6.6–8.6 [124–126], reported pI-values vary between pH 4.6 and pH 5.8 [105,127].

Insects and acari also possess several isoenzymes whith different but overlapping substrate specificity [108,119,120,128–135]. Esterase activity depends on age [68,114,124], species and strains and varies with the tissues studied [124,126,136] and may be regulated, for instance by its natural substrate juvenile hormone [117].

2.3 Glutathione S-Transferases

An important detoxifying reaction is conjugation of various foreign compounds by glutathione S-transferases (EC 2.5.1.18). These enzymes were proved in micro-organisms, plants and animals. Usually they are cytosolic and may represent 10% of total cytosolic protein (in mammalian liver). However, there were also found membrane-bound forms [137].

Biological functions of glutathione S-transferases consist of the metabolism of endogenous substrates like steroids, quinones and epoxides, the protection from lipid peroxidation products, biosynthesis of leukotrienes as well as detoxication. Exogenous compounds of different structure may be conjugated by glutathione S-transferases. In 1984 more than 3000 compounds were described to be conjugated by glutathione S-transferases [138]. Substrates (RX) only have to be lipophilic for binding at the enzyme and bear electrophilic centres for reaction with the nucleophilic glutathione (see Eq. 4).
Eq. 4:

$$RX + GSH \xrightarrow{GST} GSR + X^- + H^+$$

The structure of the tripeptide glutathione (GSH/γ-L-glutamyl-L-cysteinylglycin) is demonstrated in Fig. 2. Glutathione S-transferase reaction occurs at the sulfhydryl group of cysteine.

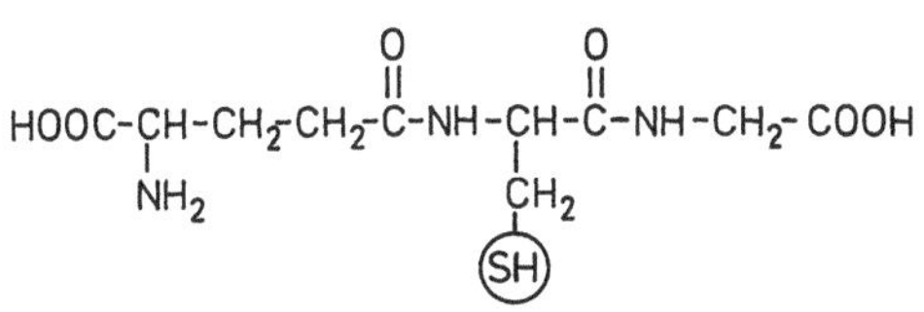

Fig. 2. Glutathione

The conjugates formed during this reaction may be further metabolized to mercapturic acids by several steps and/or excreted because of their increased water-solubility. In this way compounds may be dealkylated, dearylated, dehalogenated or cyanide may be released from organic thiocyanates, depending on the structure of R. In the last example, glutathione S-transferase reaction has toxic effects.

Note that metabolic reactions of GSH may also be non-enzymatic on the one hand, and there are other GSH-dependent enzymes than glutathione S-transferases on the other hand. An example might be DDT-dehydrochlorinase. In this case enzyme reaction with DDT does not yield a conjugate, but needs glutathione. It is still a subject of discussion, whether this enzyme is a glutathione S-transferase or not [139,140].

An important function of glutathione S-transferases is ligand binding not only to transport various substances, but also as a passive detoxication mechanism, if GSH concentration is markedly decreased in the cell [141]. Glutathione S-transferases also possess peroxidase activity [142].
(see reviews [2,143,144]).

Cytosolic glutathione S-transferases are homo- or heterodimers of several types of subunits with molecular weights of 20 to 27 kDa [143,144]. Only membrane-bound glutathione S-transferases were described as tri- and tetramers of smaller subunits with a molecular weight of 14 kDa [137].

In the cytosolic glutathione S-transferases each subunit contains a specific binding site for GSH (G-site) and a less specific site that recognizes lipophilic substrates (H-site) [143]. The subunits appear to behave independently with respect to enzyme activity [144].

Whereas some authors proposed a steady-state random mechanism for the 2-substrate-reaction of glutathione S-transferases [143,145], others suggested a hybrid mechanism of ordered sequential and ping-pong reaction [146,147]. Under physiological conditions GSH is already bound to the G-site, since normally a surplus of GSH is present in the cell (GSH-concentration of 3 to 10×10^{-10} mol/l). Therefore in-vivo reaction should be ordered sequentially [145,146].

Existence of glutathione S-transferase isoenzymes with different but overlapping substrate specificity was proved in several species. They differ in subunit composition, size, amino acid content as well as pH-optima [143,144,149]. Isoelectric points range from pH 5 to pH 10 [143].

Occurrence of the glutathione S-transferases varies with age [150-154] as well as between species [153-155], sexes and tissues [153,156,157].

Like cytochrome P-450 dependent monooxygenases glutathione S-transferases are inducible by a series of compounds, among them being phenobarbital, methylcholanthrene and β-naphthoflavone [143,153,158] and several insecticides (see table 2). Each inducer increases distinct glutathione S-transferases or subunits [164].

Several possibilities of inhibiting glutathione S-transferases by various compounds are described [165]. Since the glutathione S-transferase subunits contain two binding sites, not only blocking the specific G-site by glutathione analogues but also binding of inhibitors at the non-specific H-site influence enzyme activity [145,166].

Triphenylmethanes, for example, seem to bind at the H-site and, because they are large molecules, they also block the G-site. Tridiphane was reported to form glutathione conjugates, which inhibit glutathione S-transferase reaction [167,168].

Diethylmaleate (DEM), phenylbutenone [169], dithiocarb [170], 2,4-Dichlorophenoxyacetate (2,4-D) and 2,4,5-Trichlorophenoxyacetate (2,4,5-T) [171], sulfobromophthalein [97], mercury compounds [172], bile acids [173] and several thiol reagents [147] are inhibitors of glutathione S-transferases. Available reduced glutathione — an important factor for reaction — may be decreased by glutathione-binding reagents, like diethylmaleate and methyl iodide [174] as well as inorganic and organic mercury compounds [172] and thiol-blockers [148], like N-ethylmaleimide [147].

Although synergistic effects with several pesticides are known for diethylmaleate and N-ethylmaleimide, it is not proved that these effects originate in their

Table 2. Insecticides and synergists that induce glutathione S-transferases (in vertebrates/compounds marked with (i) also caused induction of insect glutathione S-transferases)

Compounds	Refs.
Organohalogens	
Arochlor 1254	159)
Chlordane	160)
DDT (i)	161, 162)
Dieldrin (i)	161)
Endosulfan	160)
HCH	162)
Polychlorinated biphenyls	162)
Phosphororganics	
Chlorpyriphos	160)
Etrimphos	160)
Malathion	160)
Methylparaoxon (i)	161)
Methylparathion (i)	161)
Parathion (i)	161)
Tetrachlorvinphos	160)
Carbamates	
Carbaryl (i)	161)
Pyrethroids	
Permethrin	160)
Several fumigants (i)	163)
Synergists	
Piperonyl butoxide (i)	160, 161)
Sesamex	160)

inhibition of glutathione S-transferases, since they also influence other detoxifying systems, mixed function oxidases and esterases, respectively.

Glutathione S-Transferases in Arthropods

Glutathione S-transferase activity was determined in several insect species (see Table 3).

Subunits with molecular weights of 20 to 25 kDa form dimers of 45 to 50 kDa [148, 175–180]. However, subunits of 28.5 kDa and 35 kDa were reported for *Drosophila melanogaster* [181], and subunits of glutathione S-transferase isolated from the mite *Rhizoglyphus robini* are polypeptide chains of 25 kDa [182].

Multiple forms of glutathione S-transferases, usually purified by affinity chromatography on glutathione-marked materials, were found in various insect species, e.g. *Costelytra zealandica* [178, 183], *Musca domestica* [177, 184–186], *Periplaneta americana* [187], *Wiseana cervinata* [179]. Only one single protein with glutathione S-transferase activity could be proved for *Galleria mellonella* [148], *Ceratitis capitata* [180], *Triatoma infestans* [188] and *Drosophila melanogaster* [181].

Table 3. Detoxifying enzymes in arthropods (General view of literature). In any case enzyme activity was determined with model substrates or insecticides. Since this table cannot be complete it mainly gives on overview on well established orders. A wide variety of publications on monooxygenases in about 100 insect species was summarized by Hodgson [25], where detailed information can be found

Order species	Detoxifying enzymes			Refs.
Blattodea				
Periplaneta americana	mfo			6,76)
			GST	187,200)
Heteroptera				
Triatoma infestans	mfo			114)
		esterases		99,114,201)
			GST	147,188)
Homoptera				
Myzus persicae		esterases		123,202)
Nephotettix cincticeps	mfo	esterases	GST	203)
Nilaparvata lugens		esterases		110,204,205)
			GST	110)
Coleoptera				
Antonomus grandis grandis	mfo	esterases	GST	69,93)
Costelytra zealandica			GST	178,183)
Dendroctonus terebronus	mfo			206)
Oryzaephilus surinamensis	mfo		GST	207)
Leptinotarsa decemlineata	mfo			68,208,209)
		esterases	GST	208,209)
Pitygenes chalcographus		esterases		134)
Rhizopertha dominica	mfo			210)
Sitophilus granaris			GST	211)
Tragoderma granarium			GST	163)
Tribolium spp.	mfo			212)
		esterases		113)
			GST	213)
Hymenoptera				
Apis mellifera	mfo			214)
			GST	215)
Oncophanes americanus	mfo	esterases	GST	216)
Lepidoptera				
58 species in 22 families	mfo			217)
Anticarsia gemmatalis	mfo			79)
Argotaenia citrana	mfo	esterases	GST	216)
Californica autographa	mfo			88)
Diatraea saccharalis		esterases		218)
Galleria mellonella			GST	148,219)
Heliothis spp.	mfo			57,72,84)
				85,220,221)
			GST	72,84,85)
Lymantria dispar		esterases		126)
Manduca sexta	mfo			222)
Peridroma saucia	mfo			86,87)
Platynota idaensalis	mfo			189)
Pseudaletia unipuncta		esterases		223)

Table 3 (continued)

Order species	Detoxifying enzymes			Refs.
Pseudoplusia includens		esterases		105,124)
				196)
			GST	196)
Spodoptera spp.	mfo			58,59,77)
				78,80,82)
				89,224,225
		esterases		136,225)
			GST	78,80,82)
				89,224–226)
Trichoplusia ni	mfo			88)
		esterases		227)
			GST	192)
Wiseana cervinata			GST	179)
Diptera				
Anopheles spp.	mfo			
		esterases		135,228)
			GST	229)
Ceratitis capitata		esterases		230)
			GST	180,230)
Culex spp.	mfo			231,232)
		esterases		122,130)
				232–234)
			GST	191,232)
Diploptera punctata	mfo			73)
Drosophila melanogaster	mfo			51,92,234)
				237)
			GST	181,198)
				235,236)
Encalliphora lilica	mfo			91)
Lucilia spp.	mfo			91)
		esterases		131,221)
				238,239)
Musca domestica	mfo			8,20,54)
				55,60,62)
				63,66,
				240–242)
		esterases		125,133,
				243–245)
			GST	106,119,140,
				161,168,175,
				176,184–186)
				194,200)
				246–248)
Acari				
Amblyseius fallacis	mfo		GST	249)
		esterases		108)
Boophilus microplus		esterases		127,250)
Rhizoglyphus robini			GST	182)
Tetranychus spp.	mfo			249)
		esterases		129)
			GST	182,249)
Varroa jacobsoni			GST	216)

Enzyme activity and glutathione content differ with developmental stage, between species and sexes [68,72,75,97,181,188-191]. Among several tissues glutathione S-transferase content is especially high in fatbody [80,184,187,192]. Small but still measurable amounts of microsomal glutathione S-transferase were found in *Triatoma infestans* [188].

Glutathione S-transferase activity depends on nutritional factors. Induction was shown with phenobarbital [185,193,194], 3-methylcholanthrene [185], pentamethylbenzene [195], some insecticides and fumigants (see table 2) and various plant allelochemicals. For instance, several terpenes and coumarin induced glutathione S-transferase in *Spodoptera eridania* [89], cumestrol induced the enzyme in *Pseudoplusia includens* [196]; *Anthonomus grandis* and *Heliothis virescens* showed higher enzyme activity, if larvae were fed with (+)-α-pinene, β-caryophyllene, umbelliferone or scopoletin [93]. Glutathione S-transferase activity of *Heliothis virescens* was influenced by tomato leaves, at first it was lower after 24 hours, and then increased above control level after 3 days [84].

Specific forms of glutathione S-transferases were induced in *Musca domestica* by treatment with phenobarbital, highest levels of induction were observed after 48 hours. Significant decrease of Km occurs rather than an increase of Vmax [185,193,194,197]. The induction effect varies with the strain. Usually it is lower, if basal enzyme activity is already high [185,193,198].

Catalytic mechanism and active site of insect glutathione S-transferases are intensively studied. Cytosolic or further purified preparations are taken to determine enzyme activity. An easy method for estimation of activity is incubation with model substrates and spectral determination of conjugates [149]. Note that enzyme activity may be influenced by endogenous inhibitors like quinones [199]. Contrary to vertebrate enzymes a single ligand-binding and catalytic site per dimeric enzyme molecule was suggested for the *Galleria mellonella* enzyme [141]. With the *Costelytra zealandica* enzyme it was shown that two ionization steps might be involved in the reaction [178]. A steady-state random mechanism was proposed for enzymes of *Galleria mellonella* [148], *Costelytra zealandica* [178] and *Wiseana cervinata* [179].

In general insect glutathione S-transferases resemble those of vertebrates regarding catalytic mechanism, substrate specificity and enzyme inhibition. That means that the possibility of preferentially blocking the conjugation with GSH of pesticides in insects, exploiting qualitative differences between the insect and vertebrate enzymes seems a less promising prospect [175,178,180]. Differences in reaction capacity between vertebrates and insects may primarily be a result of different isoenzyme patterns.

3 Enzyme-catalyzed Reactions in Insecticide Molecules

3.1 General Remarks

Insecticides are metabolized by various enzyme systems. Certainly monooxygenases are the most important enzymes modifying insecticide molecules, since they catalyze a wide variety of oxidative reactions at a multitude of functional groups. So they attack all three types of insecticides regarded here — organophosphorus compounds, carbamates and pyrethroids. Often monooxygenase reaction is the first step of degradation [2,10,13,25].

Since these insecticides contain ester bonds they may be hydrolyzed by esterases, however, this reaction plays a minor role in the case of carbamates [2,13,96].
Glutathione S-transferases attack mainly organophosphorus insecticides, reaction of these enzymes with carbamates and pyrethroids has not been reported until now [2,10,13,119,251]. Besides these main insecticide metabolizing enzymes other are involved in attack of special functional groups, for instance, carboxylamidases hydrolyze the amide bond of dimethoate (see Table 5).

The type of enzyme reaction depends on structure of the insecticide molecule, but also on species — this is of special interest with respect to toxicity to mammals and selectivity of an active agent — developmental stage and diet (see chapter 2; differences in enzyme content and activity) as well as on the method of administration (e.g. due to differences in transport of the insecticide and enzyme content of organs) [13].
Coadministration of insecticides may alter insecticide toxicity due to influences on their metabolism — e.g. competition at the same enzyme and inhibition of metabolizing enzymes (see Chapt. 2, organophosphates inhibit B-esterases) [227,252–254] — resulting in phenomena like synergism and antagonism.

Determination of the enzyme catalyzing a defined reaction at an insecticide molecule includes several difficulties. For instance, activity of different enzymes may yield in the same products, and therefore the pattern of metabolites may only give hints regarding the enzyme system involved. In addition to this, some metabolites may be immediately converted so that they cannot be determined, and several compounds originate from spontaneous reactions. So in any case it is necessary to ensure the involvement of enzymes. There are three main possibilities to distinguish enzymes in vitro without completely purifying them:

1) Utilization of the cellular localization of the enzymes, e.g. monooxygenases and glutathione S-transferases may be separated by preparation of microsomes and cytosolic fractions, respectively (note exceptions, see Chapt. 2)
2) Comparison of metabolism with and without cofactors, e.g. NADPH and O_2 in case of monooxygenases and glutathione in case of glutathione S-transferases (note that FAD-dependent monooxygenases also require NADPH and O_2)
3) Determination of effects of special inhibitors, e.g. piperonyl butoxide and sesamex to inhibit monooxygenases or triphenyl phosphate and S,S,S-tributylphosphorotrithioate to inhibit esterases (note that inhibitors may also affect other enzymes, and until now there is no specific inhibitor of glutathione S-transferases; see Chapt. 2)

Most reliable, of course, are results yielded with purified enzymes, but this needs extensive experimental work.

Methods to determine enzyme activity in-vitro with model substrates are relatively easy since they base on spectrophotometric determination of products. This is only seldom possible with insecticide molecules. However, because of differences in substrate specificity of the various isoenzymes (see Chapt. 2) conversion of model substrates may not be representative for the ability to attack insecticides. Involvement of enzymes in the metabolism of distinct compounds is usually investigated with radiolabelled molecules. This procedure involves four main steps, namely incubation of the radiolabelled insecticide under different conditions (cofactors, inhibitors) with enzyme preparations, partitition of reaction mixture with an organic solvent, chromatographic separation of metabolites, identification and quantitation of metabolites.

In general, all in-vitro experiments include a substantial disadvantage: they do not reflect the real situation in living organisms, since in-vivo and in-vitro enzyme activity may differ and influences of cell compartmentation and other enzymes are not considered. Therefore in-vivo experiments with special inhibitors and determination of metabolites occurring in-vivo may usefully complete and support results of in-vitro measurements.

The following chapter demonstrates important insecticide metabolizing reactions and sites of organophosphorus, carbamate and pyrethroid insecticides being sensitive to enzymatic degradation.

Literature data on proved enzyme-catalyzed reactions in these molecules are summarized and demonstrated for several well established examples. The schemes do neither reflect degree or sequence of reactions nor further metabolism of the products. Moreover, variations between species and strains are not involved. Only general differences between mammals and insects are discussed and in the tables of schemes references dealing with insecticide metabolism in insects and acari are marked with 'i' and 'ac', respectively.

See reviews on insecticide metabolism [2,10,13,25,96,119,255] for more detailed information.

3.2 Organophosphorus Compounds

Because of their structure organophosphorus compounds may be metabolized by monooxygenases [10], esterases [96,97] and glutathione S-transferases [97,119].

In mammals as well as in insects cytochrome P-450 dependent monooxygenases catalyze activation of sulfur-containing compounds, namely oxidative desulfuration

Table 4. Organophosphorus compounds that cause type I binding spectra with cytochrome P-450

Compound	Mammals	Insects	Refs.
Azinphosmethyl	×	×	260)
Azodrin		×	261)
Coumaphos	×	×	260)
Crufomate	×	×	260)
Diazinon	×	×	260)
Dimethoate	×	×	260)
Disulfoton	×		47)
Etrimfos	×		47)
Fonofos	×		47)
Malathion	×	×	222,260)
Menazon	×	×	260)
Monocrotophos	×	×	260)
Parathion	×		47)
Phorate	×		47)
Sulprofos	×		47)
Trichlorfon	×	×	260,261)

of phosphorothionates (P=S) and sulfoxidation of thioethers (C—S—C) and phosphorothiolates (P—S—C)[43,256-258]. They take part in various detoxication reactions like O- and N-dealkylation, hydroxylation, cleavage of P—O— and C—C-bonds and dehalogenation, too[10,25].

Distinct chiral isomers (e.g. in case of fonofos[259]) may be preferentially converted, and reactions with different steric isomers may have different effects. Toxicity of (—)-profenofos is greatly enhanced by sulfoxidation, whereas toxicity of (+)-profenofos is lowered, caused by different sensitivity of the target acetylcholinesterase towards the non-oxidized and oxidized isomers of this insecticide[256-258]. The situation seems to be similar with isofenphos[112].

Type I binding spectra of various organophosphorus compounds with cytochrome P-450 (for examples see Table 4) demonstrate the general possibility of cleavage by cytochrome P-450 dependent monooxygenases. Note that type I spectra with insect cytochrome P-450 were mainly found with cytochromes of resistant animals.

Probably oxidative desulfuration leads via a cyclic intermediate, monooxygenases only catalyzing the formation of the phosphooxythirane, which may spontaneously decay into two products[25] (see Eq. 5).

Eq. 5:

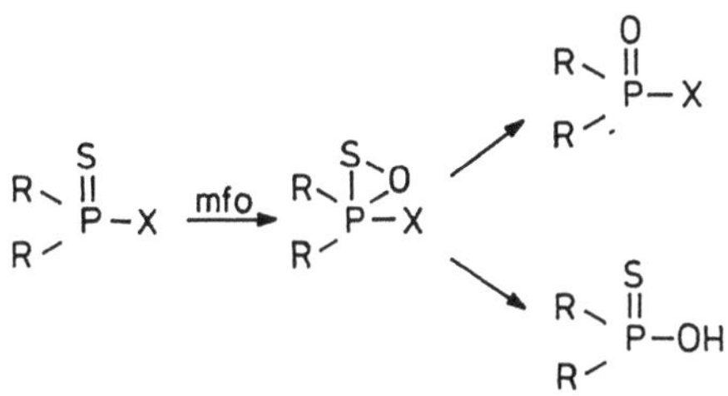

Several thioether-containing organophosphorus insecticides, e.g. demeton analogues disulfoton, fenthion, methylcarbophenothion, phorate, sulprofos, terbufos[47,49,262], were found to be substrates for sulfoxidation at their thioether-bond by FAD-dependent monooxygenases. Flavin-containing monooxygenases form only its sulfoxides[46,47], whereas cytochrome P-450 dependent monooxygenases also convert the sulfoxides to sulfones.

The phosphonothionate fonofos is desulfurated by flavin dependent monooxygenases[46,47]. Since cytochrome P-450 and flavin dependent monooxygenases yield in products of opposite chirality (see Chapt. 2), they may influence insecticide toxicity in a different manner.

Until now metabolism of insecticides by FAD-dependent monooxygenases has not been described in insects.

Carboxylesterases and phosphorotriesterases cleave C—O- and P—O-bonds both in mammals and insects. Different isoenzymes preferentially cleave P=O- or P=S-containing insecticides[13]. Carboxylamidase — a hydrolase that metabolizes dimethoate — until now has been detected only in mammals.

The most frequent reaction catalyzed by glutathione S-transferases in organophosphorus compounds is O-dealkylation. The structure of the 'leaving group' affects the rate of O-alkyl-conjugation. Both mammalian and insect enzymes prefer methoxy groups. Enzyme activity decreases with length of the alkyl-chain[119].

Table 5. Enzyme-catalyzed metabolism of organophosphorus insecticides (references dealing with insecticide metabolism in insects are marked with "i")

Compound	Reaction type	Enzymes involved	Refs.
Azinphosmethyl 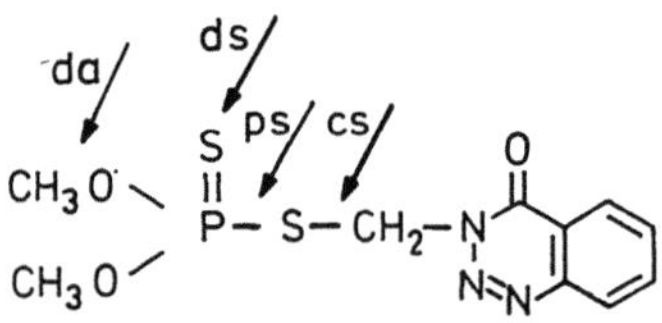	da	GST	13,97,119) 189,265(i)) 266(i))
	ps	Oxidative	267)
		Hydrolytic	267)
	ds	Oxidative	267)
	cs	GSH-dependent	267)
Diazinon 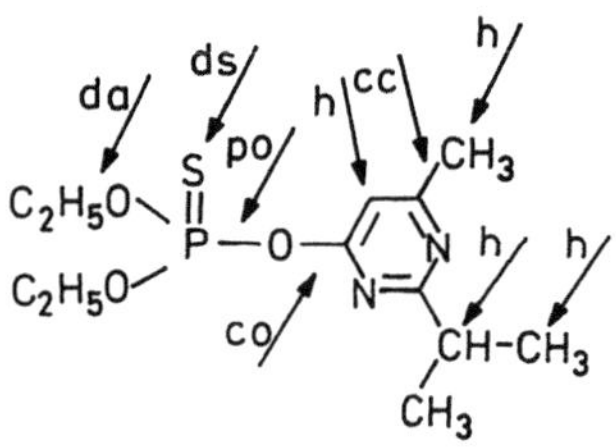	da	mfo	13,266(i))
		GST	97,119,140) 266(i))
	ds	mfo	10(i),13, 85(i))
	h	mfo	13,85(i))
	po	mfo	10(i),13)
		esterases	13,266(i))
		GST	266(i))
	co	mfo	266(i))
		GST	13,85(i), 119)
	cc	mfo	266(i))
Dichlorvos 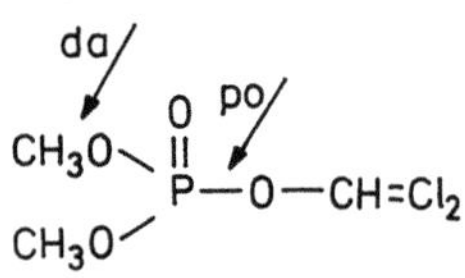	da	GST	13,119)
	po	esterases	97)
Dimethoate 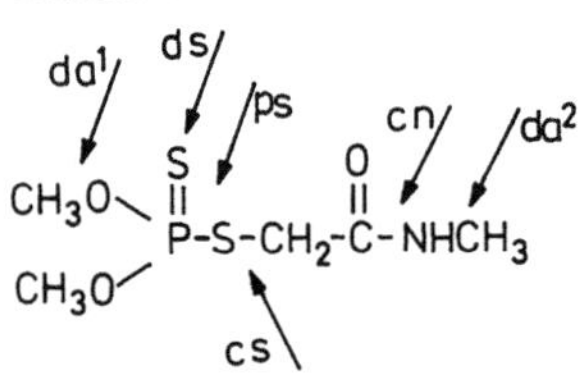	da 1	mfo	10(i),266(i))
		GST	97,266(i))
	da 2	mfo	266(i))
		hydrolysis	13)
	ds	mfo	266(i))
	ps	mfo	266(i))
		hydrolysis	13)
	cs	mfo	266(i))
	cn	carboxyl-amidase	2,96,97)
EPN 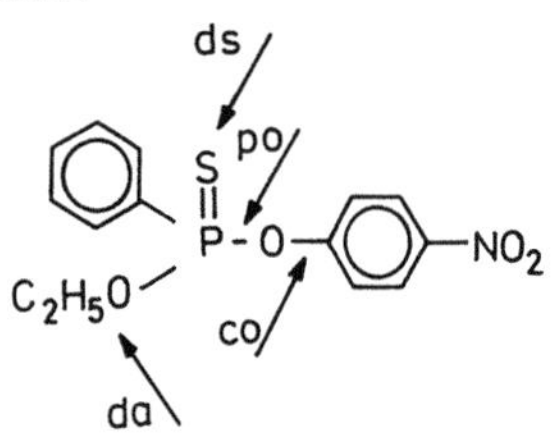	ds	mfo	263,268(i))
	po	mfo	263,268(i))
		hydrolysis	263)
		GST	119)
	co	GST	119,263) 268(i))
	da	mfo	263,268(i))

Table 5. (continued)

Compound	Reaction type	Enzymes involved	Refs.
Etrimfos 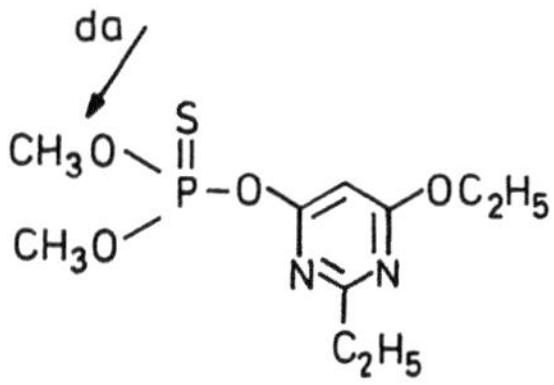	da	mfo GST	269) 119,269) 270(i))
Fenitrothion 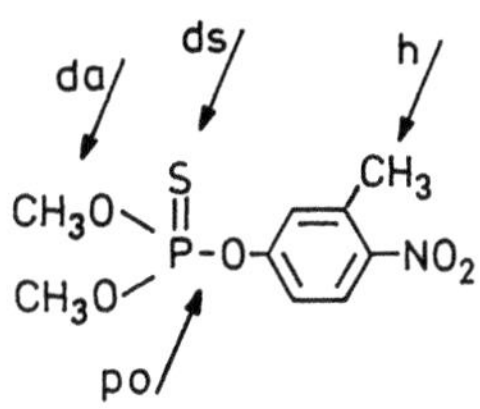	da	GST	119,271, 272(i),273(i))
	ds	mfo	272(i)) 273(i))
	po	mfo	271,272(i)) 273(i))
	h	mfo	272(i)) 273(i))
Fonofos 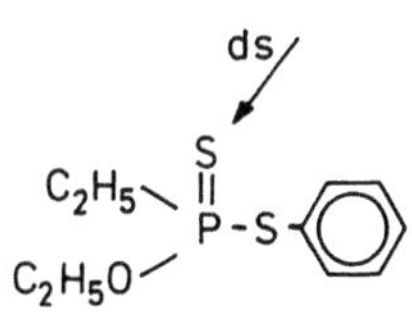	ds	mfo MFMO	43,259) 46,47)
Isofenphos 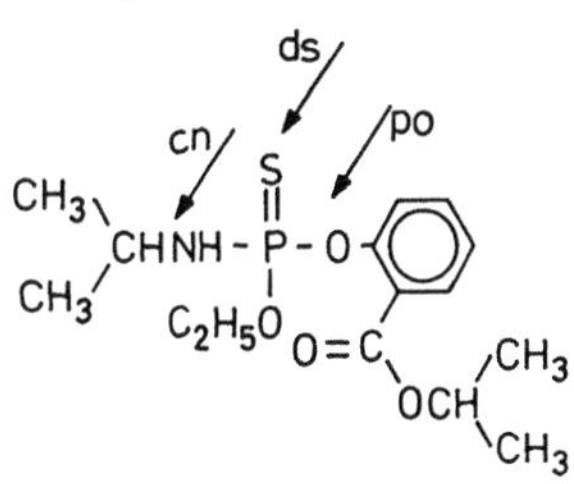	cn	mfo	112(i))
	ds	mfo	112(i))
	po	esterases	112(i))
Malathion 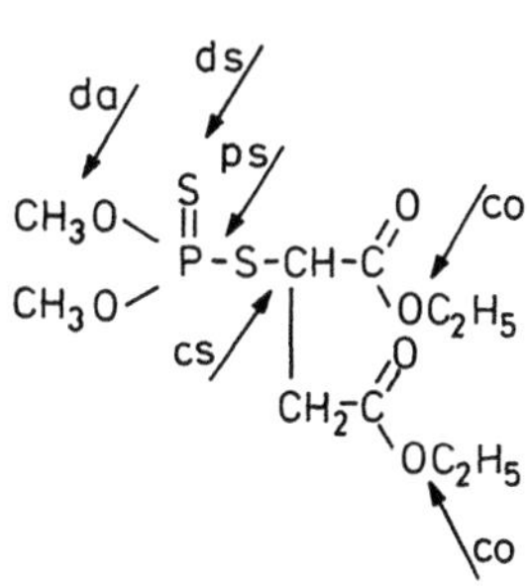	da	GST	119,274)
	ds	mfo	13,210(i)) 266(i))
	ps	mfo esterases	210(i)) 13,210(i))
	co	carboxyl-esterases	97,102,103) 114(i),201(i), 210(i),243(i), 266(i))
	cs	mfo carboxyl-esterase	266(i)) 266(i))

Table 5. (continued)

Compound	Reaction type	Enzymes involved	Refs.
Methidathion 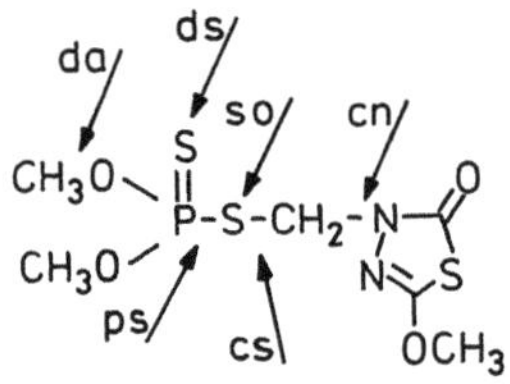	da	mfo/GST	275)
	ds	mfo	275)
	ps	enzymatic hydrolysis	275)
	cs	GSH-dependent	275)
	cn	mfo	275)
	so	MFMO/mfo	275)
Methylparathion 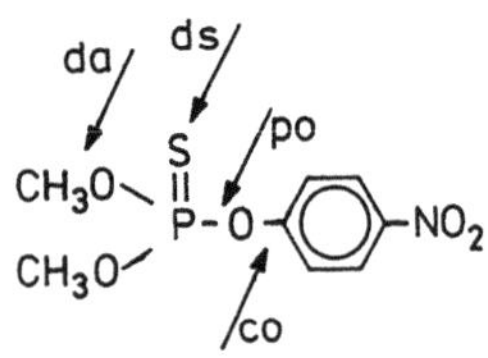	da	mfo	4(i))
		esterases	2)
		GST	4(i), 97(i)) 119)
	ds	mfo	4(i))
	co	GST	4(i), 119)
	po	mfo	4(i))
		esterases	2, 4(i))
Mevinphos 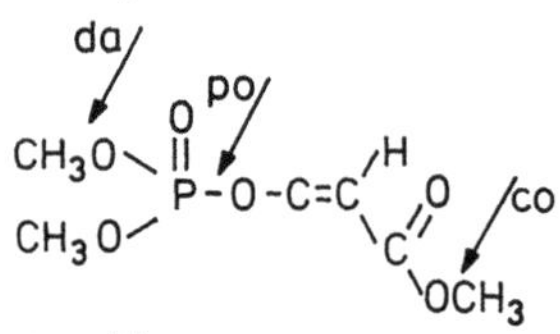	da	GST	119)
	co	esterases	13)
Parathion 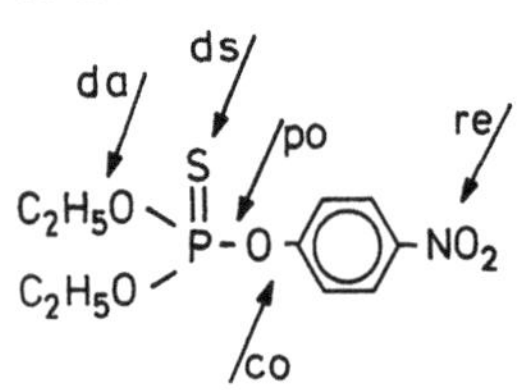	da	esterases	266(i))
		GST	97(i), 119) 266(i))
	ds	mfo	10, 13(i), 25, 268(i), 276)
	co	GST	13(i), 97(i), 119, 266(i))
	po	mfo	25(i), 43) 262, 266(i)) 276, 277)
		esterases	13, 96, 97(i), 99) 114(i), 276) 278, 279(i))
	re	nitratreductase	97(i))
Profenofos 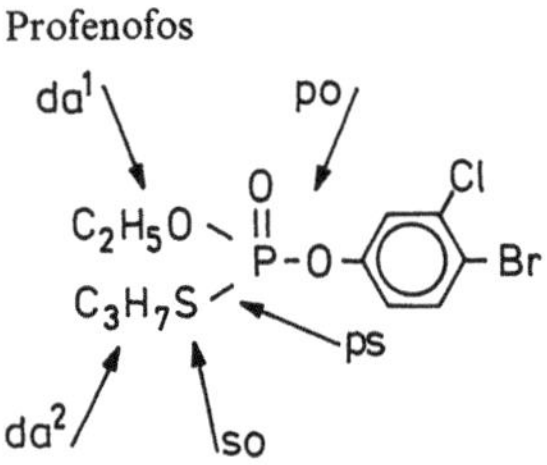	da	mfo	257)
	so	mfo	43, 256(i), 257, 258(i))
	po	mfo	257)

Table 5. (continued)

Compound	Reaction type	Enzymes involved	Refs.
Propetamphos 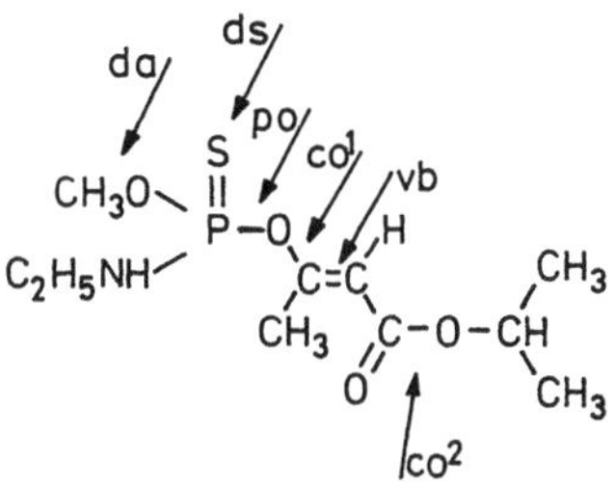	da		280(i)
	ds	mfo	280(i)
	po	mfo	280(i)
		esterases	280(i)
	co1	esterases	280(i)
	co2	GST/2 conjugates possible	264(i), 280(i)
Tetrachlorvinphos 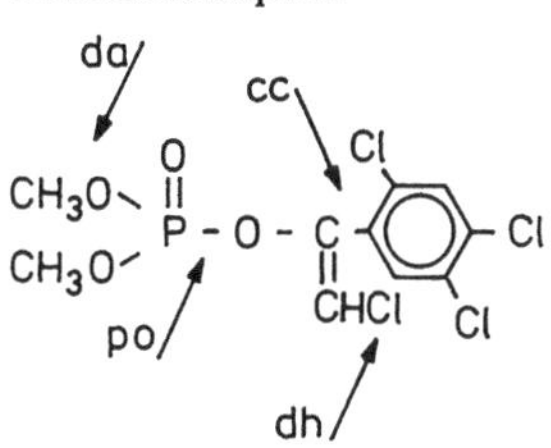	da	mfo	266(i)
		GST	119, 184, 266(i)
	po	GST	266(i)
	dh	GSH-dependent	281
	cc	Carboxylesterase	266(i)
Thiovamidothion 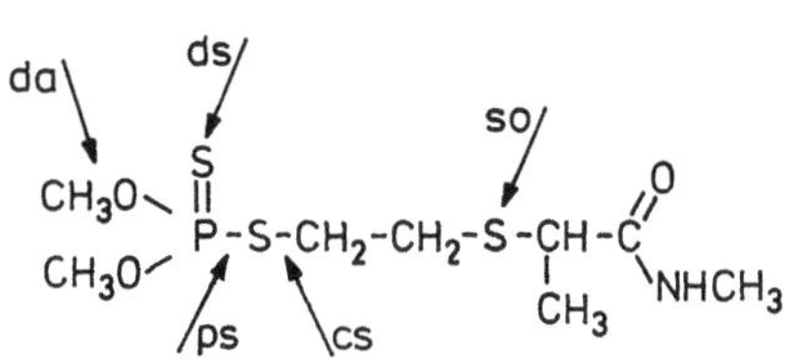	da	GST	247(i), 282
	ds	mfo	247(i), 282
	so	mfo	247(i), 282
	ps	esterases	282
		hydrolysis	247(i)
	cs	hydrolysis	247(i), 282
Trichlorfon 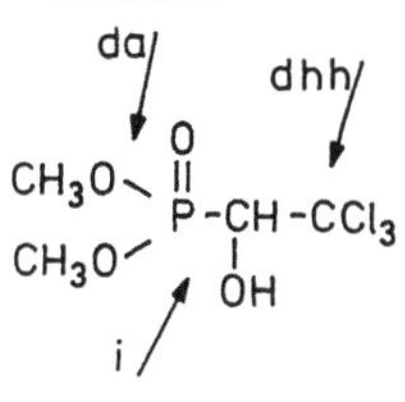	da	mfo	283(i)
		esterases	283(i)
		GST	13, 97, 283(i)
	dhh	mfo/GST	283(i)
	i	isomerization to dichlorvos	13, 283(i)

Therefore for both methylparathion and parathion monooxygenases, but only in case of methylparathion glutathione S-transferases influence detoxication via dealkylation. Phosphorothionates seem to be better glutathione S-transferase substrates than the corresponding phosphates [119]. Other glutathione S-transferase catalyzed reactions are O-dearylation, phosphonate conjugation (e.g. EPN) [119, 263] and the attacking of vinyl bonds (propetamphos) [264].

Figure 3 shows enzyme-catalyzed reactions in organophosphorus compounds and the enzymes that may be involved; in Table 5 several organophosphorus insecticides and enzymes found to be involved in their metabolism are listed.

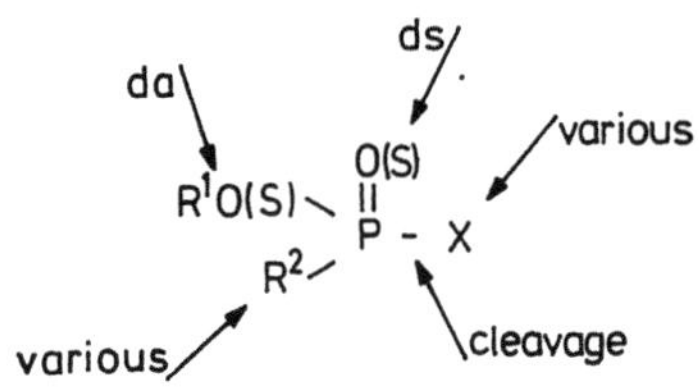

Fig. 3. Enzyme-catalyzed metabolism of organophosphorus insecticides

Abbrevation[1]	Reaction type	Enzymes catalyzing the reaction
da	dealkylation	mfo, esterases, GST
ds	desulfuration	mfo, MFMO
so	sulfoxidation	mfo, MFMO
h	hydroxylation	mfo
po	P-O-cleavage	mfo, esterases, GST
ps	P-S-cleavage	mfo, esterases
pc	P-C-cleavage	mfo
co	C-O-cleavage	esterases
cn	C-N-cleavage	carboxylamidase
cs	C-S-cleavage	mfo, esterases, GST
dh	dehalogenation	mfo, GST
vb	attack at vinyl bond	GST
dhh	dehydrohalogenation	mfo, GST
re	reduction	reductase
i	isomerization	

[1] Used in Table 5 and Fig. 3

3.3 Carbamates

Carbamates are mainly metabolized by mammalian and insect cytochrome P-450 dependent monooxygenases, which catalyze either activation due to sulfoxidation of thioether bonds (C—S—C) or detoxication by O- and N-dealkylation, epoxidation, alkyl- and ring hydroxylation [3,13].

Table 6. Carbamates that cause type I binding spectra with cytochrome P-450

Compound	Mammals	Insects	Refs.
Aldicarb	×	×	47,260)
Bufencarb	×	×	260)
Carbaryl	×	×	10,57,260,261)
Carbofuran	×	×	260)
Dimetilan	×	×	260)
Methomyl	×	×	260)
Propoxur	×	×	260,261)

Table 6 contains carbamates that cause type I binding spectra with cytochrome P-450.
Some thioether-containing carbamates, namely aldicarb, croneton, methiocarb, thiofanox, were described to be activated due to sulfoxidation by mammalian flavin-containing monooxygenases [47-49], but they are not as good substrates as organophosphorus compounds [49].

Since the esterase reaction produces less polar metabolites these enzymes are not qualified for detoxication of carbamates [13].

Glutathione-conjugation of carbamates was not found, but conjugation of hydroxylation products with glucose, glucuronide and sulfate is involved in the metabolism.

General enzyme reactions in carbamates are shown in Fig. 4; some examples are listed in Table 7.

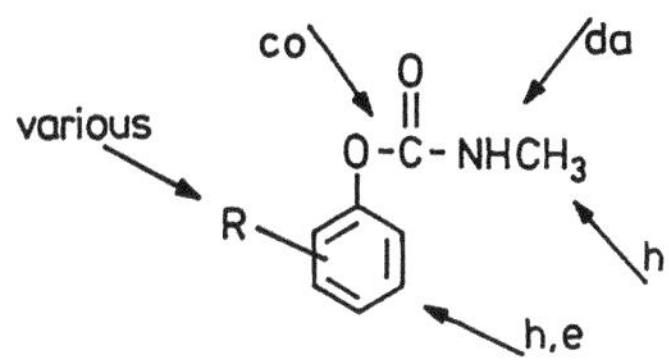

Fig. 4. Enzyme-catalyzed metabolism of carbamates

Abbreviation[1]	Reaction type	Enzymes catalyzing the reaction
co	C-O-cleavage	esterases
da	dealkylation	mfo
e	epoxidation	mfo
h	hydroxylation	mfo
so	sulfoxidation	MFMO

[1] Used in Table 7 and Fig. 4

3.4 Pyrethroids

Ester cleavage and hydroxylation are the main reactions in pyrethroid molecules both in mammals and insects [2,13,25,257,284]. Combination of these reactions with glucoside and sulfate conjugation yields in a large pattern of metabolites, for instance, permethrin is degraded into about 80 compounds [255].

Several carboxylesterases catalyze cleavage of the parent molecule into its acid (e.g. chrysanthemum acid) and alcohol (e.g. phenoxybenzol) moiety [97,255]. Although there are exceptions [285], usually trans-isomers are more susceptible to esterase attack than cis-isomers [62,86-90]. This is supported by higher toxicity of the cis-isomers. Since hydrolysis plays a minor role these isomers are mainly detoxified by oxidative reactions [290,291].

Cytochrome P-450 monooxygenases catalyze hydroxylation of methyl groups, epoxidation at double bonds (e.g. resmethrin) and ring cleavage (e.g. kadethrin) in the acid moiety [255] and hydroxylation as well as epoxidation in the alcohol moiety [2,255] being more susceptible to oxidative attack [292]. Hydroxylation of phenoxybenzoles occurs preferentially at the 5-position if they contain cyanide (Y = CN in picture 5) and at the 6-position if Y = H [255,293]. Unsubstituted phenoxybenzyl moieties are

Table 7. Enzyme-catalyzed metabolism of carbamates (references dealing with insecticide metabolism in insects are marked with "i")

Aldicarb 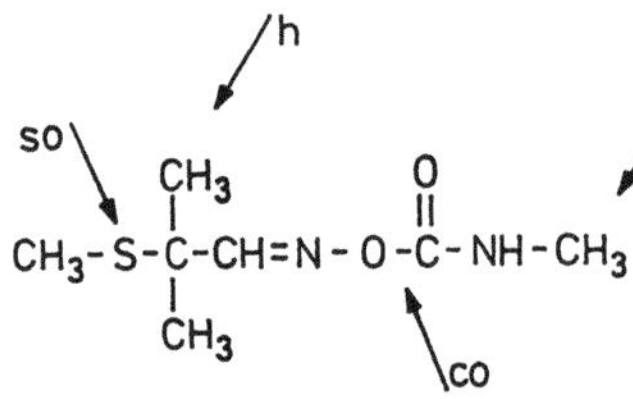	co	esterases	[4(i)]
	so	MFMO	[4,47,49]
	h	mfo	[4(i)]
Carbaryl 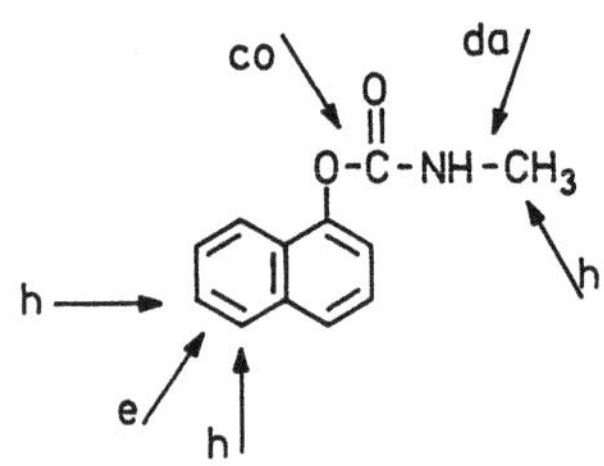	da	mfo	[2]
	h	mfo	[10(i),13]
	co	esterases	[13]
	e	mfo	[10(i),13]
Carbofuran 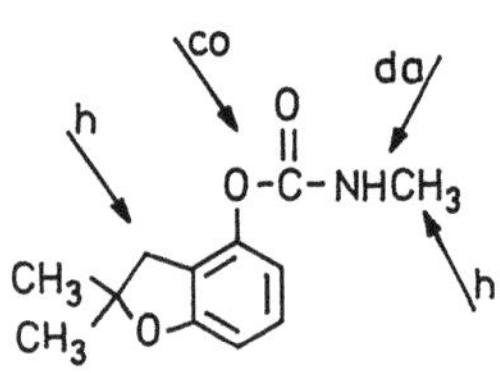	da		[2,13]
	h	mfo	[10(i),13]
	co	esterases	[13]
Propoxur 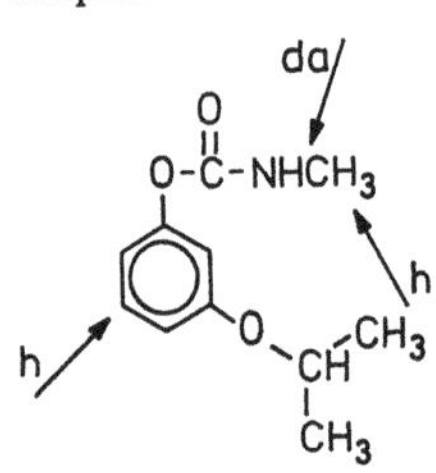	h/da	mfo	[10(i),13]

rather attacked than substituted [292]. Sites of hydroxylation also depend on species [255, 293].

Despite of cyanide release all reactions with pyrethroids lower their toxicity [255].

Incubation of cytochrome P-450 with several pyrethroids yielded type I spectra (see Table 8).

Probably neither FAD-dependent monooxygenases nor glutathione S-transferases are involved in pyrethroid metabolism.

Figure 5 demonstrates enzyme-catalyzed metabolism of pyrethroids, in Table 9 several examples are listed.

Table 8. Pyrethroids that cause type I binding spectra with cytochrome P-450

Compound	Mammals	Insects	Refs.
Pyrethrin	×	×	260)
Allethrin	×	×	222,260)
Bioallethrin	×		260)
Bioresmethrin	×		260)
Dimethrin	×	×	260)
Permethrin		×	57)

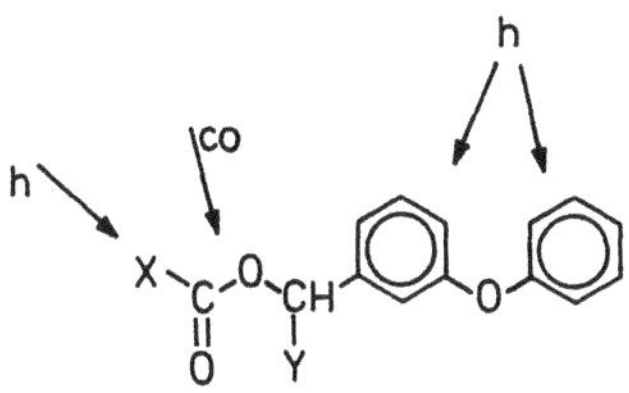

Fig. 5. Enzyme-catalyzed metabolism of pyrethroids

Abbreviation[1]	Reaction type	Enzymes catalyzing the reaction
co	C-O-cleavage	esterases
h	hydroxylation	mfo
dp	diphenylether-cleavage	

[1] Used in Table 9 and Fig. 5

Table 9. Enzyme-catalyzed metabolism of pyrethroids (references dealing with insecticide metabolism in insects and acari are marked with "i" and "ac", respectively)

Cypermethrin 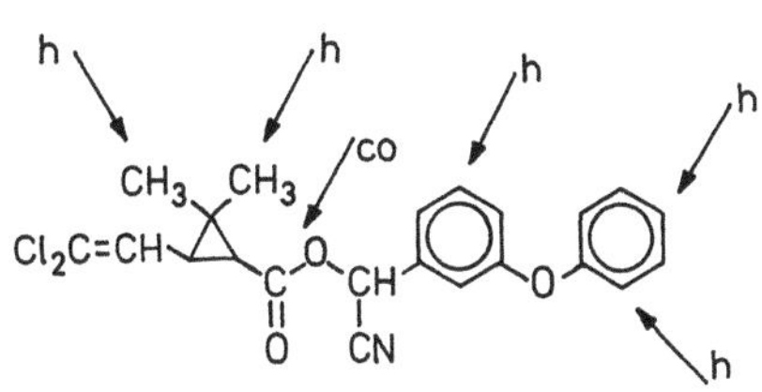	co h	esterases mfo	13,127(ac)) 227(i),255) 287,293(i))
Decamethrin 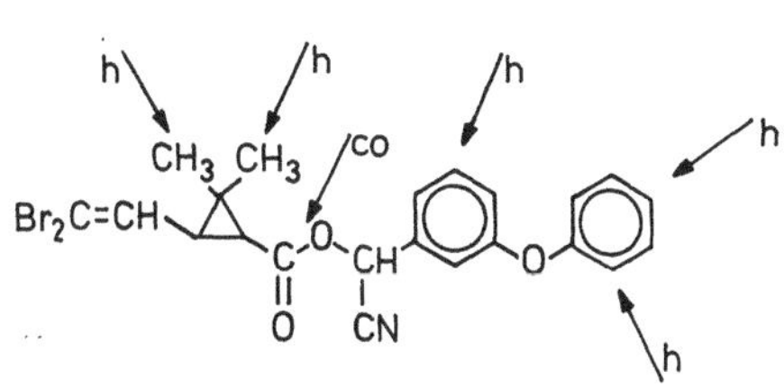	co h	esterases mfo	13,255) 13,255) 287(i))

Table 9 (continued)

Fenvalerate 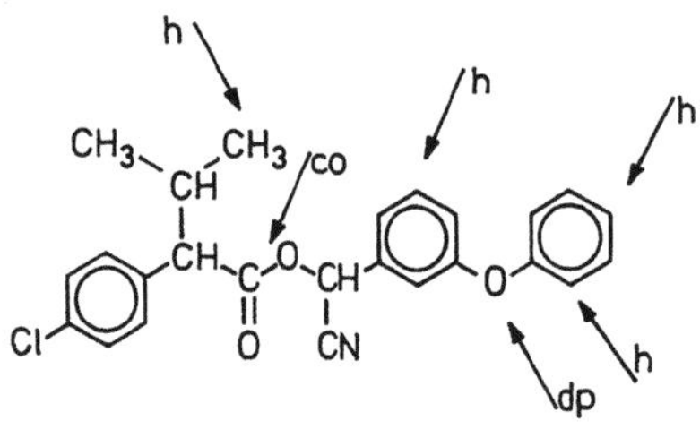	co h dp	esterases mfo	255,278,279(i), 294,295(i)) 294,295(i)) 295(i))
Permethrin 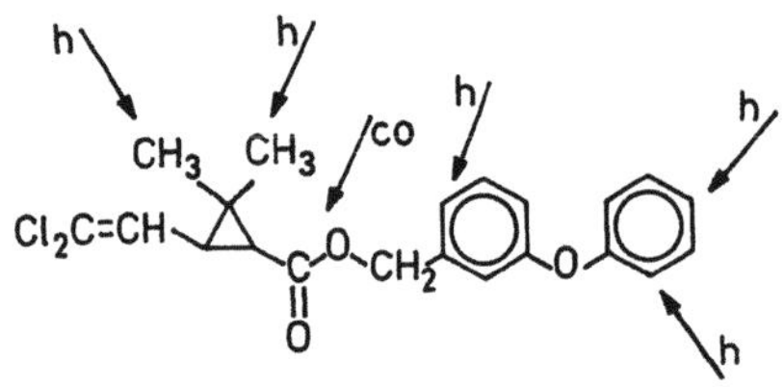	co h	esterases mfo	4(i),13,106,136(i)) 205(i),228(i),250(ac), 255) 4(i),13,255,293(i))

4 Effect of Enzyme-Catalyzed Insecticide Metabolism on Insecticide Susceptibility

Insecticides may be either activated — mainly by enzymatic oxidation — or enzymatically detoxified. Detoxication may involve production of less toxic compounds, acceleration of excretion due to increased water-solubility or rendering further metabolism possible (see Chapt. 3). In this way action of metabolizing enzymes alters insecticide toxicity and is an important factor influencing insecticide susceptibility. Variations of enzyme content and activity between sexes, developmental stages, strains and species and in connection with environmental factors (see Chapter 2) result in differences of insecticide susceptibility.

The relatively high detoxication capability of vertebrates [296] and their ability to catalyze reactions probably being absent in insects (see Chapt. 3) enable mammals to cope with a lot of compounds toxic to arthropods.

Comparing arthropod species, several authors [297–300] reported differences between species and correlations of metabolizing activity and insecticide toxicity. Usually herbivorous species dispose of higher mfo-levels than carnivorous ones [5,77]. Hence it follows that phytophagous arthropod species to be controlled by insecticides unfortunately seem to be well adapted to chemical stress. Induction of detoxifying enzymes by host plant constituents (see Chapt. 2) may play an important role in modulating insecticide toxicity. But there is also evidence that the content of glutathione S-transferases is higher in carnivorous than in herbivorous mites — possibly in order to cope with relatively high oxidative stress [5] — so carnivores might have an advantage for detoxication due to this pathway.

Among other factors, such as the decrease of target sensitivity and restricted penetration, enhanced activity of detoxication enzymes is an important resistance mechanism. For detailed information a series of reviews [2–4, 26, 94, 266, 301–305] may be used. Briefly, enzymes involved in insecticide metabolism correspond to those causing resistance. This means that monooxygenases, esterases and glutathione S-transferases were found to be involved in resistance to organophosphorus compounds, and carbamate resistance is mainly connected with monooxygenases. There is evidence that esterases and monooxygenases are involved in pyrethroid resistance, although the nerve insensitivity factor kdr and decreased penetration seem to be more important [253, 255, 306–308].

Involvement of identical isoenzymes in detoxification of several insecticides, even of different classes, is one cause for cross resistance. For instance, a distinct esterase was found to be responsible for resistance to organophosphorus and pyrethroid insecticides in *Musca domestica* [309].

The ability of species to increase catalysis of a distinct reaction may yield in relationships between insecticide structure — i.e. suitability of the compound to be a substrate for this reaction — and degree of resistance [273, 310, 311].

However, because of the occurrence of isoenzymes with different substrate specificity cross resistance patterns may vary [25]. Moreover, enhanced activity of monooxygenases causes not only detoxication but also activation of insecticides, and in this way yield in negative cross resistance [4]. Possible interaction of several resistance mechanisms complicates the picture [296, 312].

In Table 10 literature data on correlation of enzyme activity with increase of insecticide susceptibility are summarized for various compounds and arthropod species.

In these examples

— insecticide tolerance depends on enzyme action in sexes or at different ages and developmental stages (T),

— induction of enzymes caused enhanced insecticide susceptibility (I), or

— increase of enzyme action correlates with resistance (R).

It must be mentioned, that a real connection of enzyme activity with resistance was not undoubtedly proved in every case, since increase of enzyme activity may also have other causes than selection by the insecticide and differences of enzyme activity were often measured with model substrates (for disadvantages of this method remember Chapt. 3). To prove the involvement of an enzyme in resistance needs detailed examination, even overcoming resistance with specific inhibitors is no proof, these inhibitors are also effective in decreasing resistance level if concentration of insecticides is lowered by decreased penetration [25, 296]. There are examples, where enzymes, to our present knowledge not involved in insecticide metabolism, are enhanced in resistant strains. Reasons for this might be selection with several insecticides, selection of passive detoxication mechanisms — i.e. enzyme binding of xenobiotics that are not converted — or simultaneous increase of enzymes by coupled regulation of their gene activity.

Involvement of esterases in insecticide resistance was intensively studied, probably because it is relatively uncomplicated to separate, visualize and compare them by electrophoresis. Two types of changes were found: production of a mutant esterase with altered substrate specificity which preferentially converts insecticides [131, 244,

Table 10. Enhanced tolerance or resistance of arthropods towards insecticides due to activity of metabolizing enzymes (abbreviations: mfo: M, esterases: E, GST: G; R, I and T are explained in the text)

Compound	Species	Enzymes			Phenomenon			Refs.
Organophosphorus compounds in general	*Amblyseius fallacis*		E	G	R			5)
	Culex spp.		E		R			122,128,306)
		M	E		R			232)
			E		R			313)
	Lucilia illustris	M	E	G	R			238)
	Musca domestica		E		R			245)
		M			R			8)
	Myzus persicae		E		R			202)
	Nephotettix cincticeps		E		R			314)
	Tetranychus urticae			G	R			5)
Acephate	*Spodoptera frugiperda*	M				I		81)
Azinphosmethyl	*Amblyseius fallacis*			G	R			265)
	Musca domestica			G	R			265,266)
	Platynota idaensalis			G				189)
Diazinon	*Apis mellifera*	M					T	70)
	Heliothis virescens	M				I		85)
	Musca domestica			G		I		114,161)
	Nephotettix cincticeps	M	E					111)
	Spodoptera frugiperda	M				I		81)
Dimethoate	*Musca domestica*	M			R			315)
EPN	*Musca domestica*			G	R			268)
Etrimphos	*Musca domestica*			G	R			270)
Fenitrothion	*Musca domestica*			G	R			272)
			E		R			316)
	Nephotettix cincticeps	M	E		R			203)
			E		R			314)
	Nilaparvata lugens		E		R			112,132)
	Oryzaephilus surinamensis	M			R			207)
Malathion	*Aedes aegypti*		E		R			306)
	Ceratitis capitata		E		R			230)
	Culex spp.		E		R			130,233, 306,317, 318)
		M	E		R			232)
	Drosophila melanogaster	M			R			319)
	Lucilia cuprina		E		R			238)
	Musca domestica		E	G	R			320)
				G	R			272)
			E		R		T	133,243,316)
	Nephotettix cincticeps		E		R			279,317)
	Nilaparvata lugens		E		R			132,206)
	Triatoma infestans		E		R			201)
	Tribolium castaneum			G	R			213)
Methamidophos	*Spodoptera frugiperda*	M				I		81)
Methylparaoxon	*Musca domestica*		E		R			184)
Mevinphos	*Plutella xylostella*			G	R			190,321)
Monocrotophos	*Spodoptera frugiperda*	M				I		81)
	Spodoptera littoralis	M				I		37)
Paraoxon	*Lucilia cuprina*	M	E		R			277)
	Musca domestica		E		R			184)
		M			R			75)

Table 10. (continued)

Compound	Species	Enzymes			Phenomenon			Refs.
Parathion	*Lucilia cuprina*	M	E		R			277)
	Musca domestica		E		R			184)
Profenophos	*Spodoptera littoralis*	M				I		37)
Propetamphos	*Musca domestica*			G	R			264)
Temephos	*Culex spp.*		E		R			233, 318)
Tetrachlorvinphos	*Musca domestica*		E	G	R			7)
Trichlorfon	*Culex spp.*			G	R			191)
	Musca domestica		E		R			43, 283) 309, 322) 323)
	Spodoptera littoralis	M				I		81)
Carbamates in general	*Culex pipiens*		E		R			232)
	Myzus persicae		E		R			202)
Carbaryl	*Encalliphora lilica*	M					T	91)
	Heliothis virescens	M				I		88)
	Leptinotarsa decemlineata	M			R			68, 83)
	Lucilia illustris	M					T	91)
	Lymantria dispar	M					T	324)
	Phormia regina	M					T	91)
	Peridroma saucia	M				I		86)
	Spodoptera frugiperda	M			R			225)
Carbofuran	*Leptinotarsa decemlineata*	M			R			84)
Mecarbenil	*Musca domestica*	M			R			236)
Methomyl	*Anticarsia gemmatalis*	M					T	79)
	Heliothis virescens	M				I		88)
	Spodoptera frugiperda	M				I		81)
Propoxur	*Encalliphora lilica*	M					T	91)
	Lucilia illustris	M					T	91)
	Musca domestica	M				I		325)
	Phormia regina	M					T	18, 91)
	Sarcophaga bullata	M					T	18)
Pyrethroids in general	*Amblyseius fallacis*		E	G	R			5)
	Culex spp.		E		R			128)
	Heliothis armiger	M	E		R			326)
	Musca domestica		E		R			309)
	Tetranychus urticae			G	R			5)
Cypermethrin	*Spodoptera eridania*	M				I		81)
Fenvalerate	*Heliothis virescens*		E				T	106)
	Leptinotarsa decemlineata	M			R			295)
	Nephotettix cincticeps		E		R			279)
	Nilaparvata lugens		E		R			205)
	Pseudoplusia includens		E				T	106)
Permethrin	*Amblyseius fallacis*		E		R			108)
	Heliothis virescens		E				T	106)
	Leptinotarsa decemlineata	M	E	G	R			209)
	Musca domestica	M			R			236, 292)
	Myzus persicae		E		R			202)
	Nilaparvata lugens		E		R			205)
	Pseudoplusia includens		E				T	105, 124)
	Spodoptera eridania	M				I		81)
Tetramethrin	*Musca domestica*	M			R			68)

[296,327] according to the mutant 'ali-esterase theory' and enhanced content of distinct isoenzymes keeping on their catalytic properties [296]. For instance, there were proved an up to 64 fold increase of the microsomal carboxylesterase E4 in strains of *Myzus persicae* resistant to carbamate, organophosphorus and pyrethroid insecticides [123] and a 70–500 fold increase of the soluble carboxylesterases A1 and B1 in resistant strains of *Culex spp.* [122] compared to susceptible strains. Esterases E4, A1 and B1 may amount to about 3, 1–6 and 6–12% of total cell protein, respectively. The genetic basis of these increases of protein content are gene amplifications, i.e. the presence of multiple copies of structural genes [122,123,313]. In Italy there was found a mutant form of E4 — named FE4 — with altered molecular weight and an 1.5 fold hydrolytic activity compared to E4 [123].

Esterases may cause insecticide resistance due to two mechanisms: hydrolytic activity and sequestration of compounds, even if they are not detoxified [4,123,279]. This may explain resistance of *Myzus persicae* strains towards carbamates due to esterases E4. This might be similar for glutathione S-transferase isoenzymes that are described as ligand binding proteins (see Chapt. 2).

Quantitative and qualitative changes, especially with respect to substrate binding and convertion, are connected with resistance by monooxygenases [25,63,94] and glutathione S-transferases [140]. Qualitative changes support the idea that distinct cytochrome P-450s and glutathione S-transferases able to metabolize insecticides are expressed in resistant strains to a higher degree. Both enzyme systems depend on cofactors, therefore resistance due to monooxygenases and glutathione S-transferases may be connected with availability of reducing equivalents and activity of the electron-transferring reductase [292] or glutathione content as well as other glutathione-consuming processes, respectively.

It seems likely that regulatory genes control expression of structural genes at least of some isoenzymes of the various detoxifying enzymes in a balanced manner [296,328,329]. Therefore a complex of inherited genome changes may be involved in development of resistance:

- Substitution in DNA base pairs;
- amplification of preexisting genes that confer defense mechanisms;
- gene translocations, chromosome inversions and other DNA rearrangements that influence gene expression.

The first mechanism causes altered enzymes, the latter enhanced amount of enzyme protein [4].

There are several possibilities to circumvent the action of detoxifying enzymes and prevent metabolic resistance, but with each having limits and disadvantages. Inhibition of detoxifying enzymes by combination of insecticides with synergists or use of insecticide mixtures is an often practised method. For instance, pyrethroids may be combined with organophosphates or carbamates inhibiting esterases [227,284,286]. Several esterase and oxidase inhibitors (see chapt. 2) may be combined with organophosphorus compounds [330,331] and pyrethroids [40,255,284,295,332,333]. Inhibitors of glutathione S-transferases should be useful synergists for organophosphorus compounds, especially if they contain methoxy groups. However, until now only some non-specific inhibitors, namely diethylmaleate [169] and N-ethylmaleimide [147], have been described to enhance insecticide toxicity, probably via inhibition of glutathione S-transferase catalyzed detoxication.

Enzyme inhibitors eliminate the selective advantage of organisms containing high levels of detoxication enzymes, but promote development of additional resistance mechanisms [110]. Therefore this method is especially effective, if only one enzyme is involved in detoxication, e.g. carbamates combined with monooxygenase inhibitors [4, 127].

Another intention is the structural modification of insecticides to make them poorly degradable [334]. Possibly the insecticidal effect of methamidophos (O,S-dimethyl phosphoramidothioate) only slightly inhibiting the common target acetylcholinesterase is preserved by its slow metabolism [258, 335–337]. However, an oxidative activation product is also discussed as the really effective agent [338].

Promising might be protection of the alcohol moiety of pyrethroids from monooxygenase attack [292], or replacing methoxy groups in organophosphates that are degraded by glutathione S-transferases [338]. A possible danger of such derivations might be enhanced mammalian toxicity, however, for methylcarbamates an increase of insecticidal effect, but decrease of toxicity to mammals could be reached by substitution of the nitrogen proton of the carbamyl moiety [334].

5 Concluding Remarks

Our present knowledge on insecticide metabolism allows us to generalize on the main insecticide metabolizing enzymes, their sites of action and influences on insecticide toxicity. Nevertheless, there remain several interesting questions that have to be answered. Detailed investigation of the metabolism of distinct insecticides in species important in the field will be the basis for comparison of species and populations. For this purpose fast and easy but sensitive methods for enzyme preparation and determination of their activity towards insecticides are needed. Connected to this data on metabolic potency of insect species and insecticide structures sensitive to enzymatic degradation computer-based programs and pest manegement strategies should be introduced. In distinct cases biochemical parameters may be used to recognize occurrence of resistance.

Synthesis of new insecticides should take into consideration biodegradability of the compounds to find a compromise of high insecticidal effect on the one hand and low mammalian toxicity and environmental pollution on the other.

Moreover, based on detailed knowledge of metabolizing enzymes and their inhibition, structures of synergists might be further optimized.

Investigation of the regulation of gene activity with respect to resistance will be an important field, probably opening new areas to influence the development of resistance.

6 References

1. Dorough HW, Ballard SK (1982) in: Matsumura F, Krishna Murti CF (eds) Biodegradation of Pesticides. Plenum Press, New York London, p 3
2. Wilkinson CF (1985) in: Hilton JL (ed) Agricultural Chemicals of the Future (BARC Symposium 8). Totowa, p 311

3. Wilkinson CF (1983) in: Georghiou GP, Saito T (eds) Pest Resistance to Pesticides. Plenum Press, New York London, p 175
4. Brattsten LB, Holyoke Jr CW, Leeper LR, Raffa KF (1986) Science 231: 1255
5. Mullin CA, Croft BA, Strickler K, Matsumura F, Miller JR (1982) Science 217: 1270
6. Ahmad S (1979) Drug Metab. Rev. 10 (1): 1
7. Schenkman JB, Sligar SG (1981) Pharmac. Ther. 12: 43
8. Feyereisen R (1983) Pest. Bioch. 19(3): 262
9. Appleby CA (1967) Biochim. Biophys. Acta 147: 399
10. Agosin M (1985) in: Kerkut GA, Gilbert LI (eds) Comprehensive Insect Physiology Biochemistry 12. Pergamon Press, Oxford, p 647
11. Wickramsinghe RH (1975) Enzyme 19: 348
12. Hodgson E (1983) Insect Biochemistry 13(3): 237
13. Matsumura F (1985) Toxicology of Insecticides. Plenum Press, New York London 2nd ed.
14. Klingenberg M (1958) Arch. Biochem. Biophys. 75: 376
15. Garfinkel D (1958) Arch. Biochem. Biophys. 77: 493
16. Omura T, Sato R (1962) J. Biol. Chem. 237: 1375
17. Cooper DY, Levin S, Narashimhulu S, Rosenthal O, Estabrook RW (1965) Science 147: 400
18. Terriere LC, Yu SJ (1979) Pest. Bioch. 12(3): 249
19. Coon MJ (1981) Drug Metab. Dispos. 9(1): 1
20. Vincent DR, Moldenke AF, Terriere LC (1983) Insect Bioch. 13(6): 559
21. Ruckpaul K, Bernhardt R (1984) in: Ruckpaul K, Rein H (eds) Cytochrome P450. Akademie-Verlag, Berlin, p 9
22. Kulkarni AP, Hodgson E (1976) Pest. Bioch. 6(2): 183
23. Hodgson E (1979) Drug Metab. Rev. 10(1): 15
24. Agosin M (1976) Molec. Cell Biochem. 12(1): 33
25. Hodgson E (1985) in: Kerkut GA, Gilbert LI (eds) Comprehensive Insect Physiology Biochemistry and Pharmacology 11. Pergamon Press, Oxford, p 225
26. Terriere LC (1983) in: Georghiou GP, Saito T (eds) Pest Resistance to Pesticides. Plenum Press, New York London, p 265
27. Campbell MA, Gyorkos J, Leece B, Homonko K, Safe S (1983) Gen. Pharmacol. 14(4): 445
28. Fabacher DL, Kulkarni AP, Hodgson E (1980) Gen. Pharmacol. 11: 429
29. Madhukar BV, Matsumura F (1979) Pest. Bioch. 11(3): 301
30. Matkovics B, Szabo L, Ivan J, Gaal J (1983) Gen. Pharmacol. 14(6): 689
31. Moroi K, Ushiyama S, Satoh T, Kuga T (1976) Toxicol. Appl. Pharmacol. 37: 377
32. Habazin-Novak V, Plestina R (1984) Period. Biol. 86(4): 315
33. Guengerich FP, Dahn GA, Wright ST, Martin MV, Kaminsky LS (1983) Xenobiotica 12(11): 701
34. Terriere LC (1984) Ann. Rev. Entomol. 29: 71
35. Mico BA, Branchflower RV, Pohl LR (1983) Biochem. Pharmacol. 32(15): 2357
36. Dowd PF, Smith MC, Sparks TC (1983) Insect Bioch. 13(5): 453
37. Dittrich V, Gisin D, Luetkemeier N (1981) J. Appl. Entomol. 5: 474
38. Gil DL, Wilkinson CF (1976) Pest. Bioch. 6(4): 338
39. Nedel'kina SV, Leonova IN, Salganik RI, Zenkova TY, Weiner LM, Popova VI (1985) Biokhimiya 50(8): 1284
40. Chen SJ, Sun CN (1986) J. Econ. Entomol. 79(1): 22
41. Marcus CB, Murray M, Hetnarski K, Wilkinson CF (1987) Pest. Bioch. 28(1): 140
42. Kulkarni AP, Hodgson E (1978) Pest. Bioch. 9(1): 75
43. Levi PE, Hodgson E (1985) Toxicol. Lett. 24(2/3): 221
44. Feyereisen R, Langry KC, Ortiz De Montellano PR (1984) Insect Bioch. 14(1): 19
45. Poulsen LL, Ziegler DM (1979) J. Biol. Chem. 254: 6449
46. Smyser BP, Hodgson E (1985) Biochem. Pharmacol. 34(8): 1145
47. Tynes RE, Hodgson E (1985) J. Agric. Food Chem. 33: 471
48. Tynes RE, Hodgson E (1985) Arch. Biochem. Biophys. 240(1): 77
49. Smyzer BP, Sabourin PJ, Hodgson E (1985) Pest. Bioch. 24(3): 368
50. Smyzer BP, Levi PE, Hodgson E (1986) Biochem. Pharmacol. 35(10): 1719

51. Mitchell MJ, Smith SL (1986) Insect Bioch. 16(3): 525
52. Vincent DR, Moldenke AF, Farnsworth DE, Terriere LC (1985) Pest. Bioch. 23(2): 171
53. Vincent DR, Moldenke AF, Farnsworth DE, Terriere LC (1985) Pest. Bioch. 23(2): 182
54. Feyereisen R, Vincent DR (1984) Insect Bioch. 14(2): 163
55. Vincent DR, Terriere LC (1985) Insect Bioch. 15(2): 299
56. Hällström I, Blanck A, Atuma S (1984) Biochem. Pharmacol. 33(1): 13
57. Collins PJ, Hooper GHS (1984) Comp. Biochem. Physiol. B 77(4): 848
58. Brattsten LB, Price SL, Gunderson CA (1980) Comp. Biochem. Physiol. C 66: 231
59. Crankshaw DL, Hetnarski K, Wilkinson CF (1981) Insect Bioch. 11(5): 515
60. Moldenke AF, Vincent DR, Farnsworth DE, Terriere LC (1984) Pest. Bioch. 21(3): 358
61. Capdevila J, Ahmad N, Agosin M (1975) J. Biol. Chem. 250(3): 1048
62. Stanton RH, Plapp Jr FW, White RA, Agosin M (1978) Comp. Biochem. Physiol. B 61: 297
63. Yu SJ, Terriere LC (1979) Pest. Bioch. 12(3): 239
64. Moldenke AF, Terriere LC (1980) Pest. Bioch. 16(3): 222
65. Collins PJ, Insect Bioch. (1985) 15(4): 551
66. Terriere LC, Yu SJ, Farnsworth DE, Rose HA (1980) Pest. Bioch. 13(3): 274
67. Terriere LC, Yu SJ (1976) Pest. Bioch. 6(3): 223
68. Leonova IN, Nedel'kina SV, Salganik RI (1986) Biokhimiya 51(3): 426
69. Brattsten LB (1987) Pest. Bioch. 27(1): 1
70. Smirle MJ, Winston ML (1987) J. Econ. Entomol. 80(1): 5
71. Wells DS, Rock GC, Dauterman WC (1983) Pest. Bioch. 20(2): 238
72. Gould F, Hodgson E (1980) Pest. Bioch. 13(1): 34
73. Feyereisen R, Farnsworth DE (1985) Insect Bioch. 15(6): 755
74. Hällström I, Sundvall A, Rannug U, Grafström R, Ramel C Chem.-Biol. Interactions 34: 129
75. Oppenoorth FJ, Van Der Pas LJT (1986) Pest. Bioch. 25(1): 40
76. Pratt GE, Jennings RC, Weaver RJ (1984) Insect Bioch. 14(6): 609
77. Brattsten LB, Wilkinson CF (1977) Science 196: 1349
78. Yu SJ (1984) Pest. Bioch. 22(1): 60
79. Christian MF, Yu SJ (1986) Comp. Biochem. Physiol. C 83(1): 23
80. Gunderson CA, Brattsten LB, Fleming JT (1986) Pest. Bioch. 26(3): 238
81. Yu SJ (1982) Pest. Bioch. 17(1): 59
82. Yu SJ (1983) Pest. Bioch. 19(3): 330
83. Rose RL, Brindley WA (1985) Pest. Bioch. 23(1): 74
84. Riskallah MR, Dauterman WC, Hodgson E (1986) Insect Bioch. 16(3): 491
85. Riskallah MR, Dauterman WC, Hodgson E (1986) Pest. Bioch. 25(2): 233
86. Yu SJ, Berry RE, Terriere LC (1979) Pest. Bioch. 12(3): 280
87. Moldenke AF, Berry RE, Terriere LC (1983) Comp. Biochem. Physiol. C 74(2): 365
88. Farnsworth DE, Berry RE, Yu SJ (1981) Pest. Bioch. 15(2): 158
89. Brattsten LB, Evans CK, Bonetti S, Zalkow LH (1984) Comp. Biochem. Physiol. C 77(1): 29
90. Marcus CB, Murray M, Wang C, Wilkinson CF (1986) Pest. Bioch. 26(3): 310
91. Rose HA, Terriere LC (1980) Pest. Bioch. 14(3): 275
92. Zijlstra JA, Vogel EW, Breimer DD (1984) Chem.-Biol. Interactions 48(3): 317
93. Brattsten LB (1987) Pest. Bioch. 27(1): 13
94. Hodgson E, Kulkarni AP (1983) in: Georghiou GP, Saito T (eds) Pest Resistance to Pesticides. Plenum Press, New York London, p 207
95. Aldridge WN (1953) Biochem. J. 53: 110
96. Dauterman WC (1982–83) J. Tox. Clin. 19(6–7): 623
97. Dauterman WC (1985) in: Kerkut GA, Gilbert LI (eds) Comprehensive Insect Physiology Biochemistry and Pharmacology 12. Pergamon Press, Oxford, p 713
98. Junge W (1984) in: Bergmeyer HU (ed) Methods of Enzymatic Analysis 4. Verlag Chemie Weinheim Deerfield Beach, Florida Basel, p 2
99. De Malkenson NC, Wood EJ, Zerba EN (1984) Insect Bioch. 14(4): 481
100. Mc Ilvain JE, Timoszyk J, Nakatsugawa T (1984) Pest. Bioch. 21(2): 162
101. Robbi M, Beaufay H (1983) Eur. J. Biochem. 137: 293
102. Lin PT, Main AR, Motoyama N, Dauterman WC (1983) Pest. Bioch. 20(2): 232

103. Lin PT, Main AR, Motoyama N, Dauterman WC (1984) Pest. Bioch. 22(1): 110
104. Yu SJ (1985) Environ. Entomol. 14: 512
105. Dowd PF, Sparks TC (1986) Pest. Bioch. 25(1): 73
106. Dowd PF, Sparks TC (1987) Pest. Bioch. 27(2): 237
107. De Vries DH, Georghiou GP (1981) Pest. Bioch. 15(3): 242
108. Chang CW, Whalon ME (1986) Pest. Bioch. 25(3): 446
109. Noppun V, Miyato T, Saito T (1987) Appl. Entomol. Zool. 22(1): 98
110. Georghiou GP (1983) in: Georghiou GP, Saito T (eds) Pest Resistance to Pesticides. Plenum Press, New York London, p 793
111. Hama H, Hosoda A (1983) Appl. Entomol. Zool. 18(4): 475
112. Gorder GW, Kirino O, Hirashima A (1986) J. Agric. Food Chem. 34(6): 941
113. Mackness MJ, Walker CH, Rowlands DG, Price NR (1983) Comp. Biochem. Physiol. C 74(1): 65
114. Fontan A, Zerba E (1984) Comp. Biochem. Physiol. C 19(1): 183
115. Yu SJ, Terriere LC (1977) J. Agric. Food Chem. 25(5): 1076
116. Yu SJ, Terriere LC (1977) J. Agric. Food Chem. 25(6): 1333
117. Jones G, Jones D, Hirematl S (1987) Insect Bioch. 17(6): 896
118. Beranek AP, Oppenoorth FJ (1977) Pest. Bioch. 7(1): 16
119. Motoyama N, Dauterman WC (1986) Rev. Biochem. Toxic. 2: 49
120. Ahmad S (1976) Insect Bioch. 6(5): 541
121. Hughes PB, Green PE, Reichman KG (1984) J. Econ. Entomol. 77(6): 1400
122. Fournier D, Bride JM, Mouches C, Raymond M, Magnin M, Berge JP, Pasteur N, Georghiou GP (1987) Pest. Bioch. 27(2): 211
123. Devonshire AL, Searle LM, Moores GD (1986) Insect Bioch. 16(4): 659
124. Dowd PF, Sparks TC (1984) Pest. Bioch. 21(2): 275
125. Kao LR, Motoyama N, Dauterman WC (1985) Pest. Bioch. 23(2): 228
126. Kapin MA, Ahmad S (1980) Insect Bioch. 10(3): 331
127. De Jersey J, Nolan J, Davey PA, Riddles PW (1985) Pest. Bioch. 23(3): 349
128. Georghiou GP, Pasteur N (1978) J. Econ. Entomol. 71: 201
129. Šula J, Weyda F (1983) Experentia 39: 78
130. Miyata T, Saito T, Yasutomi K (1984) Jap. J. Sanit. Zool. 35(3): 245
131. Hughes PB, Raftos DA (1985) Bull. Entomol. Res. 75(3): 535
132. Hasui H, Ozaki K (1984) Appl. Entomol. Zool. 19(1): 52
133. Kao LR, Motoyama N, Dauterman WC (1984) Pest. Bioch. 22(1): 86
134. Ritzengruber O, Führer E (1986) J. Appl. Entomol. 101(2): 187
135. Tadano T (1987) J. Am. Mos. Con. Ass. 3(2): 137
136. Abdel-Aal YA, Soderlund DM (1980) Pest. Bioch. 14(3): 282
137. Morgenstern R, De Pierre JW (1983) Eur. J. Biochem. 134: 591
138. Das M, Bickers DR, Mukthar H (1984) Biochem. Biophys. Res. Commun. 120(2): 427
139. Clark AG, Shamaan NA (1984) Pest. Bioch. 22(3): 249
140. Clark AG, Shamaan NA, Sinclair MD, Dauterman WC (1986) Pest. Bioch. 25(2): 169
141. Clark AG, Carrol N (1986) Biochem. J. 233: 325
142. Tan KH, Meyer DJ, Belin J, Ketterer B (1984) Bioch. J. 220(1): 243
143. Mannervik B (1985) Adv. Enzymol. Relat. Areas Mol. Biol. 57: 357
144. Ketterer B (1986) Xenobiotica 16(10/11): 957
145. Chen WJ, Boehlert CC, Rider K, Armstrong RN (1985) Biochem. Biophys. Res. Commun. 128(1): 233
146. Jakoby WB (1978) Adv. Enzymol. Relat. Areas Mol. Biol. 46: 383
147. Wood EJ, De Villar MIP, Melgar FJ, Zerba EN (1986) Pest. Bioch. 26(2): 170
148. Chang CK, Clark AG, Fieldes A, Pound S (1981) Insect Bioch. 11(2): 179
149. Habig WH, Pabst MJ, Jakoby WB (1974) J. Biol. Chem. 249: 7130
150. Spearman ME, Leibman KC (1983) Life Sci. 33: 2615
151. Spearman ME, Leibman KC (1984) Drug Metab. Dispos. 12(5): 661
152. Spearman ME, Leibman KC (1984) Biochem. Pharmacol. 33(8): 1309
153. Gregus Z, Varga F, Schmelas A (1985) Comp. Biochem. Physiol. C 80(1): 85
154. Gregus Z, Klaassen CD, Schmelas A (1983) Zentralbl. Pharm. Pharmakother. Laboratoriumsdiagn. 122(12): 1273

155. Tu CPD, Matsushima P, Li NQ, Rhoads DM, Spikumar K, Reddy AP, Reddy CC (1986) J. Biol. Chem. 260(20): 9540
156. Stohs SJ, Al-Turk WA, Angle CR, Heinicke RJ (1982) Gen. Pharmacol. 13: 519
157. Tu CPD, Weiss MJ, Li NQ, Reddy CC (1983) J. Biol. Chem. 258(8): 4658
158. Felton JS, Ketley JN, Jakoby WB, Aitio A, Bend JR, Nebert DW (1980) Molec. Pharmacol. 18: 559
159. Hammock B, Ota K (1983) Toxicol. Appl. Pharmacol. 71(2): 254
160. Kulkarni AP, Fabacher DL, Hodgson E (1980) Gen. Pharmacol. 11: 437
161. Hayaoka T, Dauterman WC (1982) Pest. Bioch. 17(2): 113
162. Hennighausen G (1984) Z. gesamte Hyg. 30: 603
163. Shivandandappa T, Rajendran S (1987) Pest. Bioch. 28(1): 121
164. Awasthi YC, Partridge CA, Theodore C, Dao DD (1984) Comp. Biochem. C 78(1): 39
165. Askulöf P, Guthenberg C, Jakobson I, Mannervik B (1975) Bioch. J. 147: 513
166. Ong LK, Clark AG (1986) Biochem. Pharmacol. 35(4): 651
167. Lamoureux GL, Rusness DG (1986) Pest. Bioch. 26(3): 323
168. Lamoureux GL, Rusness DG (1987) Pest. Bioch. 27(3): 318
169. Welling W, De Vries JW (1985) Pest. Bioch. 23(3): 358
170. Dierickx PJ (1984) Pharmacol. Research. Comm. 16(2): 135
171. Vessey DA, Boyer TD (1984) Toxicol. Appl. Pharmacol. 73(3): 492
172. Dierickx PJ (1985) Pharmacol. Res. Comm. 17(5): 489
173. Hayes JD, Mantle TJ (1986) Biochem. J. 233(2): 407
174. Gregus Z, Varga F (1985) Acta Pharmacol. Toxicol. 56(5): 398
175. Motoyama N, Dauterman WC (1977) Insect Bioch. 7: 361
176. Motoyama N, Dauterman WC (1978) Insect Bioch. 8: 337
177. Clark AG, Shamaan NA, Dauterman WC, Hayaoka T (1984) Pest. Bioch. 22(1): 51
178. Clark AG, Dick GL, Smith JN (1984) Biochem. J. 217(1): 51
179. Clark AG, Drake B (1984) Bioch. J. 217(1): 41
180. Yawetz A, Koren B (1984) Insect Bioch. 14(6): 663
181. Cochrane BJ, Morissey JJ, LeBlanc GA (1987) Insect Bioch. 17(5): 731
182. Cohen E, Gerson U (1986) Insect Bioch. 16(3): 449
183. Clark AG, Dick GL, Martindale SM, Smith JN (1985) Insect Bioch. 15(1): 35
184. Oppenoorth FJ, Van Der Pas LJT, Houx NWH (1979) Pest. Bioch. 11(2): 176
185. Ottea JA, Plapp Jr FW (1984) Pest. Bioch. 22(2): 203
186. Clark AG, Dauterman WC (1982) Pest. Bioch. 17(3): 307
187. Usui K, Fukami JI, Shishido T (1977) Pest. Bioch. 7(3): 249
188. Wood E, Casabe N, Melgar F, Zerba E (1986) Comp. Biochem. Physiol. B 84(4): 607
189. Wells DS, Rock GC, Dauterman WC (1983) Pest. Bioch. 20(2): 238
190. Cheng EY, Chou TM, Kao CH (1984) Chung-hua Nung Yeh Yen Chin 33(1): 73
191. Jiang JL (1984) Acta Entomol. Sin. 27(3): 248
192. Ottea JA, Hammock BD (1986) Insect Bioch. 16(2): 319
193. Ottea JA, Plapp Jr FW (1981) Pest. Bioch. 15(1): 10
194. Hayaoka T, Dauterman WC (1983) Pest. Bioch. 19(3): 344
195. Cochrane BJ, LeBlanc GA (1986) Biochem. Pharmacol. 35(10): 1679
196. Dowd PF, Rose RL, Smith M, Sparks TC (1986) J. Agric. Food Chem. 34(3): 444
197. Abd-Elroaf TK, Dauterman WC (1981) Insect Bioch. 11(6): 649
198. Yu SJ, Terriere LC (1973) Pest. Bioch. 3: 141
199. Motoyama N, Kulkarni AP, Hodgson E, Dauterman WC (1978) Pest. Bioch. 9(3): 255
200. Clark AG, Smith JN, Speir TW (1973) Bioch. J. 135(3): 385
201. Wood EJ, De Villar MIP, Zerba EN (1985) Pest. Bioch. 23(1): 24
202. Devonshire AL, Moores GD (1982) Pest. Bioch. 18(2): 235
203. Hama H (1984) Jpn. J. Appl. Entomol. Zool. 28(3): 143
204. Dai S, Sun CN (1984) J. Econ. Entomol. 77(4): 891
205. Chang CK, Whalon ME (1987) Pest. Bioch. 27(1): 30
206. White Jr RA, Franklin RT, Agosin M (1979) Pest. Bioch. 10: 233
207. Rose HA, Wallbank BE (1986) J. Econ. Entomol. 79(4): 896
208. Nedel'kina SV, Solomennikova JV, Volkotrub EH, Leonova IN, Roslavceva CA, Salganik RI (1987) Agrochimija 24(4): 103

209. Soderlund DM, Hessney CW, Jiang M (1987) J. Agric. Food Chem. 35(1): 100
210. Matthews WA (1980) Pest. Bioch. 13(3): 303
211. Starrat AN, Bond EJ (1981) Pest. Bioch. 15(3): 275
212. Cohen E (1982) Comp. Biochem. Physiol. C 71: 123
213. Wool D, Noiman S, Manheim O, Cohen E (1982) Bioch. Genet. 20(7–8): 621
214. Yu SJ, Robinson FA, Nation JL (1984) Pest. Bioch. 22(3): 360
215. Baars AJ, Driessen OMJ (1984) J. Apic. Res. 23(1): 37
216. Croft CA, Mullin CA (1984) Environ. Entomol. 13(5): 1330
217. Rose HA (1985) Ecol. Entomol. 10: 455
218. Vines RC, Reagan TE, Sparks TC, Pollet DK (1984) J. Econ. Entomol. 77(4): 857
219. Clark AG, Letoa M, Wong ST (1977) Life Sci. 20: 141
220. Collins PJ, Hooper GHS (1984) J. Aust. Ent. Soc. 23: 249
221. Hollinghaus JG, Sander G, Little RJ (1984) J. Econ. Entomol. 77(4): 1393
222. Kulkarni AP, Hodgson E (1976) Insect Bioch. 6: 385
223. Tanada Y, Hess R, Omi EM (1980) Insect Bioch. 10(2): 125
224. Brattsten LB, Gunderson CA, Fleming JT, Nihbahlet KN (1986) Pest. Bioch. 25(3): 346
225. McCord E, Yu SJ (1987) Pest. Bioch. 27(1): 114
226. Wadleigh RW, Yu SJ (1987) Insect Bioch. 17(5): 759
227. Ishaaye I, Casida JE (1980) Pest. Bioch. 14(2): 178
228. Poehling HM, Meyer W (1980) Insect Bioch. 11(1): 189
229. Hemingway J, Malcolm CA, Kissoon KE, Boddington RG, Curtis CF, Hill N (1985) Pest. Bioch. 24(1): 68
230. Koren B, Yawetz A, Perry AS (1984) J. Econ. Entomol. 77(4): 864
231. Raymond M, Fournier D, Bride JM, Cuany A, Berge J, Magnin M, Pasteur N (1986) J. Econ. Entomol. 79(6): 1452
232. Villani F, Hemingway J (1987) Pest. Bioch. 27(2): 218
233. Maryama K (1984) Insect Bioch. 14(2): 181
234. Houk EJ, Hardy JL (1981) Insect Bioch. 11(1): 97
235. Baars AJ, Jansen M, Breimer DD (1979) Mutat. Res. 62(2): 279
236. Jansen M, Baars AJ, Breimer DD (1984) Biochem. Pharmacol. 33(22): 3655
237. Holwerda BC, Morton RA (1983) Pest. Bioch. 20(2): 151
238. Raftos DA, Hughes PB (1986) J. Econ. Entomol. 79(3): 553
239. Raftos DA (1986) Pest. Bioch. 26: 302
240. Robacker KM, Hodgson E (1982) Insect Bioch. 12(4): 427
241. Vincent DR, Moldenke AF (1985) Pest. Bioch. 23(2): 182
242. Scott JC, Georghiou GP (1986) Pest. Sci. 17: 195
243. Niwa Y, Miyata T, Saito T (1977) J. Pest. Sci. 2(2): 151
244. De Villar MIP, Van Der Pas LJT, Smissaert HR, Oppenoorth FJ (1983) Pest. Bioch. 19(1): 60
245. Yeoh CL, Kuwano E, Eto M (1981) Appl. Entomol. Zool. 16(3): 247
246. Ugaki M, Shono T, Tsukamoto M, Fukami JI (1985) Appl. Entomol. Zool. 20(1): 73
247. El-Oshar MA, Kodama O, Motoyama N, Dauterman WC (1987) Pest. Bioch. 27(1): 132
248. Kulkarni AP, Motoyama N, Dauterman WC, Hodgson E (1978) Bull. Environ. Contam. Toxicol. 20(2): 227
249. Mullin CA, Matsumura F, Croft BA (1985) Comp. Biochem. Physiol. 79(1): 85
250. Riddles PW, Davey PA, Nolan J (1983) Pest. Bioch. 20(2): 133
251. Zolotova TB, Roslavceva CA (1981) Chimija v sel'skom chozjajstve 11(30): 3
252. Cohen SD (1984) Fundam. Appl. Toxicol. 4: 315
253. Chadwick PR, Slatter R, Bowron MJ (1984) Pestic. Sci 15(2): 112
254. Gaughan LC, Engel JL, Casida JE (1980) Pest. Bioch. 14(1): 81
255. Casida JE, Ruzo LO (1980) Pestic. Sci. 11: 257
256. Wing KD, Glickman AH, Casida JE (1983) Science 219: 63
257. Glickman AH, Wing KD, Casida JE (1984) Toxicol. Appl. Pharmacol. 73: 16
258. Wing KD, Glickman AH, Casida JE (1984) Pest. Bioch. 21(1): 22
259. Lee PW, Allahyari R, Fukuto TR (1978) Pest. Bioch. 8(2): 158
260. Kulkarni AP, Mailman RB, Hodgson E (1975) J. Agric. Food Chem. 23(2): 177

261. Hodgson E, Motoyama N (1984) in: Evered D, Collins GM (eds) Origins and development of adaptation. Pitman Publishing Ltd., London p 167
262. Osimitz TG, Kulkarni AP (1985) Drug Metab. Dispos. 13(2): 197
263. Nomeir AA, Dauterman WC (1979) Pest. Bioch. 10(2): 190
264. Wells DS, Motoyama N, Dauterman WC (1986) Pestic. Sci. 17(6): 631
265. Motoyama N (1982) J. Pest. Sci. 7(3): 415
266. Roslavceva S (1983) Agrochimija 2: 126
267. Levine BS, Murphy SD (1977) Toxicol. Appl. Pharmacol. 40: 393
268. Nomeir AA, Hajjar NP, Hodgson E, Dauterman WC (1980) Pest. Bioch. 13(2): 112
269. Ioannou YM, Dauterman WC (1978) Pest. Bioch. 9(2): 190
270. Ioannou YM, Dauterman WC (1979) Pest. Bioch. 10(1): 31
271. Kawamura Y, Takeda M, Uchiyama M (1982) Eisei Kagaku 28(2): 65
272. Ugaki M, Shono T, Fukami J (1985) Pest. Bioch. 23(1): 33
273. Konno Y, Shishido T (1987) J. Pest. Sci. 12: 469
274. Isono H, Miyaura S, Kubota T (1984) Eisei Kagaku 30(6): 363
275. Chopade HM, Dauterman WC (1981) Pest. Bioch. 15(2): 105
276. Sultatos LG, Minor LD (1985) Toxicology 36(2/3): 159
277. Hughes PB, Devonshire AL (1982) Pest. Bioch. 18(3): 289
278. Kao LR, Motoyama N, Dauterman WC (1985) Pest. Bioch. 23(1): 66
279. Motoyama N, Kao LR, Lin PT, Dauterman WC (1985) Pest. Bioch. 21(2): 139
280. Wells DS, Dauterman WC, Afifi LM (1986) J. Agric. Food Chem. 34(1): 79
281. Akthar MH, Foster TS (1979) J. Agric. Food Chem. 27(1): 113
282. El-Oshar MA, Motoyama N, Dauterman WC (1987) J. Agric. Food Chem. 35(1): 138
283. Lange C (1981) Untersuchungen zu Trichlorfon-Metabolismus und -Resistenz bei Insekten. Thesis, Akademie der Landwirtschaftswissenschaften der DDR, Berlin
284. Sawicki RM, Devonshire AL, Farnham AW, O'Dell KE, Moores GD, Denholm I (1981) Bull. Ent. Res. 74: 174
285. Ishaaya J, Casida JE (1981) Environ. Entomol. 10: 681
286. Gaughan LC, Unai T, Casida JE (1977) J. Agric. Food Chem. 25: 9
287. Shono T, Ohsawa K, Casida JE (1979) J. Agric. Food Chem. 27(2): 316
288. Schnitzerling HJ, Nolan J, Hughes S (1983) Pestic. Sci. 14(1): 64
289. Suzuki T, Miyamoto J (1978) Pest. Bioch. 8(2): 186
290. Soderlund DM, Casida JE (1977) Pest. Bioch. 7(4): 391
291. Scott JG, Georghiou GP (1984) Pest. Bioch. 21(1): 53
292. Scott JG, Georghiou GP (1986) Pestic. Sci. 17(3): 195
293. Shono T, Casida JE (1978) J. Pest. Sci. 3: 165
294. Mumtaz MM, Menzer RE (1986) J. Agric. Food Chem. 34(6): 929
295. Soderlund DM, Hessney CW, Jiang M (1987) J. Agric. Food Chem. 35(1): 100
296. Oppenoorth FJ (1985) in Kerkut GA, Gilbert LI (eds) Comprehensive Insect Physiology Biochemistry and Pharmacology 12. Pergamon Press, Oxford, p 731
297. Novoshilov KV, Smirnova IM (1984) Agrochimija 9: 94
298. Novoshilov KV, Federova SP, Smirnova IM (1986) Agrochimija 5: 91
299. Bigley WS, Plapp FW (1978) J. Agric. Food Chem. 26: 1128
300. Mallipudi NM, Fukuto TR (1979) Pest. Bioch. 12(1): 55
301. Oppenoorth FJ (1984) Pest. Bioch. 22(2): 187
302. Georghiou GP (1983) in: Georghiou GP, Saito T (eds) Pest Resistance to Pesticides. Plenum Press, New York London, p 1
303. Magnin M, Fournier D, Pasteur N (1985) Cah. Orstom. Ser. Entomol. Med. Parasitol. 23(4): 273
304. Taylor CE (1985) Biol. J. Limnean Soc. 27: 103
305. Dauterman WC (1983) in: Georghiou GP, Daito T (eds) Pest Resistance to Pesticides. Plenum Press, New York London, p 229
306. Brealey CJ, Crampton PL, Chadwick PR, Pickett FE (1984) Pest. Sci. 15(2): 121
307. Farnham AW, O'Dell KE, Denholm I, Sawicki RM (1984) Bull. Ent. Res. 74: 581
308. Sawicki RM, Farnham AW, Denholm I, Church VJ (1986) Pest. Sci. 17(5): 483
309. Sawicki RM, Devonshire AL, Farnham AW, O'Dell KE, Moores GD, Denholm I (1984) Bull. Ent. Res. 74(2): 197

310. Konno Y, Shishido T, Tanaka F (1986) J. Pest. Sci. 11(3): 393
311. Konno Y, Shishido T (1985) J. Pest Sci. 10(2): 285
312. Pereguda TA (1985) Agrochimija 22(8): 123
313. Mouches C, Pasteur N, Berge JB, Hyrien O, Raymond M, De Saint Vincent BR, De Silvestri M, Georghiou GP (1986) Science 233 (4765): 778
314. Miyata T, Saito T, Fukamachi S, Kiritani K, Kawahara S, Yoshioka K, Ozaki K, Sasaki Y, Tsuboi A, Hama H, Iwata T (1981) Jpn. J. Appl. Entomol. Zool. 25(3): 150
315. Hammock BD, Mumby SM, Lee PW (1977) Pest. Bioch. 7: 261
316. Ugaki M, Abe T, Fukami JI, Shono T (1983) Appl. Entomol. Zool. 18(4): 447
317. Miyata T, Saito T, Yasutomi K (1984) Jap. J. Sanit. Zool. 35(2): 103
318. Tang ZH, Wood RJ (1986) Bull. Entomol. Res. 76(3): 505
319. Morton RA, Holwerda BC (1985) Pest. Bioch. 24: 19
320. Motoyama N, Hayaoka T, Nomura K, Dauterman WC (1981) J. Pest. Sci. 5(3): 393
321. Cheng EY, Chou TM, Kao CH (1983) Chung-hua Nung Yeh Yen Chiu 32(4): 373
322. Pfeiffer G, Otto D, Pilz R, Slonina C (1979) Arch. Phytopathol. Pflanzenschutz 15: 149
323. Leonova IN, Nedel'kina SV, Naumova NB, Salganik RI (1986) Biokhimiya 51(3): 420
324. Ahmad S, Forgash AJ, Das YT (1980) Pest. Bioch. 14(3): 236
325. Terriere LC, Yu SJ (1974) J. Agric. Food Chem. 22(3): 366
326. McCaffery AR, Ahmad M (1986) British Crop Prot. Conf. Abstracts 2 4c–14: 591
327. Oppenoorth FJ (1965) Ann. Rev. Entomol. 10: 185
328. Plapp Jr FW (1984) Pest. Bioch. 22: 194
329. Plapp Jr FW, Wang TC (1983) in: Georghiou GP, Saito T (eds) Pest Resistance to Pesticides. Plenum Press, New York London, p 47
330. Baker SR, Mc Gilliard AD, Dahn PA (1979) Pest. Bioch. 11(1): 20
331. Otto D, Pfeiffer G, Lange C (1983) Arch. Phytopathol. Pflanzenschutz 19: 175
332. Liu MY, Chen JS, Sun CN (1984) J. Econ. Entomol. 77(4): 851
333. Kamel EB, McKee MJ, Knowles CO (1987) J. Econ. Entomol. 80(2): 332
334. Fukuto TR, Mallipudi NM (1983) in: Georghiou GP, Saito T (eds) Pest Resistance to Pesticides. Plenum Press, New York London, p. 557
335. Rose RL, Sparks TC (1984) Pest. Bioch. 22(1): 67
336. Khawasinah AMA, March RB, Fukuto TR (1978) Pest. Bioch. 9(2): 211
337. Suksayretrup P, Plapp Jr FW (1977) J. Agric. Food Chem. 25(3): 481
338. Eto M, Okabe S, Ozoa Y, Mackawa K (1977) Pest. Bioch. 7(4): 364

Insecticide Compounds, Acting on the Acetylcholine Receptor of the Insect Nervous System

Rainer Richter, Dieter Otto, and H.-J. Mengs
Institute for Plant Protection Research of the Academy of Agricultural Sciences of GDR, Stahnsdorfer Damm 81, DDR-1532 Kleinmachnow

Today some insecticides are known for their effectivity on the acetylcholine receptors of the insects. Nicotinic and muscarinic acetylcholine receptors have been investigated intensively in vertebrates. However there are some differences in the mode of action and in drug specifity of these receptors in insects when compared to those in vertebrates.

Nicotine has been described as the first compound with this mode of action followed by nicotine derivates. Currently, there are some commercial used insecticides with interaction on the acetylcholine receptor, developed as nereistoxine derivates such as cartap, thiocyclam or bensultap.

The review summarizes the knowledge of the use of insecticides to interact with the acetylcholine receptor in insects, describes these substances in relation to their pharmacological properties and gives an overview about the synthesized and patent covered derivatives, the structure-activity-relations, as well as the importance of these pesticides in plant protection. The possibility of attacking this receptor in insects in step with pesticide efficiency indicated that the acetylcholine receptor may be further exploited as one new target for the nereistoxins and some other insecticides.

Chemistry of Plant Protection 2

1 Introduction

The majority of the known insecticides of practical importance act on the insect nervous system.

Two main targets are involved in the mechanisms of action of these insecticides: the acetylcholinesterases, which will be inhibited by compounds of the large group of organophosphorus esters and carbamates [1] and the sodium channel protein in the neuronal axon membranes, which will be affected by pyrethroids and DDT by pertubing the kinetics of the action potential [2,3]. In both targets insecticide insensitive mutants occur as a suitable basis for the development of resistance in a pest population by selection processes, when only the one type of action of an insecticide is frequently used repetitively [4,5].

Supporting by the pesticide developing industries and institutes it is a world wide aim to have available a broader spectrum of insecticides with quite different mechanisms of action to prevent selection of resistance by cyclic rotation of different active compounds and to meet the differentiated and increasing demands for effectiveness, toxicological properties and environmental safety [6]. Some few examples are known, in which the active compound produces insecticidal effects by interfering with other neurochemical types of synaptic signal transmission, e.g. at octopaminergic synapses by chlordimeforme and other formamidines or at gamma aminobutyric acid mediated (GABAergic) synapses by the avermectines from the fungus *Streptomyces avermytilis* [7,8,9].

The high effectiveness of organophosphates and carbamates shows that interference in the acetylcholine (ACh) mediated (cholinergic) transmission process is a potent insecticide mechanism. Beside the inhibition of the ACh-splitting acetylcholinesterases (AChE) by these agents, other possibilities of disturbing this synaptic process exist. Hexachlorocyclohexane (HCH) and cyclodiene [3] enhance the release of ACh from the presynaptic membrane, augmenting the ACh-level in the synaptic cleft. The acetylcholine receptor protein itself (AChR) can serve as a target for insecticides, too. The AChR is known as the site of action of the insecticide nicotine and described as the target of cartap hydrochloride, thiocyclam oxalate and some related insecticidal compounds.

The aim of this review is to summarize the available information about the properties, occurrence, localization and possible function of the AChR in insects and the action of different groups of insecticides at this site. Considering AChR as an insecticide target is encouraged by the fact that the insecticides known are related to either naturally occurring agents (nicotine) or synthetic derivatives from a natural toxin (Nereistoxin), because it is a successful concept to take a natural product with considerable potency and a novel mode of action against insects as a prototype and base structure for the development of new insecticides. With regard to the demand to search for insecticides with high toxicity to insect pests but with low toxicity to vertebrates, it is necessary to compare the properties of AChR from insects and vertebrates according to function and pharmaca binding specifities and to point out differences as a base for further concepts in biorational design of new insecticides.

2 Structural, Functional and Pharmacological Properties of Cholinergic Receptor Proteins

In neural signal transmission from neuron to neuron a chemical neurotransmitter substance is released from the presynaptic neuron into the synaptic cleft, which diffuses to the postsynaptic membrane where it reacts with the receptor protein and triggers a cellular response [10,11]. Either a rapid, transient change in ionic permeability of the postsynaptic membrane results, which may cause a depolarization (excitation) or a hyperpolarization (inhibition) of the postsynaptic cell depending on the ionic species involved, or alternative metabolic steps may be interposed between the transmitter recognition and the cellular response. Secondary messengers such as cyclic GMP may be involved in these slower, long-term physiolocigal responses to the transmitter [12].

Acetylcholine is an excitatory neurotransmitter in many vertebrate [13] and invertebrate [14] peripheral and central synapses. From vertebrate pharmacology it is known that the AChRs differ in their responses to drugs. There are two major types: the nicotinic AChR and the muscarinic AChR.

2.1 The Nicotinic Acetylcholine Receptor of Vertebrates (nAChR)

This kind of AChR is activated by ACh and nicotine. This gives the receptor its common name. The nAChR is an intensively studied receptor protein in vertebrate skelett muscle, brain, nerve cord and ganglia [11,15,16,7]. High concentrations of it are also found in the electric organ of some fishes (*Torpedeo* and *Electrophorus* spp.) [12]. Both functionally and biochemically, this receptor closely resembles the AChR of the mammalian neuromuscular junction [18]. It has a molecular mass of about 260000 [18,19,20]. The nAChR is an integral membrane protein composed of five glycosylated polypeptide chains with apparent molecular mass of 40000 (alpha 2×), 50000 (beta), 60000 (gamma) and 65000 (delta) around the channel porus in alpha, beta, alpha, gamma, delta arrangements (Fig. 1) [18,21,22].

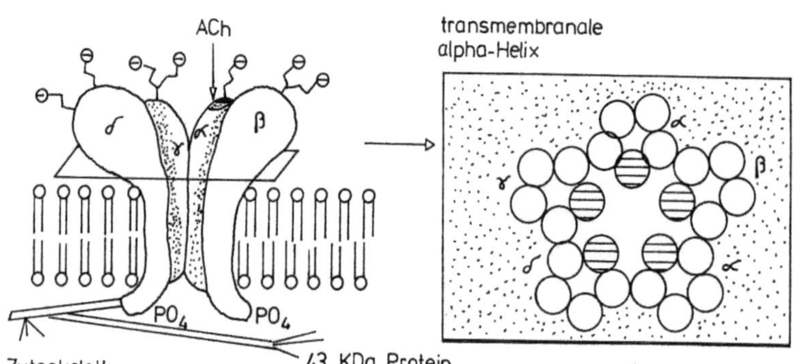

Fig. 1. Structural model of the nAChR of *Torpedo californica*. All of the subunits involve 4 hydrophobic and 1 amphipatic alpha helicae, which are constructing the ion channel of the receptor [25]

Its complete amino-acid-sequence has been identified from both vertebrates [23, 24, 25] and insect cells [26].

Two ACh-molecules have to bind simultaneously to the two alpha-subunits of the receptor protein in order to trigger the opening of the channel for sodium ions. This process leads to the depolarization of the postsynaptic membrane as a reaction to presynaptical transmitter release, thus exciting the postsynaptic neuron or muscular cell. The opening and closing of the single channel complex can be recorded by high sensitive electrophysiological methods [27, 12, 28].

Pharmacological studies of the nAChR have been greatly assisted by the use of alpha-neurotoxins, particularly of radioactive labelled alpha-bungarotoxin (BTX) [29]. This snake venom is one of the well known poisons which irreversibly inhibits the nAChR of the vertebrate neuromuscular junction [30, 31]. The alphabungarotoxin binding at nAChRs of the central nervous system of vertebrates is, however, questionable [11].

Besides the inhibiting effects of alpha-neurotoxins d-tubocurarin a former plant-derived arrow poison of some South American Indians, also acts as an antagonist at the n-AChR. It binds to the same protein like acetylcholine and the tobacco alkaloid nicotine, but in contrast to these two agonists, it prevents the ion channel from opening [32, 12].

By recent cloning studies the AS-sequence of the alpha subunits of the nAChR of *Torpedo californica* electroplax could be analyzed [33, 34, 35]. An investigation by McCormick and Atassi [36] led to the determination of that part of the AS-sequence, which is responsible for ACh binding and alpha BTX binding, respectively. The binding site is in a high degree connected with the sequence between Cys-128 and Cys-142 and could be elaborated as followed:

```
        125            130
        Lys-Ser-Tyr-Cys-Glu-Ile-Ile-Val-Thr-His
                    |                       |
                    S                       Phe 135
145                 |  141                  |
Gly-Leu-Lys-Met-Tyr-Cys-Asn-Glu-Glu-Asp-Phe-Pro
```

The 125–147 peptide was synthesized, and the affinity to [^{125}J]-alpha BTX and [^{3}H]-ACh was compared with that of natural AChR-protein and with unspecific binding of non relevant proteins (Table 1). In the synthetic peptide Met-144 was substituted by norleucine (Nle) to avoid problems caused by the oxidation of the

Table 1. Binding of BTX and ACh to synthetic alpha-subunit peptide 125–147 of the AChR [36]

Sepharose adsorbent	^{125}J-alpha BTX bound (cpm)	^{3}H ACh bound (cpm)
Native AChR	51130 ± 1150	12150 ± 1440
Peptide 125–147	19090 ± 670	3593 ± 960
Bovine serum albumin	1200 ± 220	903 ± 110
Haemoglobin A	1830 ± 330	780 ± 70

thioether side chain of methionine, which might occur during isolation and disulphide bond formation.

Another group of natural poisons does not bind to the two alpha units but to ionic channel sites of the nAChR of vertebrate motor endplates and *Torpedo* electric organs [37,38,39,40]. This is the spiro-alkaloid histrionicotoxin (HTX), which inhibits neuromuscular transmission in a non-competitive fashion [41,42]. Local anaesthetics and the hallucinogen phencyclidine are also thought to bind directly to a site associated with the ionic channel of the nAChRs [43,44,45,46]. Some other AChR-binding substances are known. They are discussed in connection with the presentation of AChRs of the insects.

2.2 The Muscarinic Acetylcholine Receptor of Vertebrates (mAChR)

Whereas the nAChR is closely coupled to an ion-channel, a second messenger type of transduction operates in the case of the muscarinic receptor, which is activated by ACh and muscarine respectively. The mAChR is present in neurons in both the peripheral and central nervous system, in many exocrine glands, and in smooth muscle. The mAChR have a molecular mass of about 70000–80000 depending on preparation [49] and they are membrane proteins. Pharmacological and electrophysiological investigations have shown that mAChR in different tissues may react in different ways. So the activation of mAChRs causes depolarization in brain neurons or the reverse by causing hyperpolarization of cardiac pacemaker and atrial cells by increasing potassium permeability [47,48]. There is some evidence that the mAChR is linked to guanylate cyclase as a second messenger. Activation of this enzyme leads to an increase in intercellular cyclic guanidinmonophosphate (cGMP) which effects the internal ion composition and therefore initiates a cellular response [12] (Fig. 2). The muscarinic receptor is inhibited antagonistically by atropin, scopolamine, and quinuclidinyl benzilate (QNB).

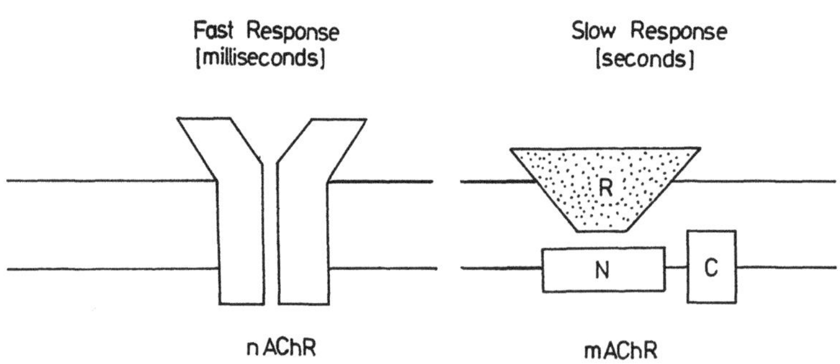

Fig. 2. Structural model of both acetylcholine receptors [15] R = receptor; N = nucleotide binding protein; C = nucleotide cyclase

2.3 The Acetylcholine Receptors of the Insect Nervous System[1]

Nicotinic and muscarinic AChRs have been identified in the invertebrate central nervous system, too. But there are some differences in drug binding specifities and sensitivities to active reagents between receptors in insects and in vertebrates. Furthermore, the insect brain contains more nicotinic than muscarinic receptors, while the reverse is found in the mammalian brain [15]. It is very interesting that the insect central nervous system (CNS) contains a third kind of putative AChR. This receptor has been identified in homogenates of insect CNS, mainly based on the specifity pattern of their cholinergic drug binding properties in biochemical experiments. The discovery of these receptors dates back to the binding studies of [^{3}H] muscarine to extracts of the housefly (*Musca domestica*) [50]. The third AChR type binds nicotinic as well as muscarinic drugs [50] (mixed affinity type, multifunctional type) (Table 2).

The knowledge about the AChR's of insects is summarized in an instructive review by Sattelle [51].

The cholinergic transmission in insects can be localized in special parts of the CNS by histochemical methods, especially by help of radiolabelled alpha-bungarotoxin. Using these methods, Schmidt-Nielson et al. (1977) [53] showed that a alpha-bungarotoxin binding AChR in *Drosophila* is confined to synaptic areas of the neuropile in the brain. Hildebrandt and co-workers [54] demonstrated the toxin

Table 2. Inhibition of the specific pharmaca binding at different types of AChRs from *Musca domestica* heads by other compounds [52]

Type Standard:	nAChR alpha-BTX (= alpha bunga-rotoxine) 10^{-5} M	mAChR QNB (= Quinuclidi-nylbenzilate) 10^{-4} M	mixed AChR Decamethonium 10^{-5} M
Compounds, active at nAChR of vertebrates:			
nicotine	95	5	41
alpha BTX	100	2	3
dihydro-β-erythroidin	98	8	3
curare	74	77	21
succinylcholin	15	3	61
decamethonium	14	19	94
Compounds, active at mAChR of vertebrates:			
QNB	3	100	17
dexetimide	10	96	95
atropine	43	91	45
scopolamine	5	91	25
pilocarpine	5	81	71

[1] See also the review "Molecular properties and function of insects acetylcholine receptors" which was published by Breer and Sattelle in J. Insect Physiol. 1987, 33, 771 after we had finished preparing this paper

Table 3. Voltage-dependence of antagonist suppression of ACh-induced current in the cell body membrane of the fast coxal depressor motoneurone of the metathoracic ganglion of *Periplaneta americana* [64]

Concentration of Antagonist	Voltage-independent block of ACh-response	Voltage dependent block of ACh-response
5.0×10^{-7} M	alpha-BTX	
1.0×10^{-6} M	Dihydro-β-erythroidine	
5.0×10^{-6} M	Mecamylamine	
1.0×10^{-4} M		d-Tubocurarine
1.0×10^{-5} M		Atropine

binding to nAChR in frozen sections of the brain of the moth *Manduca sexta*. It could be shown that sensoric axons from the antennae terminate in conduced regions of neuropile (glomeruli) containing primary afferent synapses. Toxinbinding is highly concentrated in the glomeruli.

A few binding sites were also detected in nonsynaptic regions such as the non-synaptic neuropile, the neuronal cell-body regions and antennal nerves. Studies on the localization of nAChR in the 6th abdominal ganglion of the cockroach (*Periplaneta americana*) reveal dense binding in the central neuropile on either side of the middleline in the frozen sections [55]. This is the region in which cercal afferent sensoric fibres make synaptic contacts with the dentritic branches of the giant interneurons.

All of the listed synaptic loci of bungarotoxin binding are contact regions of sensoric afferent nerves with interneurons. Studies on the antennal afferent input to the antennal lobes of the moth *Manduca sexta* [56,57,58] and the cercal afferent input to the giant interganglionic interneurons in the abdominal ganglion of the cockroach *Periplaneta americana* [55] suggest the hypothesis that ACh is a major sensoric neurotransmitter in insect CNS. In contrast to vertebrates cholinergic transmission is not involved in the neuromuscular junction of insects.

New findings in nonsynaptical regions of the CNS of the insects admit a conclusion and promote the hypothesis that besides the membrane-bound AChRs of the synaptic region of central neuropil another receptor population is coexistent. Autoradiographic and immunohistochemic investigations at the 6. abdominal ganglion of *Periplaneta americana* indicated that non-synaptic membranes of cell bodies in the region of neurons and glia cells in the periphery of the ganglion around the neuropil posses facilitating drug accessibility being sensitive to locally or in vivo applied neurotransmitters, alpha-BTX, and functionally related compounds [55,59]. Physiological dates about this ACh- and alpha-BTX sensible voltage-dependent AChR-ion channel complex in the neuron cell body membrane, especially of the fast coxal depressor motoneuron are provided by Sattelle, 1983 [64] (Table 3).

2.4 Pharmacological Characterization of the AChR Binding Properties

The initial step of ligand-receptor-interaction can be studied by biochemical-pharmacological techniques. Putative AChR from the CNS and from peripheral zones have been characterized by radiolabelled ligand binding methods [60] using at first [^{3}H]-muscarine [51] later [^{125}J]-alpha bungarotoxine (BTX) or [^{3}H]-quinuclidinyl benzilate (QNB). A saturation of specific binding indicates a limited number of binding sites [61,62]. These studies took their pattern from studies of vertebrate tissue. [^{125}J]-BTX binding in insect tissue was demonstrated in frozen sections of *Drosophila melanogaster* by Hall and Teng (1975) [63] and [^{3}H)-QNB binding to mAChR by Yamamura and Snyder (1974) [65–69]. After these initial studies several laboratories began to characterize these toxin binding components from particulate extracts from *Drosophila melanogaster* heads, *Periplaneta americana* terminal ganglions, *Musca domestica* homogenates and *Locusta migratoria* neuronal tissues with a row of different pharmaca, competing with the radiolabelled BTX and QNB. All these investigations result in biochemical data, which can be compared with those from vertebrate AChRs.

Alpha-bungarotoxin (alpha-BTX)
Antagonist of the AChR of vertebrates and invertebrates
Molecular mass: 8000, Polypeptide

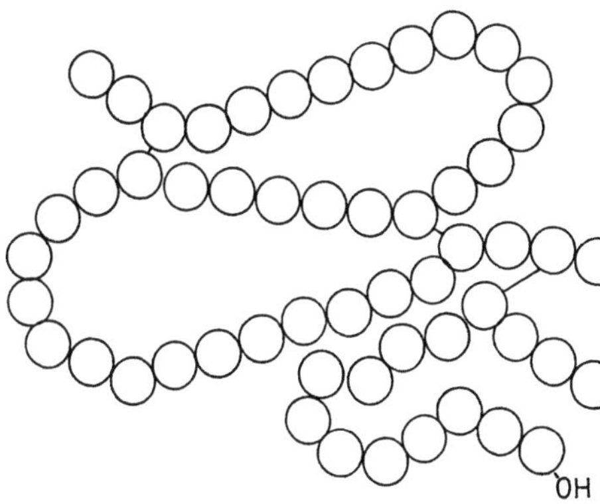

Fig. 3. Alpha-Bungarotoxin

Specific binding:

1.3 +/—0.1 pM/mg protein of 105000 × g supernatant of *Periplaneta americana* 6. abdominalganglion
(Richter, 1987) [59]

0.9 +/— 0.4 pM/mg protein of 500 × g supernatant of *Drosophila melanogaster* heads
(Dudai, 1977) [70]

1.5–1.8 nM/mg protein of enriched membranes from Electroplaxs of *Torpedo californica*
(Solar et al., 1983) [71]

Dissoziation constant:

2.2×10^{-8} M	in 105000 × g supernatant of *Periplaneta americana* 6. abdominalganglion (Richter, 1987) [59]
1.8×10^{-10} M	in 40000 × g pellet of *Drosophila melanogaster* heads (Jimenez and Rudloff, 1980) [27]

	$0.6–1.9 \times 10^{-9}$ M	in 500 × g supernatant of *Drosophila melanogaster* heads (Dudai, 1978) [73]
	0.3×10^{-8} M	membrane preparation of *Musca domestica* (Eldefrawi, 1980)
Kd1	4.0×10^{-10}	on high degree of purity AChR of membrane fragments of *Electrophorus* (Maelicke et al., 1977) [74]
Kd2	2.0×10^{-11}	
Kd	3.7×10^{-9}	(Lummis and Sattelle, 1985) [75]

Alpha-BTX and all other alpha-toxins, isolated from the venoms of the elapide snakes, have very high affinities for, and bind quasi-irreversibly to the nicotinic AChRs of motor endplates and the electric organes of fish. Alpha-BTX does not bind to some AChRs in the CNS of the vertebrates.

In insects we can find a higher dissociation rate than in vertebrates tissue, nevertheless the affinity of alpha-BTX is relatively strong in comparison to other ligands of nAChR. The nAChR of insects binds alpha-BTX more reversibly than that of mammalian muscle [76, 39]. Alpha-BTX blocks responses to applied ACh to nerve cells in the nerve cord of *Periplaneta americana* [77].

It is very interesting to know that this strong poison for vertebrates has no toxic effect on insects. The conclusion may be that antagonists which exclusively react with the nAChR of insects are probably no insecticides. We have investigated the insecticide activity of alpha-BTX to *Periplaneta americana* by injection of 20 μl of 10^{-2} M toxin intersegmentally. The animals did not show any reaction to this poison, though alpha-BTX had bound on the nAChR of target tissue as could be demonstrated by immunhistochemical methods in connection with injection tests.

d-Tubocurarin

Molecular mass: 681.67

Alcaloid

$C_{37}H_{42}Cl_2N_2O_6$

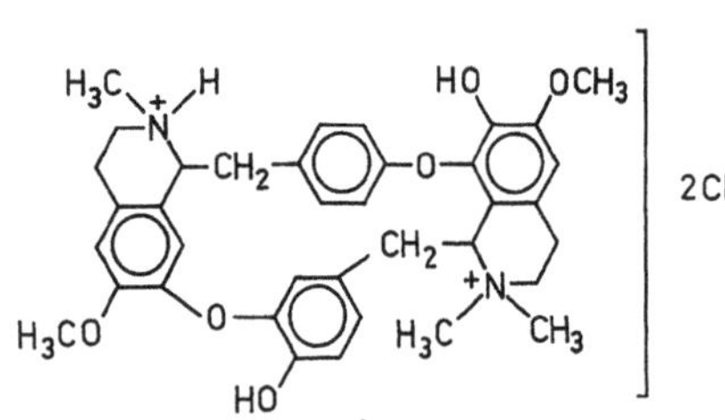

1

D-tubocurarin (1) inhibits, the neuromuscular transmission in vertebrates by antagonistically competing with ACh at the binding site. It is a strong poison for vertebrates. D-tubocurarin inhibits the nAChR in invertebrates, too. The inhibition-constant (k_i) has been measured by comparison with alpha-BTX in a preincubation test. The inhibition constant is defined as $K_i = IC_{50}/(1 + [L]/K_d)$.

Inhibition-constant

1.0×10^{-7} M	for 105000 × g supernatant of *Periplanota americana* (Richter, 1987) [59]

8.5×10^{-7} M	for 40000 × g supernatant of *Periplaneta americana* (Lummis and Sattelle, 1985)[75]
1.0×10^{-7} M	purified AChR from electroplaxs *Electrophorus electricus* (Maelicke and Reich, 1977)[78]

After injection of d-tubocurarin *Periplaneta americana* does not react to this poison like to alpha-bungarotoxin; so the hypothesis emphasizes that antagonists of the nicotinic AChR are not potent insecticides.

There is some evidence that d-tubocurarin has also some binding affinity at the insect mAChR too[75, 79].

Physiostigmin (Eserin)
Molecular mass: 275.34
Alkaloid
$C_{15}H_{21}N_3O_2$

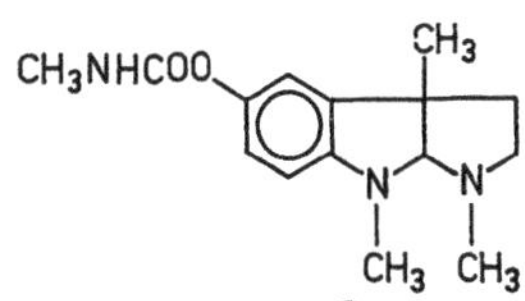

2

The binding of [^{125}Iod] alpha-BTX to acetylcholine receptors of the ganglionic homogenate of the molusc *Aplysia* and of the fruitfly *Drosophila* is blocked by the anticholinesterase eserin.

Inhibition constant

3.2×10^{6} M	in 500 × g pellet of *Aplysia dactylomela* (Carpenter et al., 1976)[80]
1.0×10^{5} M	in Triton-x-100 extracts of *Drosophila melanogaster* (Schmidt-Nielson et al., 1977)[53]

Therefore, it might be that certain anticholinesterase insecticides owe some of their toxicity not only to inhibition of acetylcholinesterases, but also to interactions with AChRs of insects.

Quinuclidinyl benzilate (QNB)
molecular mass: 337.42
$C_{21}H_{23}NO_3$

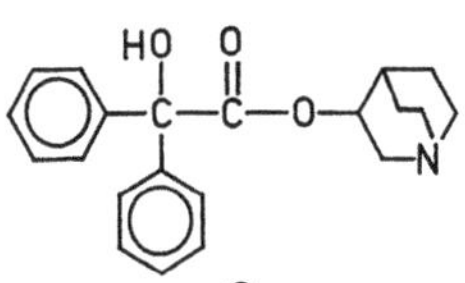

3

Dissoziation constant

2.0×10^{-9} M	in 500 × g supernatant of *Drosophila melanogaster* heads (Haim et al., 1979)[81]
0.2×10^{-9} M	in Guinea pig brain extract (Eldefrawi M. E, 1983)[15]

[^{3}H]-QNB (3), as a potent muscarinic cholinergic antagonist, has proved to be an effective substance in the characterization of the CNS receptors of vertebrate brain extracts[69]. Dudai and Ben-Barak have pointed out QNB (3) also characterizing a saturable, specific component of binding in *Drosophila* [82]. By the aid of the binding measurement of [^{3}H]-QNB the ratio between nicotinic and muscarinic AChR in insects and vertebrates could be determined.

Whereas in vertebrates the ratio of the receptors in the CNS shifts to the majority of muscarinic receptors, in *Drosophila* head extracts for example in the CNS only 0.02 muscarinic receptors of 1 nAChR have a ratio 1:50 [15]. The affinity of QNB to mAChR of insects, however, is one order of magnitude lower than for muscarinic receptors of vertebrates. This is an interesting differentiating property between insects and vertebrates.

Injection-tests demonstrated a insecticidal activity of QNB (20 µl, 10^{-2} M) on *Periplaneta americana.*

Atropin

Molecular mass: 289.36
Tropan-Alkaloid
$C_{17}H_{23}NO_3$

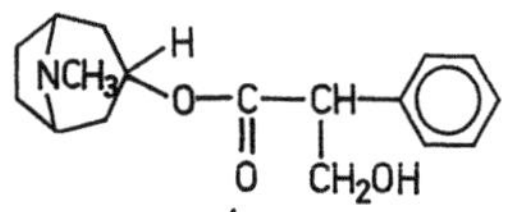

4

Inhibition constant

1.7×10^{-7} M	in *Periplaneta americana* (nerve cord) (Lummis and Sattelle, 1985) [75]
4.0×10^{-7} M	in *Locusta migratoria* (head ganglia) (Breer, 1981) [83]
4.0×10^{-7} M	in *Drosophila melanogaster* (heads) (Haim et al., 1979) [81]
4.2×10^{-10} M	in rat brain extracts (Gilbert et al., 1979) [84]

Atropin (4) is an antagonist of muscarinic AChR of vertebrates and invertebrates. It could be shown that atropin in the injection-test (20 µl; 10^{-2} M) has no insecticidal efficacy. But it inhibits the nicotine binding on the "mixed functional" receptor of the 100000 × g supernatant of housefly head extracts [79]. This is a hint to an allosteric influence at two binding sites of one and the same receptor molecule.

Muscarine chloride

Molecular mass: 208.71
$C_9H_{20}ClNO_2$

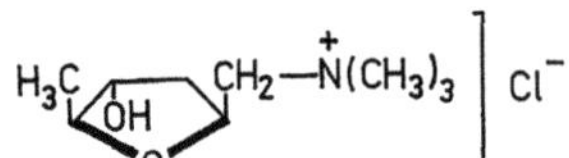

5

Muscarine chloride (5) is the natural product of the toadstools (fungi) *Amanita muscarina* and *Amanita pentharina.* It reacts like the natural transmitter acetylcholine and is an agonist of the mAChR. Muscarine, which had no effect on the binding of alpha-BTX to *Drosophila* putative nicotinic receptors at concentrations up to 1.0×10^{-3} M [70] inhibited [^{3}H]-QNB binding with the IC 50 6.0×10^{-5} M. O'Brien, R. and Eldefrawi, A. [50] characterized the binding of [^{3}H]-muscarine to extracts of housefly (*Musca domestica*) heads, using equilibrium dialysis. Most of the binding was present in a high-speed (100000 × g) supernatant. Since binding was inhibited equally well by nicotinic and muscarinic ligands, muscarine (5) was suggested to bind at AChR of mixed affinity (= Third type "multifunctional" receptor). Extent and affinity of the binding to fly head homogenates was comparable to that found for binding of this reversible ligand to electroplax of *Torpedo fish* [85].

3 The Action of Nicotine and its Derivatives on the Acetylcholine Receptor (AChR) and its Insecticide Properties

3.1 The Potence and Importance of Nicotine as an Insecticide

Molecular mass: 162.23
Pyridine-Alkaloid
$C_{10}H_{14}N_2$

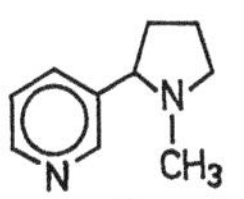

6

Nicotine (S)-3-(1-methylpyrrolidine-2-yl)-pyridine) (6), an ingredient of the plant *Nicotiana tabacum*, displays an agonistic efficacy when applied to vertebrate and insect cholinergic neurons [12]. Nicotine is one of the oldest known insecticides [86, 11, 15]. The LD 50 values for a lot of insect species are reported by Negherborn [87] and evaluated for topical application, e.g. for *Periplaneta americana* with 650 mg/kg, for *Apis mellifica* with 315 mg/kg (for comparison: LD 50 DDVP 6.5 mg/kg). The *Bombyx mori* silkworm is extremely sensitive (LD 50 topical = 4 mg/kg), the yellow mealworm *Tenebrio molitor* extremely insensitive (LD 50 topical = 3200 mg/kg). Nicotine possesses a remarkable vaporous activity. The median lethal concentration of vapors (in µg/l) ranged from 3–8 for aphids and thrips, from 150 to 300 for the Japanese beetle and Mexican bean beetle. Nicotine kills insects rapidly, often within an hour. Intensive tremors and convulsions are followed by paralysis. As early as 1746 the insecticidal activity of crude extracts of tobacco leaves was described (88, 89). But despite this long history, nicotine has never been an important insecticide. Metcalf reported [90] that the equivalent of 1.2 million pounds of free nicotine was used in agriculture in the USA in 1944. With the development of synthetic organic insecticides nicotine has assumed an even less important role. The formerly used extracts of tobacco leaves have been replaced by technical nicotine with 95–98% or by nicotine sulphate with 40% alkaloids. They are non-persistent, non-systemic contact insecticides with some ovicidal properties. In the case of nicotine sulphate the addition of an alkali is necessary to liberate the free nicotine base. A fumigation formulation is also available. Today two commercial preparates of nicotine are registered in France and Great Britain [91, 92, 93].

Nicotine has a fairly high mammalian toxicity and is toxic to man by inhalation and by dermal contact (LD 50 s.c. rat: 16 mg/kg and acute oral: 50–60 mg/kg) [87]. This is one of the reasons why nicotine has been limited in utilization. Besides this nicotine has a narrow insecticidal spectrum (aphids, thrips and certain species of mites) and a very short life span and low residual activity under field conditions [94]. So nicotine cannot compete with modern organosynthetic insecticides. Otherwise, nicotine can cope with the resistance to organic phosporus compounds and carbamates, and insect strains which develop resistance to such kinds of insecticides are not likely to posses cross resistance to nicotine due to the different targets [94].

3.2 Mechanism of Insecticidal Action of Nicotine

Nicotine represents the class of "nicotinoids" with a unique mechanism of action. Nicotine mimics ACh and some of the AChR types fail to distinguish between ACh and nicotine. A direct interaction with the nAChR is demonstrated by its effectiveness in competitively displacing [^{125}Iod]-alpha BTX from both housefly heads and *Torpedo* membranes [39,44]. It is also displacing [^{3}H]-decamethonium binding from the cholinergic binding protein in housefly brain high speed centrifugation fraction [79].

Inhibition constant (due to alpha BTX binding)

Ki =	8.0×10^{-8} M	*Drosophila melanogaster* (heads) (Dudai, 1978) [73]
	3.5×10^{-6} M	*Periplaneta americana* (nerve cord) (Lummis and Sattelle, 1985) [75]
	4.6×10^{-8} M	*Periplaneta americana* (6. abdominal ganglion) (Richter, 1987) [59]
	1.2×10^{-7} M	Chick optic lobe (Wang et al., 1978) [95]

As an agonist of the nAChR, nicotine produces, in vertebrates, effects on the function of the autonomic nervous system and the effector organs receiving both symphatic and parasymphatic innervation. It has a strong activity on the neuromuscular junction in vertebrates.

In insects nicotine binds to the nAChR and to the "mixed functional", decamethonium sensitive receptor. The direct interaction between nicotine and the acetylcholine binding site of the receptors has been analyzed by biochemical methods [15].

Nicotine is believed to act at cholinergic synapses in the insect central nervous system on afferent sensoric function. It has no direct effect on the skeletal muscles of insects, which are non-cholinergic, but receive innervation from excitatoric neurons [94].

3.3 Structure-Activity-Relation in Nicotine Derivatives

Some work has been done to correlate the chemical structure with biological activity of nicotine and related compounds in mammals [96], much less attention has been devoted to their effects on insects. Eldefrawi et al. [97,98] found that such nicotinoids, which have insecticidal potency, inhibit the binding of the ligand [^{3}H]-muscarine at

Table 4. Comparison of the toxicity and the relative inhibition on the mAChR to the housefly by some selected nicotinoids

Drug	Toxicity to housefly LD 50 (μg/fly) (1)	Relative toxicity	% Inhibition of [3H]-muscarin binding (10^{-6}) (2)
Nicotine [6]	5	1	50
Anabasine [7]	4	0.75	25
N-(3 Pyridyl-methyl)dimethylamine [8]	16	0.46	45
N-(3 Pyridylmethyl)-diethylamine [9]	11	0.10	61
Coramine [10]	100	—	0
N-(3-Pyridyl-methyl)morpholine [11]	100	0.038	0

(1) Yamamoto et al., [99] based on LD 50 to house fly by topical application method. Corrected on molar basis;
(2) Eldefrawi et al., [98]

the "soluble" AChR of housefly brains. Some results, summerized in Table 4, show a positive correlation between the degree of toxicity and the inhibition of muscarine binding. There was reversible binding of [^{3}H]-nicotine and [^{3}H]-muscarine at the same protein. But the correlation is only weak, because toxicity is depending on many factors, only one of which is the affinity of the drug to the target.

Because both the ACh-receptor and the ACh-esterase posses binding sites for ACh, these proteins must have similar parts of affine structures. Therefore, it is to suppose that also other ligands of the AChR can affect the AChE and reversely. But there is no correlation between the insecticide potency of nicotinoids or 2-pyridyl-methylamines to houseflies and I 50 values for ACh splitting of AChE (Table 5) [99].

It has to be taken into consideration, too, that the concentration required to kill houseflies is far lower than the concentration required for the inhibition of AChE [99]. So the mechanism of action of nicotine and its derivatives is not primarily connected with antiacetylcholinesterase activity.

Detailed requirements by Yamamoto et al. [99] led to a good understanding of the structure-activity-correlations. Necessary is the pyridine ring, its nitrogen possesses a relatively low basicity. A ring without nitrogen or a cyclohexal ring in the molecule is not active. The second nitrogen must not be involved in a five-membered pyrrolidine ring, but it must have a constellation resulting in a high basicity (higher than 6, best of all between pka 8 and 9 at 37 °C), and the distance between the two nitrogens may be about 4.2 Å. So the essential part of the molecule,

Table 5. Absence of correlation between the insecticide potency of nicotinoids to houseflies and I_{50} value for ACh, splitting of AChE [99]

Drug	Relative toxicity	Basicity pKa at 37 °C	Anticholinesterase activity/ACh 2mM I_{50}
dihydronicotyrine [12]	1.16	7.21	0.35
nicotine [6]	1.00	7.66	2.60
n-methylanabasine [13]	1.00	7.66	0.72
anabasine [14]	0.80	6.41	3.50
nicotyrine [15]	0.18	4.51	2.30

Yamamoto et al. [99]: Determination of anticholinesterase activity was done at substrate level 2 mM ACh (pH 7.4 and 37 °C).

responsible for high nAChR affinity and high insecticidal potency is far lower than the concentration required for inhibition of acetylcholinesterases [99].

Altering the position of the side chain at the pyridine ring (Table 6) and the chain length between the 3-pyridyl and basic nitrogen (Table 7) had significant effects on toxicity. The secondary and tertiary amines were found to have insecticidal potency whereas the primary amino derivates have not (Table 8).

In general, in tertiary amino types (nicotinoids and N,N-disubstituted 3 pyridylmethylamines) the tendency can be outlined, the higher the pKa the higher the inhibition of AChR. In contrast primary and secondary amines show lowered inhibition, though they have a high basicity and they are ionized in a high degree.

Because of the high basicity of the second nitrogen in comparison with the pyridine nitrogen, this pyrrolidine or related nitrogen becomes protonated in a physiological medium at pH 7, e.g. inside the insect body. In this way the molecule acquires the active form, which is able to compete with ACh on the receptorprotein. From this point of view, nicotine (6) or structurally related compounds are only pre-insecticides, which enable the compound to penetrate easily into the body and into the nervous tissue to the site of action. The protonization converts this

Table 6. Effects on toxicity altering the position of the side of chain at the pyridine ring

Structure	Relative toxicity	Basicity/37 °C
Diethylaminomethylpyridine		
[9]	0.085	8.42
[9 a]	0.450	8.12
[9 b]	0.10	7.96

Table 7. Effects of toxicity altering the chain length between the 3-pyridyl and the basic nitrogen

Structure	Relative toxicity
3(N-Piperidylmethyl)pyridine (16)	0.79
3(N-Piperidylethyl)pyridine (17)	0.14

Table 8. Dependency of toxicity from the methylation of the basic nitrogen

Structure	Name	Relative toxicity	Basicity/37 °C
primary (18)	3-Aminomethyl-pyridine	0.029	8.0
secondary (19)	3-Methylamino-methylpyridine	0.300	8.0
tertiary (8)	3-Dimethylamino-methylpyridine	0.460	7.76

pre-insecticide to the active form with the quarternary nitrogen, which is essential for the binding of the molecule to the anionic site of the receptor.

Several natural analogs of nicotine are found in tobacco or other plants which have similar physiological effects in mammals. These include nornicotine (20) and anabasine (7), which are toxic to insects, too [94].

Coniine (21), the principal poison of hemlock; cytisine (22), a alkaloid of laburnum; anatoxin-a (23), the exotoxine of the freshwater blue green algae *Anabaena flosquae*,

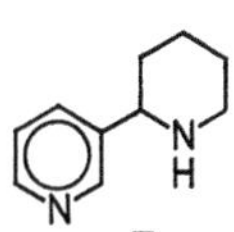
7

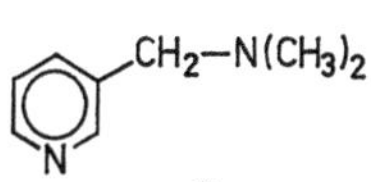
8

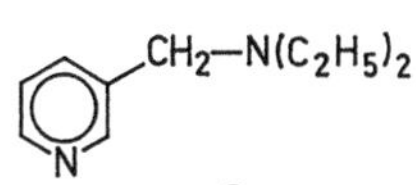
9

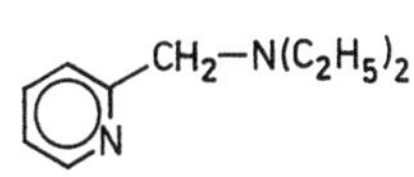
9a

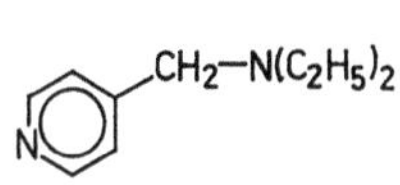
9b

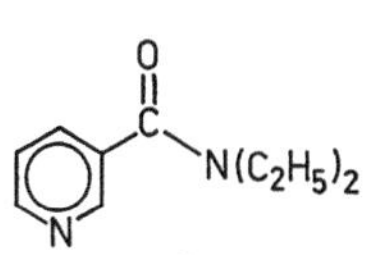
10

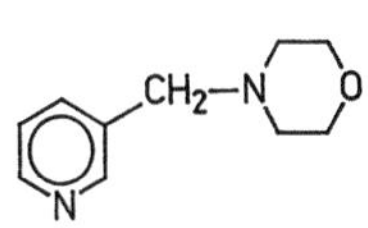
11

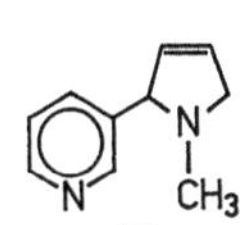
12

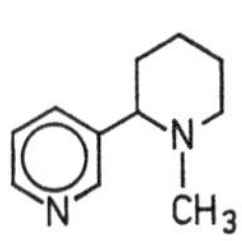
13

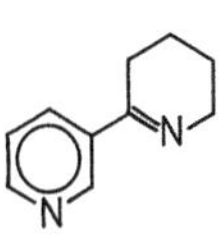
14

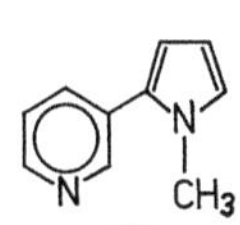
15

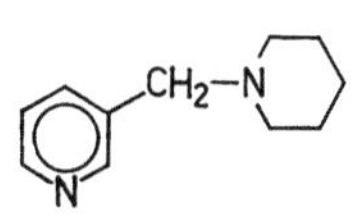
16

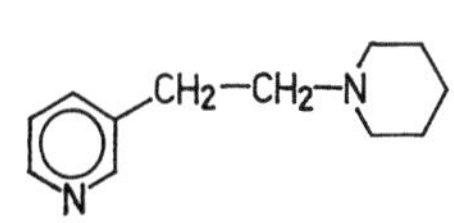
17

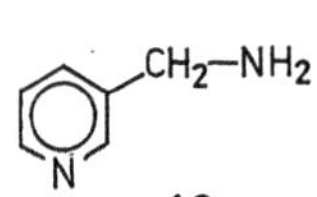
18

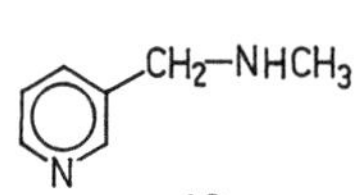
19

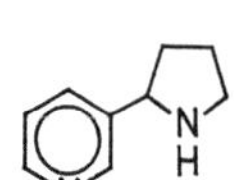
20

and lobeline (24), the active ingredient of the Indian tobacco show similar efficacy like nicotine. The structural variations of these compounds offer in comparison with nicotine activity on mammals that the nicotine structure can be modified drastically without loss in activity [94].

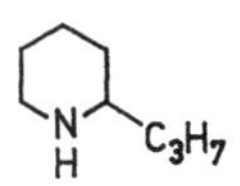

21

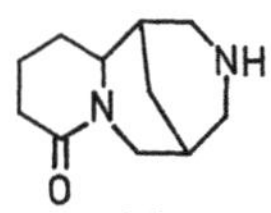

22

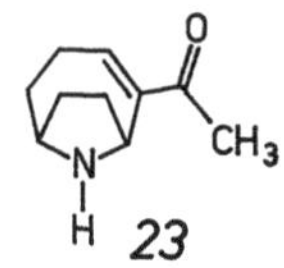

23

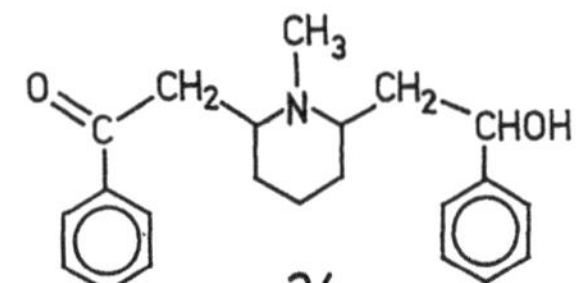

24

Although nicotine has been in use for a long time, development of better nicotinoid insecticides which act as receptor poisons has not progressed. This may emphasize that drug receptors remain relatively unexploited as targets of insecticides.

4 The Action of Nereistoxin and its Derivatives on the Acetylcholine Receptor (AChR) and its Insecticide Properties

4.1 Synthetic Derivatives of Nereistoxin as Commercial Insecticides in Plant Protection

Nereistoxin is 4-dimethylamino-1,2-dithiolane (25), naturally occurring in the marine worm *Lumbriconereis heteropoda* MARENZ (Polychaeta, Annelida). It was at first isolated and pharmacologically characterized by Nitta in 1934 [100]. Determination of the chemical structure and toxic action was carried out by Okaichi and Hashimoto [101]. In 1964 Sakai demonstrated the high insecticidal activity of nereistoxin, in particular to larvae of the rice stem borer (*Chilo suppressalis*, one of the most serious insect pests in rice in Asia) and some other lepidopterous and coleopterous insects [102]. This was the starting point of an intensive chemical and biological research programme for the derivation of the nereistoxin molecule, as a basis for developing new insecticides. Today three synthetic derivatives of nereistoxin are commercially available: Cartap hydrochloride (26), Thiocyclam oxalate (27) and Bensultap (28). They have acquired remarkable importance in the field of plant protection.

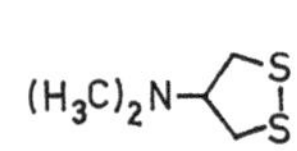

25

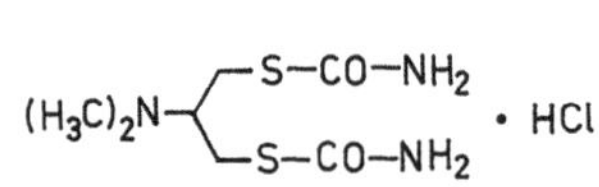

26

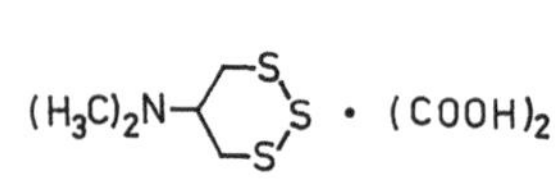

27

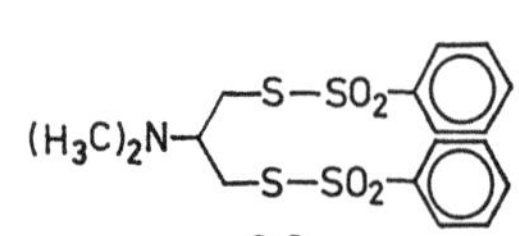

28

The general advantages of these insecticides are

- the activity on a broad spectrum against harmful insect pests, especially to coleoptera and lepidoptera,
- the action as a contact and stomach poison with ovicidal, larvicidal and adulticidal effects [103, 104, 105], against sucking and leaf-biting pests.
- The systemic transportability within the water stream in plants [104, 106].
- Full toxicity to such pest populations which have already developed resistance to all other classes of insecticides (chlorinated hydrocarbons, carbamates, phosphoric esters, pyrethroids) [107–110].
- Only moderate toxicity to beneficial insects and honey bees [111–118].
- Without persistance in the environment [119–121].
- With only moderate (Cartap, Thiocyclam) or weak (Bensultap) toxicity to mammals,
- control of main pests in main crops like rice, citrus, sugar cane, alfalfa, potatoes, rape, sugar beet, cereals, cotton, vegetables and fruits (Table 9).

Some special properties of these active agents and their formulated preparations are summerized as follows. (For spectrum of biological activity and toxicological properties see Table 10 and 11).

CARTAP, as cartaphydrochloride
2-dimethylamino-1,3-bis(thiocarbamoyl)propane hydrochloride (26)
Code name: Ti 1258.
Patents: priority Japan 1964 Takeda Chemical Industries, Ltd.

GB P 1126204
US P 3332943
FR P 1452338

former review see [162].
Chemical and physical properties:
colourless crystals, Fp: 179 to 181 °C
stable in acidic medium; hydrolysis in neutral or alkaline solution; soluble in water (c. 200 g/l) at 25 °C), slightly soluble in ethanol and methanol, not soluble in acetone, ether, ethyl acetate, chloroform, benzene and n-hexane.

Table 9. Literature about reommendated application in crops

Culture	
Rice	.122–130, 113, 104, 131–137
Citrus	.107
Sugar cane	.138, 139
Alfalfa	.103
Potatoes	.105, 107, 110, 120, 131, 140–146
Rape	.147, 148
Sugar beet	.149–151
Cereals	.131, 139, 152–154
Cotton	.155, 156
Vegetables/fruits	.114, 120, 131, 157–161

Table 10. Proofs of efficacy of Padan, Evisect and Bancol in pest control, published in literature of applied entomology

Pest	Padan	Evisect	Bancol
Acari			
Tetranychus urticae		163	
Polichrosis botrana	160		
Thysanoptera			
Scirtothrips dorsalis			164
Stenochaetothrips biformis	122		
Saltatoria			
Gryllotalpa africana	164		
Gryllotalpa unispira	165		
Aeneolamia varis saccarina		138	
Homoptera, Cicadoidea			
Nephotettix vincticeps	113, 123		
Nephotettix varescens		132	
Nilaparvata lugens	113	132	
Homoptera, Aphidae			
Myzus persicae	151	163	
Aphis fabae	151		151
Brevicoryne brassicae			164
Macrosiphoniella sanbornii		163	
Heteroptera, Lygaeidae			
Dimorphopterus spinolae	166		
Coleoptera, Nitidulidae			
Meligethes aeneus	148	164	164
Coleoptera, Coccinellidae			
Henosepilachna vigintioctopuncta			120
Epilachna varivestis	164		
Coloptera, Anobiidae			
Anobium punctatum	167		
Coleoptera, Chrysomelidae			
Leptinotarsa decemlineata	105, 109, 140–144	131, 146	110, 120
Oulema melanopus	154	153, 154	110, 168
Oulema oryzae	124, 125		
Coleoptera, Curculionidae			
Bothynoderes punctiventris	149, 164		164
Psalidium maxilosum	149		
Anthonomus grandis	164		164
Tanymecus palliatus			110, 164
Tanymecus dilatocillis			110
Ceutorhynchus macula alba	169	109	
Ceutorynchus quadridens		147	
Sitophilus granarius			170
Otiorrhynchus ovatus	159		
Otiorrhynchus juglandiformis	161		
Otiorrhynchus albidus	161		
Sciaphobus balcanicus	161		
Mecinus pyraster	161		
Lepidoptera, Plutellidae			
Plutella xylostella	108, 158, 164	108	120, 164
Lepidoptera, Gracilariidae			
Lithocollis blancardella			164

Table 10. (Continued)

Pest	Padan	Evisect	Bancol
Lepidoptera, Bucculatricidae			
Bucculatrix thurberiella	164		164
Lepidoptera, Torticidae			
Laspeyresia pomonella	164		
Laspeyresia funebrana	164		
Lobesia botrana	164		
Lepidoptera, Pyralidae			
Chilo suppressalis	126, 129, 130		164
Chilo infuscatellus		164	
Ostrinia nubilalis	164	131	
Tryporyza incertulas	128, 126, 133	131, 132	131, 132
Tryporyza innotata	126, 128		
Diatraea saccharalis	164		
Lepidoptera, Nycteolidae			
Earias insulana		156	
Lepidoptera, Arctiidae			
Hyphantria cunea	164		
Lepidoptera, Noctuidae			
Agrotis ypsilon	164	164	
Sesamia inferens	126	164	
Spodoptera littoralis	155		
Heliothis armigera	155		
Laphygma frugiperda	164		
Alabama argillacea	164		
Lepidoptera, Lasiocampidae			
Malacosoma neustria	164		
Lepidoptera, Lymantriidae			
Euproctis melania		172	
Lepidoptera, Hesperidae			
Parnara guttata			164
Lepidoptera, Pieridae			
Pieris rapae	164		164
Pieris brassicae		131	
Hymenoptera			
Mesoneura rufononta		172	
Hoplocampa minuta	164		
Hoplocampa testudinea	164		
Diptera			
Hydrellia sasuki		132	
Myiopardalis pardalina	157		
Rhagoletis cerasi	164		
Orseolia oryzae		133	
Aedes vexans	174		
Anopheles sinensis	173		
Liriomyza trifolii		163	

Formulations (Takeda Chemical Industries, Ltd.):

Padan	50 SP,	water soluble powder	50% a.s.
Padan	25 SP,	water soluble powder	25% a.s.
Padan	10 G,	granulate	10% a.s.
Padan	4 G,	granulate	4% a.s.
Padan	DD,	dust	2% a.s.

THIOCYCLAM, as thiocyclam hydrogen oxalate
= N,N-dimethyl-1,2,3-trithian-5-ylamine hydrogen oxalate (27)
Code name: SAN 155 I.
Patents: priority Sandoz AG, Swiss, 1977

IP PS 7 8142529
DE OS 2 03555

former review see [131)]
Chemical and physical properties:
The hydrogen oxalate is a colourless crystalline solid, mp: 125 to 128 °C;
Stability: Half life (protected from light, a 40 mg/l water, 22 to 24 °C): pH 3.5 = 42 d; pH 7.9 = 11 d
practically insoluble in cerosin
Formulation (Sandoz AG):

Evisect WP	wettable powder	50% a.s.
Evisect G	granulate	5% a.s.
Evisect 90SP	water soluble	90% a.s.

Table 11. Toxicological properties of nereistoxin derivatives [92, 164)]

Parameter	Cartap	Thiocyclam	Bensultap
LD 50 acute oral mg/kg			
rat	325–345	195–310	1100–1120
mouse,	225	270	580
LD 50 acute dermal mg/kg			
rabbit			>2000
rat		1000	
mouse	>1000		
LD 50 i.v. mg/kg			
rat		12.5	
LD 50 i.p. mg/kg			
rat		30	503
no effect level, 90 d			
rat mg/kg diet	10	100	250
LD 50 carp mg/l; 48 h	1.3	1.12	13.2
dangerous for honey bees			15 µg/bee
half life (d)			
in plants		0.9–4	3–7 lab.
in soil	3	1–4	30 field

BENSULTAP
= 2-dimethylamino-1,3-bis(phenylsulfonylthio)propane (28)
Code name: Ti 78, Ti 1671.
Patents: priority Japan 1968 Takeda Chemical Industries, Ltd. DEOS 1 917246,
Chemical and physical properties:
light yellow powder, Fp: 82 to 83 °C, stable under acid conditions, under neutral and alkaline condition hydrolysis occurs, not hygroscopic, solubility: in water 0.7 to 0.8 mg/l, in methanol 25 g/l, in ethanol 13 g/l, in n-hexane 0.12 g/l, in acetone, DMF and chloroform very good solubility.
Formulation (Takeda Chemical Industrie, Ltd.):

Bancol	WP	wettable powder	50% a.s.
Vicetenon	WP	wettable powder	50% a.s.

4.2 The Mode of Insecticidal Action of Nereistoxin and its Derivatives

In 1941 Nitta [100] reported that adult flies fed on fresh dead annelid worm *Lumbriconereis heteropoda* were knocked down immediately after feeding, but they soon recovered from the intoxication and flew away. Further investigations followed [102] and generally an anaesthetic like effect could be observed after application.

First proofs about the mechanism of action of nereistoxin at the cholinergic synaptic transmission were detected in investigations of vertebrate pharmacological preparations [100]. At first nereistoxin was found to block cholinergic transmission in frog rectus abdominalis nerve-muscle preparations by decreasing the release of ACh from presynaptic nerve terminals and by reducing sensitivity of the postsynaptic membrane to applied ACh [174]. In further detailed investigations at amphibian and mammalian nerve-muscle-junctions, it could be shown that the reduced sensitivity of postsynaptic AChR is the most important part in the nereistoxin mechanism of action, and that there are, indeed, no remarkable effects on presynaptic ACh release [44]. The results led to the conclusion that nereistoxin is an antagonistic inhibitor of postsynaptic AChR. During occupation of receptor protein with nereistoxin the transmitted ACh is not able to trigger excitation at the postsynaptic membrane.

The specific binding site of nereistoxin at the receptor protein is not yet clear. Nereistoxin seems to be a weak inhibitor of alpha-bungarotoxin binding on the electric organ of *Torpedo* spec. [44]. It failed to inhibit the binding of [^{3}H]-perhydrohistrionicotoxine and, therefore, failed to block the ion-channel site of the AChR of the *Torpedo* electroplax membranes. There is a good evidence that nereistoxin acts as an AChR blocking agent in insects, too. In contrast to nicotine, which has an ACh mimetic excitatory effect on the postsynaptic neuron, nereistoxin and its derivatives have an antagonistic effect. Intoxicated insects do not show exciting effects, they die without convulsive symptomes, like in narcotic state. The antagonist nereistoxin competes on the nAChR the agonist ACh.

The nereistoxin binding is reversible. So, revival may occur when intoxication is not sufficient. Cockroaches (*Periplaneta americana*) treated with 23 µg/g of the toxin were inactivated for about 15 minutes, and regained normal behaviour 5 hours after treatment [102]. Generally, however, insects are rapidly paralysed.

Electrophysiological investigations were able to establish a blocking action at cercal afferent-giant interneurone nicotinic cholinergic synapses in *Periplaneta americana*[175–177]. A concentration above 2×10^{-8} M nereistoxin partly suppressed the electrophysiological response in the postsynaptic membrane after transmitter release in response to stimulation of cercal nerve XI[177,178].

Also in insects nereistoxin is a weak inhibitor of alpha-bungarotoxin binding, for instance in housefly head preparations[44]. Special investigations[59] confirmed that a concentration of about 10^{-4} M nereistoxin occupied 50% of alpha-bungarotoxin binding at the nAChR of the 6th. abdominal ganglion of *Periplaneta americana.* Nereistoxin is also a weak inhibitor of Quinuclidinyl benzilate binding. That means that nereistoxin does not act at the muscarinic AChR of the housefly head[179]. A slight anticholinesterase activity was also detected[180], but this may not have any importance for the mode of action.

Analogous with an agonistic action, at higher concentrations, up to 10^{-5} M, a postsynaptic depolarization was induced by nereistoxin (25)[1,177]. Despite the inhibitory effect at the AChR in the central nervous ganglion a stimulation of the heart and the gut of *Periplaneta* by nereistoxin can be observed. The stimulatory action on the visceral muscle can be partially antagonized by atropine and d-tubocurarine, both AChR active-compounds.

The different effects, antagonizing and depolarizing, may be due to the dosis dependency of the reaction and to the presence of delicate differences between organs in the structure and sensitivity of the AChR.

However, it seems that more than one type of action of nereistoxin (25) can be detected, one or more of which may be involved in the insecticidal action of this natural invertebrate toxin[12].

The synthetic derivatives of nereistoxin (25) and the insecticides cartap (26), thiocyclam (27) and bensultap (28) were transformed to nereistoxin inside the insect body[59,181]. So they ought to be pre-insecticides with good properties for handling, stability, transport and uptake[182]. Nereistoxin is in all cases the effective structure.

Toxins with similar structure, like nereistoxin have been isolated from the algae *Chara globularis*[183], namely 4-methylthio-1,2-dithiolane and 5-methylthio-1,2,3-trithiane propan-1,3-dithiolane with insecticidal properties, and from the mangrove *Bruguiera conjugata* stem and bark, which are named brugierol (cis I) and isobrugierol (trans I)[184].

4.3 Chemistry of Nereistoxin Derivatives in Relation to the Insecticidal Activity

A synthesis of Nereistoxin (25) (4-dimethylamino-1,2-dithiolane) (185) from dimethylamino-dichloro-propanes (36, 37) or bis(benzylthio)derivatives (29, 30) was described by Hagiwara and Konishi et al.[186–188].

The nucleophilic substitutions take place with intermediary formation of cyclic sulfonium (31) and immonium (35), which caused isomeric by-products, because of the two reaction centres of the cyclic intermediate. An improved synthesis of 4-dimethylamino-1,2-dithiolane and related compounds has been achieved in a three step sequence, starting from 1,3-bis(benzylthio)propan-2-one (38) [189] (or 2-phenyl-1,3-dithian-5-one) without being accompanied by unfavourable by-products but with unsatisfactory yields [188]. Konishi [185,190] reported the synthesis of various insecticidally active thiolester derivatives of 2-dimethylamino-1,3-bis(mercapto)-propane (dihydronereistoxin) (33) in a one step sequence, starting from 2-dimethyl-amino-1,3-dichloropropane (36) or 1-dimethylamino-2,3-dichloropropane (37) by a reaction with different sulfur nucleophiles RS_-, e.g. via (39, 40).

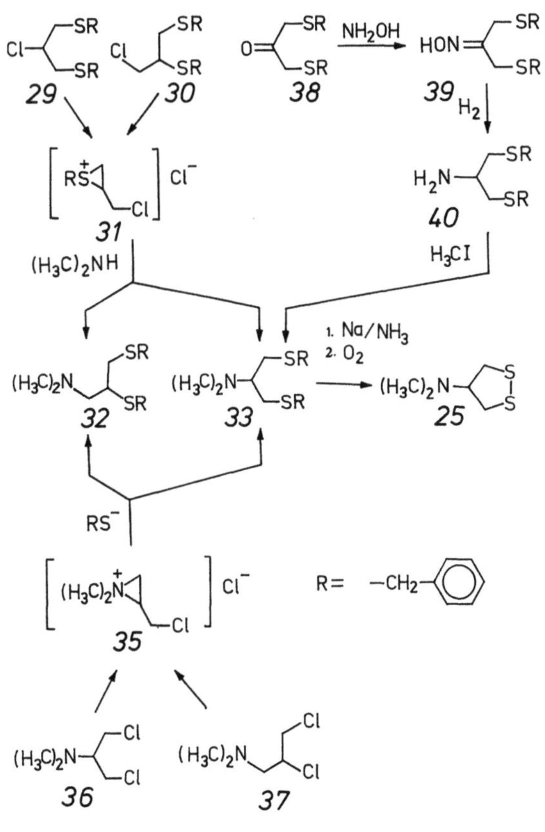

Fig. 4. Synthetic routes to nereistoxin

Biological tests showed that most of the thiolesters have contact toxicity to insects, especially against the larvae of the rice stem borer (*Chilo suppressalis*). Among the 2,3-isomers only dithiol and dithiocyanate had effects in insects.

The salt formation of the free base stabilized the compounds, but weakened their biological activities.

The treatment of 2-dimethylamino-1,3-bis(thiocyanato) propane with hydrochloric acid as well as the reaction of dihydronereistoxin, with two equivalents of alcali cyanides produces in the thiolcarbamate (26) good yields [189-195].

2-dimethylamino-1,3-bis(carbamoylthio)propane hydrochloride Cartap (26) is rapidly hydrolyzed in physiological conditions to 2-dimethylamino-1,3-bis(mercapto)propane, which was oxidized to nereistoxin (25) [121, 189].

A similarly easy conversion to nereistoxin can be expected from a large number of disulfane compounds [196, 197].

2-dimethylamino-1,3-bis(tert.-butylthio)propane (41) shows a high contact- and feeding activity against insects. Like other dithio-derivatives it is relatively stable and biological degradation is favourable.

The fish activity and the toxicity to warmblooded animals are also low [162].

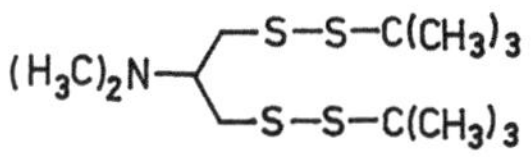

41

Some 2-dialkylamino-1,3-bis(arylsulfonylthio)propanes [197-201] represented by Bensultap (28) [121], have distinguished insecticidal properties with the lowest mammalian toxicity [121, 198].

An other group of derivatives with similar activity are the 2-dimethylamino-1,3-bis(alkoxyalkylthio)propanes (42), they have been prepared by the reaction of 2-dimethylamino-propane-1,3-thiol with chlorinated ethers [196, 203]

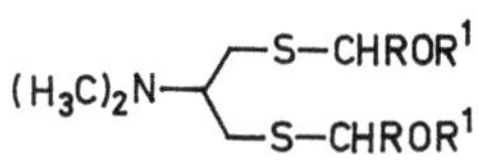

$R=R^1=H; C_{1-3}$-alkyl

42

The S,S′-diacyl derivatives of propane-1,3-dithiol (43) respectively exhibit a various activity against insects [199, 204].

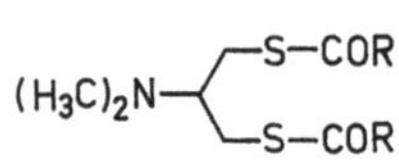

$R=-CH_3$
$-C_6H_5$

43

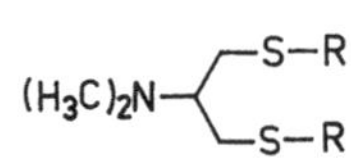

44

In a screening test insecticidal activity has been tested on adults of the adzuki bean weevil (*Callosobruchus chinensis*) and also larvae of the rice stem borer (*Chilo suppressalis*), because these insects are highly sensitive to the activity of the original compound, nereistoxin [185, 205, 206]. Typical acyclic insecticidal derivatives are listed in Table 12.

The activity of the 1,3-propane-dithioles, the derivatives of the dihydronereistoxin, is likely to depend on the stability of the S-R linkage.

More active compounds, the thioacetates, thiocyanates, thiocarbamates and thiosulfonates would be easily hydrolyzed at the linkage than alkyl- or arylthioethers, compounds which are composed of more stable S-R linkages.

It is presumed that active 1,3-dithiols are biologically converted into dihydronereistoxin from the cleavage of the S-R linkage. The conversion of cartap and dithiocyanates in insect, mammalian and plant tissues has been detected [205].

Tab. 12. Insecticidal activity and mammalian toxicity of acyclic derivatives (44)

Compound (44) R =	Base/Salt	*C. ch.* LD 50 (μg/dish)[a]	*Ch. s.* (μg/g)[b]	Mice (mg/kg)[c]	Rats (mg/kg)[c]
Nereistoxin	$(COOH)_2$	7.9	1.7	118	420
H	$(COOH)_2$	5.9	20		
CH_3	Base	1000	50		
$CH_2—C_6H_5$	Base	1000	50		
$COCH_3$	$(COOH)_2$	20.3	3.1		
COC_6H_5	$(COOH)_2$	1000	0.86		
$CONH_2$	HCl	8	5	165	250
$CSN(CH_3)_2$	Base	1000	4.2		
$C(NH)N=_2$	2 HCl	1000	50		
CN	$(COOH)_2$	4	3	23.2	125
$SCH(CH_3)_2$	$(COOH)_2$	9.9	3		
$S(p—Cl—C_6H_4)$	$(COOH)_2$	14	3		
SO_2H	inner	48	20		
SO_2CH_3	$(COOH)_2$	50	2.4	331	
$SO_2—C_2H_5$	$(COOH)_2$	60	50		
$SO_2—C_6H_5$	$(COOH)_2$	5000	50	415	1120
$SO_2—CH_2C_6H_5$	$(COOH)_2$	1000	50		
$SO_2—(p—CH_3—C_6H_4)$	Base	1100	50		
SO_3Na	Na	10.6	6.1	170	333–400
$P(S)(OC_2H_5)_2$	$(COOH)_2$	990	35	6.0	

[a] Dry film contact method (Petri-dish of 9 cm in diameter);
[b] Topical application;
[c] Oral.
C. ch. = *Callosobruchus chinensis*
Ch. s. = *Chilo suppressalis*

The reaction of thiolsulfonic acid derivatives, especially the 2-dialkylamino-1,3-bis(benzenesulfonylthio)propanes or the sodium thiosulfonate with sodium sulfides leads to the 1,2-dithiolanes, the derivatives of nereistoxin.

These compounds were oxydized to 4-dialkylamino-1,2-dithiolane-1-oxides by treatment with peracids or hydrogen peroxide. It was found that 1,2-dithiolane-1-oxides react with sodium sulfide, thereby one sulfur atom is inserted into the ring and the sulfoxide bond is reduced forming the 1,2,3-trithianes (45) in high purity and high yield [207–209].

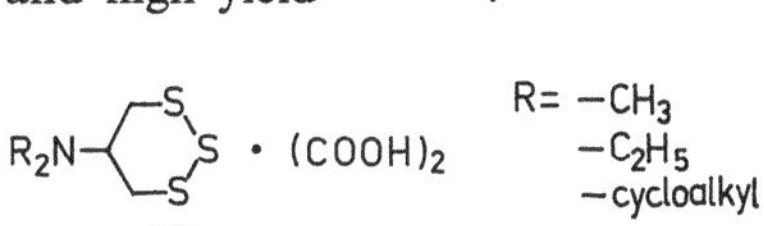

45

5-dimethylamino-1,2,3-trithian (45) ($R=CH_3$) Thiocyclam (Evisect R) SAN 155, is also prepared by the treatment of 2-dimethylamino-1,3-dimercaptopropane with sulfurdichloride, is used ay an insecticide and fungicide with low toxicity in warm-blooded animals [210].

2-Imino-1,3-dithiane (46) show also some insecticidal activity.

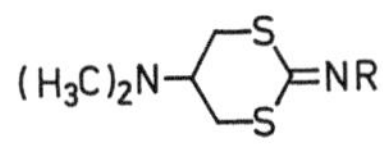

46

R= $-OH, -COOR$
$-COR, -CONR_2, -CSNR_2$
$-CN, -OCONR_2$

2,2-disubstituted 1,3-dithianes (47) with specific action towards various groups of insects are synthesized as a large number of derivatives [136].

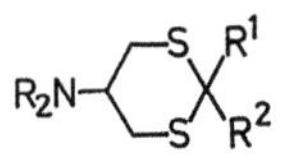

47

R= $-CH_3$; $-C_2H_5$; -cycloalkyl
$R^1 = -H, -CN, -COOR, -COR^4, -SO_2R^5$
$R^2 = -H, -CN, -COOR^3, -COR^4, -SO_2R^5$

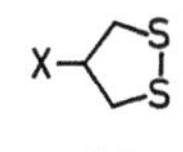

48

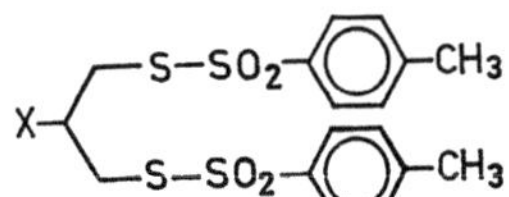

49

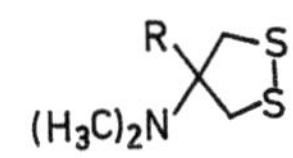

50

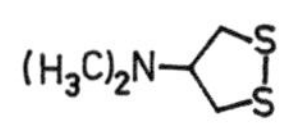

51

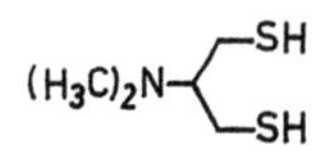

52

Table 13. Synaptic blocking and insecticidal activity of some N-substituted derivatives [198,205]
4-amino-1,2-dithiolanes (48)

Compound X=	Minimum syn. blockingdose (μg/ml)	*C. ch.* LD50 contact (μg/dish)	*Ch. s.* LD50 topical (μg/g)
H_2N- [a]	75	1000	50
$(CH_3)NH-$ [a]	7.5	66	6.3
$(CH_3)_2N-$ [a]	2.5	7.9	1.7
$(CH_3)_3N-$ [b]	75	1000	40
$(CH_3)(C_2H_5)N-$ [a]	2.5	10	10
$(CH_3)(C_2H_5)N-$ [a]	2.5	10	10
$(C_2H_5)_2N-$ [c]	7.5–25	9.9	18
N-Pyrridin- [c]	25	29	28
N-Piperidin- [c]	7.5–25	26	17
N-Morpholin- [c]	7.5–25	14	23

2-amino-1,3-bis-(toluolsulfonylthio)-propanes (49)

Compound X=	*C. ch.* LD50 contact (μg/dish)	*Ch. s.* LD50 topical (μg/g)
$(CH_3)_2N-$ [d]	1100	50
$(CH_3)(C_2H_5)N-$ [d]	210	<50
$(C_2H_5)_2N-$ [a]	1500	50
$(C_3H_7)_2N-$ [a]	1000	50
$(C_4H_9)_2N-$ [d]	1000	50
N-Pyrilidin- [d]	1000	50
N-Piperidin- [d]	1000	50
N-Morpholin- [d]	1000	50

[a] oxalate; [b] hydroiodide; [c] picrate; [d] free base;
C. ch. = *Callosobr. chinensis*
Ch. s. = *Chilo suppr.*

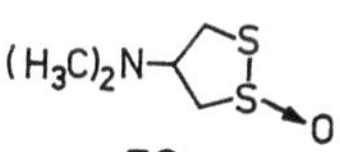

53

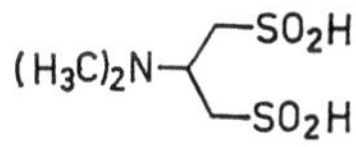

54

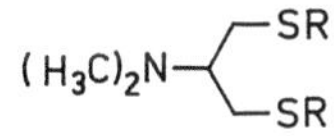

55A

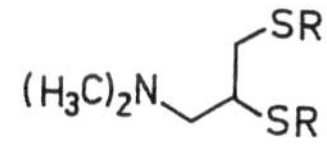

55B

In all classes of derivatives, as in dithiolanes, di- and trithiones, the dimethylamino group has been substituted. The effect of the amino-substitution in dithiolanes and thioltosylates is shown in Table 13.

The substitution of the amino group considerably influences the activity. Activity appears in monomethylamine derivative, but it decreases with substitution of the more bulky alkylamine than with dimethylamine. Dithiolanes substituted by diethylamine, pyrrolidine, piperidine and morpholine are almost equally active. Quarternary ammonium is inactive presumably owing to its high ionizable characteristic, which will result in the inability to penetrate a certain barrier of the target area.

Substitution of the hydrogen on the nitrogen neighbouring carbon atom in the skeletal structure generally weakens insecticidal activity (Table 14).

The activity of the oxidized products of nereistoxin and acyclic 2-dialkylamino-1,3-propanedithiols are explained by the assumption that these compounds are converted into nereistoxin or dihydronereistoxin by biological redoxreactions (Table 15).

It has been recognized that the active compounds belong to the 4-alkylamino-1,2-dithiolanes, 5-dialkylamino-1,2,3-trithianes or 2-dialkylamino-1,3-propanedithiols. In contrast, the non sulfurated amino-propanes and 1-dimethylamino-2,3-propanedithiol derivatives have no or less insecticidal activity (Table 16).

Table 14. Insecticidal activity of 4-substituted-4-dimethylamino-1,2-dithiolanes (50)

Compound R=	*C. ch.* LD50 contact (μg/dish)	*Ch. s.* LD50 topical (μg/g)
—H	7.9	1.7
$-CH_3$	26.0	50.0

Table 15. Insecticidal activity of nereistoxin, dihydronereistoxin and their oxidized products

Compound	*C. ch.* LD50 contact (μg/dish)	*Ch. s.* LD50 topical (μg/g)
(51)	7.9	1.7
(52)	5.9	20.0
(53)	16.3	5.3
(54)	48.0	20.0

Table 16. Insecticidal activity of 2-dimethylamino-1,3-propanedithiols and 1-dimethylamino-2,3-propanedithiol-Derivatives [55A/B]

Compound A/B R=	*C. ch.*	LD50 contact (µg/dish)	*Ch. s.*	LD50 topical (µg/g)
	A	B	A	B
$-H$	5.9	1000	20	50
$-CN$	4.0	1000	3	50
$-CONH_2$	8.0	1000	5	50
$-COCH_3$	20.3	1000	3.1	50

C. ch. = *Callosobruchus chinensis;*
Ch. s. = *Chilo suppressalis*

5 The Action of Other Insecticides at the Acetylcholine Receptor

For insecticides, such as chlorinated hydrocarbons, phosphororganic esters, carbamates, and pyrethroids, the main relevant mechanism of action is well known and different from the action at the AChR. But there is evidence that some of these types of insecticides have effects at the AChR too, and it cannot be decided at the present level of knowledge whether these interactions and side effects are of any importance in the mode of action.

Besides commercial insecticides some other groups of active compounds have also been reported to have interactions with the AChR, such as ryanodine [216], isothiocyanate [217,218,177,219], organic mercuriates [11,88].

5.1 Chlorinated Hydrocarbons

Chlorinated hydrocarbons of the DDT-type act at the tetrodotoxin sensitive Na-channel in excitative axon membranes of the neurones by prolonging the open state of this channel protein during action potential transfer [220].

However, DDT and DD-OH-palmitic acid give rise to an increased (^{3}H)QNB-binding on muscarinic AChR of mice cerebral cortex [221]. This is connected with a shift in the relation of the frequency of different receptor types by decreasing high affinity carbachol binding sites and increasing low affinity QNB binding sites [16]. Regional effects of organochlorine insecticides on cholinergic muscarinergic receptors of rat brain are also reported [222]. Some derivatives of DDT such as dicofol and chlorbenzilate have an acaricide effect (see Fig. 5) (56)–(60). Chlorbencilate inhibits (^{3}H)QNB binding to bovine brain *corpus striatum* receptors, and this compound inhibits, like dicofol, the (^{3}H)-HTX binding to the *Torpedo* electric organ (see Table 17) [15]. Chlorbencilate reduces [^{3}H]-QNB binding to housefly heads with the same Ki values like to bovine brain extracts [001) 179]. Whether the acaricidal effects of chlorbenzilate are related to its action on muscarinic receptors in mites, has not been decided. Yet three muscarinic agonists (oxotremorine, arecoline and pilocarpin) were reported to have an acaricidal effect [223].

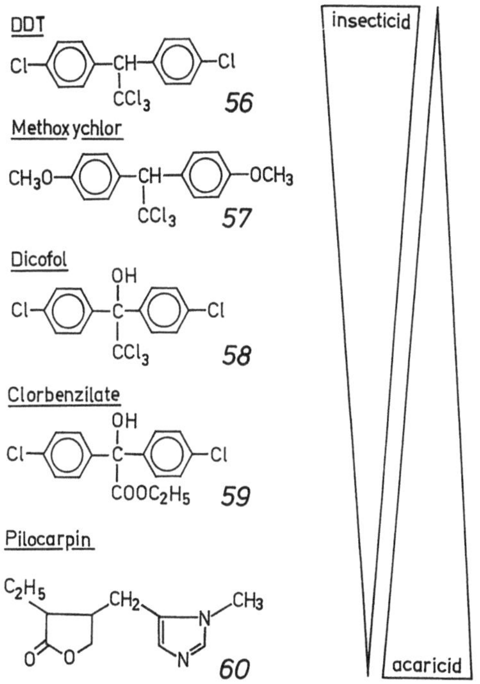

Fig. 5. Insecticidal and acaricidal activity of chlorinated hydrocarbons

Table 17. Comparison of the inhibition properties from chlorbenzilat to HTX and QNB binding AChR's

Drug (10^{-4} M)	% Inhibition of binding +/— S.D.	
	[^{3}H]-H12-HTX	[^{3}H]-QNB
DDT	0	35 +/— 15
Methoxychlor	0	0
Dicofol	22 +/— 6	0
Chlorbenzilat	61 ×/— 2	45 +/— 10

A zero value is used where % inhibition is not significant ($P < 0.05$) in three experiments (15).
HTX ... *Torpedo* electric organ;
QNB ... Bovine Brain, Corpus striatum

The toxic potency of muscarine ligands on ticks and mites indicates that the mAChR in mites may possibly be a target for acaricides.

5.2 Phosphororganic Insecticides and Carbamates

These insecticides are known as strong inhibitors of the ACh-esterase (86). Their ability to compete with the ACh at the binding site of the ACh-esterases may support the conclusion that they can also interact with the ACh binding site of the AChR. An interference of AChE-Inhibitors with the nAChR-channel complex has been demonstrated with homogenates from *Torpedo* electroplax (Table 18)[224, 11, 15] and from the molusc *Aplysia*[79].

Table 18. Inhibition properties from some AChE blocking pesticides on a AChR in *Torpedo* electroplax[15]

Pesticide 100 μM	% Inhibition in *Torpedo* electroplax	
	AChR against [^{3}H]-ACh 1 μM	AChE (Ellmann method)
Aminocarb	80 +/— 4	90 +/— 2
Aldicarb	74 +/— 16	87 +/— 2
Monocrotophos	93 +/— 5	99 +/— 1
Dichlorvos	88 +/— 1	99 ×/— 1

The alpha-BTX binding at the AChR of the 105000 × g supernatant of the *Periplaneta americana* 6. abdominal ganglion was effected by DDVP and Aldicarb[59]. Anticholinesterase compounds were shown to inhibit (^{3}H)-decamethionium binding on putativee AChR of high-speed homogenates in housefly head extracts[225]. In order of the effectiveness these were Tetram > Guthoxon > R 16661 (3-ethyl-5-methyl-1,3-oxazolidine) > DFP. Carbamate anticholinesterases eserin, neostigmin, pyridostigmine and Physiostigmine also inhibit the alpha-bungarotoxin-binding to the receptor sites in *Torpedo* membranes[224] and *Drosophila* head extracts[53].

Neostigmin, pyridostigmine and physiostigmin at low concentrations inhibit acetylcholinesterase whereby indirectly potentiating ACh enhancement of (^{3}H)-HTX binding to the channel sites of nAChR. However, at higher concentrations they inhibited ACh-action due to their direct binding to AChR[2224]. Preincubation of ACh-receptors with any of these three carbamates reduced the rates of binding of [^{125}Iod]-BTX and increased the inhibitoric effectivity of carbamylcholine against alpha-BTX.

Whereas aldicarb and monocrotophos are able to displace (^{3}H)-ACh from the *Torpedo* electroplax AChR in a remarkable degree, they fail to inhibit (^{3}H)-QNB binding[15] in housefly head homogenates (Table 19). That indicates that the muscarinic AChR of insects does not seem to be a target for anticholinesterase insecticides.

Table 19. Documentation of the inhibition properties on the muscarinic AChR of housefly [15] in comparison to table 18

Pesticide 100 μM	% Inhibition of [^{3}H]-QNB binding (1 nM) housefly head
Aldicarb	4 +/− 6
Monocrotophos	17 +/− 3

5.3 Pyrethroids

The well established knowledge of the mechanism of action of pyrethroids indicates the very high effectivity at the axon membrane Na-ion-channel protein, like DDT. The interactions of natural pyrethrins and nine synthetic pyrethroids with the nicotinic acetylcholinic receptor/channel complex of *Torpedo* electric organ membranes were studied by Abbassy et al., 1983 [226].

None caused significant reduction in [^{3}H]-ACh binding to the receptor sites, but all inhibited [^{3}H]-H12-HTX binding to the channel sites. Due to this interaction with the H12-HTX sensitive Na-ion-channel site of the nAChR two different types of pyrethroid molecules could be observed. Natural pyrethrins, allethrin, bioallethrin, tetramethrin and resmethrin are more potent and act rapidly and their highest level of inhibition against HTX binding can be reached in <30 seconds. On the other hand inhibition of [^{3}H]-H12-HTX binding by permethrin, fluvalinate, cypermethrin, and fenvalrate is increased slowly and takes more time. There is no positive correlation between the insecticidal action (LD50 *Periplaneta americana*) and the action on channel site of the nicotine AChR in the postsynaptic membrane (IC50-[^{3}H]-H12-HTX(2nM)) (Table 20). That means that the effects of pyrethroids at the HTX sensitive ion channel site of the nAChR are not involved in the mechanism of high effectivity of pyrethroids in knocking down and killing insects.

Table 20. Comparison of insecticidal and inhibition properties of two types from pyrethroids against HTX sensitive ion channel site of the AChR of the american cockroach

Pyrethroid	IC50 HTX μM	LD50 mg/kg *Periplaneta americana*
Type I		
Allethrin	2	1.20
Type II		
Cypermethrin	1000	0.02

6 Concluding Remarks

Interaction with the acetylcholine (ACh) mediated neural signal transmission from a presynaptic neuron to a postsynaptic neuron by insecticides is a well known and frequently exploited possibility of killing injurious insects. So the cholinergic synaptic process is a sensitive target for effective chemical interventions. There are a lot of highly effective compounds acting as inhibitors of the ACh-esterase, but at present only a few which act at the acetylcholine receptors (AChR).

Though the anticholinesterases are mimetica of ACh at the cholinesterases, they cannot imitiate ACh at the binding site of the AChR in a relevant level.

The effectivity and the properties of Cartap, Thiocyclam and Bensultap show that the AChRs can be suitable targets for insecticidal attacks, calling for attention in conceiving and developing new insecticides. The known insecticides, acting at the AChR, are derived from naturally occurring toxin compounds, the agonistic nicotine and inhibitoric nereistoxin. Both have been taken as model structures for intensive derivatization and structure — insecticidal effectivity — correlation analysis. Therefore in literature, especially in patent literature, a large range of synthesized and tested derivatives can be found. Only a few have become economically important, but the field has not been exploited yet, and it is not impossible that new structures can be developed which act as insecticides at this target protein and possess new advantages. The review has pointed out some remarkable differences between AChR proteins from vertebrate and insect neurons due to distribution in peripheric or central nervous loci, the physiological functions which they are involved in and the affinity spectrum to ligands. The biochemical differences can be the base for efforts to develop effective insecticides with moderate or low toxicity to mammals [227]. Another strategy which can be used to develop selective compounds is given in the case of nereistoxin derivatives, i.e. the possibility of choosing a selective additive at the two sulfur atoms in the dimethylamino-propane structure. These substituents can increase the insecticidal effects and remarkably decrease the toxicity to mammals. Bensultap is an outstanding example of this.

The knowledge of the molecular properties of the relevant insect AChR proteins is, up to now, not sufficient for a direct "biorational design" of new insecticidal molecules. But there are a lot of suitable assertions about essential parts and properties of the known lead structures which can stimulate and provoke new purpose-directed synthesis.

7 References

1. Eldefrawi AT (1985) in: Kerkut GA, Gilbert LI (eds) Comprehensive Insect Physiology Biochemistry and Pharmacology 12. Pergamon Press, Oxford, p 115
2. Ruigt GSF (1985) in: Kerkut GA, Gilbert LI (eds) Comprehensive Insect Physiology Biochemistry and Pharmacology 12. Pergamon Press, Oxford, p 183
3. Osborne MP (1985) in: Kerkut GA, Gilbert LI (eds) Comprehensive Insect Physiology Biochemistry and Pharmacology 12. Pergamon Press, Oxford, p 131
4. Oppenoorth FJ (1985) in: Kerkut GA, Gilbert LI (eds) Comprehensive Insect Physiology Biochemistry and Pharmacology 12. Pergamon Press, Oxford, p 731
5. Brattsten LB, Holyoke CWJr, Leeper JR, Raffa KF (1986) Science 231. p 1255
6. Geissbühler H (1984) Chimia 38. p 307
7. Keyserlingk HCv (1986) Agricult. Zool. Rev 1. p 195
8. Keyserlingk HCv, Jäger A, Szepanski Cv (1985) Approaches to New Leads for Insecticides Springer, Berlin, Heidelberg p 159
9. Lund AE (1985) in: Kerkut GA, Gilbert LI (eds) Comprehensive Insect Physiology Biochemistry and Pharmacology 12. Pergamon Press, Oxford, p 9
10. Katz B (1969) Release of Neuronal Transmitter Substances. Univ. Press New York

11. Eldefrawi AT, Monsour NA, Eldefrawi ME (1982) Pharmac. Ther. 16. p 45
12. Sattelle DB (1985) in: Kerkut GA, Gilbert LI (eds) Comprehensive Insect Physiology Biochemistry and Pharmacology 12. Pergamon Press, Oxford, p 395
13. Krnjevic K (1974) Physiol. Rev. 54. p 418
14. Gerschenfeld HM (1973) Physiol. Rev. 53. p 1
15. Eldefrawi ME, Eldefrawi AT (1983) J. Envir. S. B18. p 65
16. Raftery MA (1983) Cold S. Harb. 48. p 21
17. Einarson BW, Gullick B, Conti-Tronconi B and Lindstrom J (1982) Biochemistry 21. p 5295
18. Reynolds J, Karlin A (1978) Biochemistry 17. p 2035
19. Hucho F, Bandini G, Suraz-Isla BA (1981) Europ. J. Bioch. 83. p 335
20. Conti-Tronconi BM, Raftery MA (1982) Annu. Rev. Biochem. 51. p 491
21. Lindstrom J, Merlie J, Yogeeswaran G (1979) Biochemistry 18. p 4465
22. Raftery MA, Hunkapiller M, Strader CD, Hood LE (1980) Science 208. p 1454
23. Numa S, Noda M, Takahashi H (1983) Cold S. Harb 38. p 57
24. Finer-More J, Stroud RM (1985) Proc. Nat. Acad. Sci. 91. p 4532
25. Repke H, Liebmann C (1987) Membrane Receptors and their Effektor Systems. VCH Weinheim
26. Hermans-Borgmeyer I, Zopf D, Rysek RP, Hovemann B, Betz H (1986) EMBO J. 5. p 1503
27. Hille B (1977) Biosystems 8. p 195
28. Harrow ID, Hue B, Pelhate M, Sattelle (1979) J. Physiol. 295. p 63
29. Morrison M, Bayse GS (1970) Biochemistry 9. p 2995
30. Lee CY (1972) Ann. Rev. Pharmacol. Toxicol. 12. p 265
31. Albuquerque EX, Eldefrawi AT, Eldefrawi ME (1979) in: Lee CY (eds) Snake venoms. Springer Berlin, Heidelberg, New York, p 377
32. Lee CY, Chang SL, Kau ST, Luh SH (1972) J. Chromatogr 72. p 71
33. Noda M, Takashashi H, Tanabe T (1982) Nature 299. p 793
34. Noda M, Takashashi H, Tanaba T, Toyosoto M, Kikyotani S (1983) Nature 301. p 251
35. Claudio T, Balliot M, Patrick J, Heinemann S (1983) Proc. Nat. Acad. Sci. U.S.A. 80. p 1111
36. Mc Cornick DJ, Atassi M (1984) Bioch. J. 224 p 995
37. Albuquerque E, Adler M, Spivak CE, Aguayo L (1980) Ann. N.Y. Acad. Sci. 358. p 204
38. Eldefrawi ME (1978) Fed. Proc. 37. p 2823
39. Eldefrawi AT, Eldefrawi ME (1980) in: Sattelle LM, Hall LM, Hildebrand JG (eds) Receptors for Neurotransmitters, Hormones and Pheromones in Insects. Elsevier, North Holland, Biomedical Press, Amsterdam, p 59
40. Betz H (1982) Neurosci. L. 33. p 153
41. Eldefrawi ME, Warnick JG, Schofield GG (1982) Bioch. Pharmac. 27. p 135
42. Daly JW, Karle J, Myers CW, Tokuyama T (1971) Proc. Nat. Acad. Sci. U.S.A. 68. p 1870
43. Eldefrawi ME, Aronstam RS, Barky NN (1980) Proc. Nat. Acad. Sci. U.S.A. 77. p 2309
44. Eldefrawi AT, Barky NM, Eldefrawi ME (1980) Mol. Pharmacol. 17. p 172
45. Elliot J, Dunn SMJ, Blanchard SG, Raftery MA (1979) Proc. Nat. Acad. Sci. U.S.A. 76. p 2576
46. Karlin A (1980) in: Cohman W, Poste G, Nicolson GL (eds) The Cell Surface and Neuronal Function. Elsevier, New York, p 191
47. Nathanson NM (1982) Trends Neur. 5. p 401
48. Hartzell HC (1981) Nature 291. p 539
49. Peterson GL, Scott-Herron G, Schinerlik J (1984) Proc. Nat. Acad. Sci. U.S.A. 81. p 4993
50. O'Brien RD, Eldefrawi AT (1970) J. Neurochem. 17. p 1287
51. Sattelle D (1980) Adv. Insect 15. p 215
52. Triphati RK, Tripathi HL, O'Brien RD (1978) Biochem. Biophys. A. 586. p 624
53. Schmidt-Nielson BK, Gepner JI, Teng NN, Hall IM (1977) J. Neurochem. 29. p 1013
54. Hildebrand JG, Hall LM, Osmond BC (1979) Proc. Nat. Acad. Sci. U.S.A. 76. p 499
55. Satelle DB, Harrow JD, Hue B, Pelhate M, Gepner JI, Hall L (1983) J. Exp. Biol. 107. p 473
56. Sanes JR, Hildebrand JG (1976) Develop. Biol. 51. p. 282
57. Sanes JR, Hildebrand JG (1976) Develop. Biol. 52. p 105
58. Sanes JR, Prescott DJ, Hildebrand JG (1977) Brain Res. 119. p 389

59. Richter R (1987) An alpha-bungarotoxin binding AChR of insects as a target of insecticides Diss., Academy of Agricultural Science of GDR, Berlin
60. Snyder SH, Bennett JP (1976) Ann. Rev. Physiol. 38. p 153
61. Birdsall NJ, Hulme EC (1976) J. Neurochem. 27. p 7
62. Burt D (1978) in: Yamamura HI, Enna SJ, Kuhar MJ (eds) Neurotransmitter receptor binding. Raven press, New York
63. Hall M, Teng NNH (1975) in: Mc Mahon D (ed) Development Biology-pattern Formation gene Regulation, ICN-UCLA symposia on Molecular and Cellular Biology 2. Benjamin, Menlo Park, p 282
64. Sattelle DB (1980) in: Borcovec AB, Kelly TJ (eds) Insect Neurochemistry and Neurophysiology. Plenum Press, New York, p 51
65. Yamamura HI, Snyder SH (1974) Brain Res. 78. p 320
66. Yamamura HI, Snyder SH (1974) Mol. Pharmacol. 10. p 861
67. Yamamura HI, Kuhar MJ, Greenberg D, Snyder SH (1974) Brain Res. 66. p 541
68. Yamamura HI, Kuhar MJ, Snyder SH (1974) Brain Res. 80. p 170
69. Snyder SH, Chang KJ, Kuhar MJ, Yamamura HI (1975) Fed. Proc. 34. p 1915
70. Dudai Y (1977) FEBS Letter 76. p 211
71. Soler G, Farach HA, Mattlingly JR, Martinez M (1983) Arch. Biochem. Biophys. 225. p 872
72. Jimenez F, Rudloff E (1980) FEBS Letter 113. p 183
73. Dudai Y (1978) Biochem. Biophys. 539. p 505
74. Maelicke A, Fulpius BW, Klett RP, Reich E (1977) J. Biol. Chem. 252. p 4811
75. Lummis SCR, Sattelle DB (1985) Comp. Biochem. Physiol. 80. p 75
76. Hall LM (1980) in: Sattelle DB, Hall LM, Hildebrand JG (eds) Receptors for Neurotransmitters, Hormones and Pheromones in Insects. Elsevier, North Holland, Biomedical Press, Amsterdam p 111
77. Sattelle DB, David JA, Harrow ID, Hue B (1980) in: Receptors for Neurotransmitters, Hormones and Pheromones in Insects. Elsevier, North Holland, Biomedical Press, Amsterdam, p 125
78. Maelicke A, Reich E (1976) Cold S. Harb. 40. p 232
79. Monsour NA, Eldefrawi AT, Eldefrawi ME (1977) Biochemistry 16. p 4126
80. Carpenter DO, Greene LA, Shain W, Vogel Y (1976) Mol. Pharmac. 12. p 999
81. Haim N, Nahum S, Dudai Y (1979) J. Neurochem. 32. p 543
82. Dudai Y, Ben-Barak J (1977) FEBS Letter 81. p 134
83. Breer H (1981) Neurochem. 3. p 155
84. Gilbert FRT, Hanley MR, Iverson LL (1979) Br. J. Pharmacol. 65. p 451
85. O'Brien RD, Gilmour LP, Eldefrawi ME (1969) Proc. Nat. Acad. Sci. U.S.A. 65. p 438
86. Corbett JR (1974) The Biochemical Mode of Action of Pesticides Academic Press, New York p 1
87. Negherborn WO (1959) in: Spector WS (ed) Handbook of Toxicology; Insecticides 4. Sounders, Philadelphia, p 854
88. Shepard HH (1951) The Chemistry and Action of Insecticides Mc Graw Hill, New York
89. Waite MB (1925) Year. Agr. 6. p 453
90. Metcalf RL, McKelvey JJ (1976) in: J. Wiley (ed) The Future for Insecticides: Needs and Prospects. Wiley, New York, p 112
91. Koll. Index Phytosanitaire [Acta Service lutte antiparasitaire, Fr] (1988) p 72
92. Worthing CR (1979) The Pesticide Manual, 6th edn. Lovenham Press, London, p 382
93. Koll. Agriculture Chemicals Approval Scheme, Approved products for Farmers and Growers, GB p 193
94. Eldefrawi ME (1984) in: Kerkut GA, Gilbert LI (eds) Comprehensive Insect Physiology Biochemistry and Pharmacology 12. Pergamon Press, Oxford, p 263
95. Wang GA, Molinaro S, Schmidt J (1978) J. Biol. Chem. 253. p 8507
96. Barlow RB, Hamilton JT (1962) Brit. J. Pharmacol. 18. p 510
97. Eldefrawi AT, O'Brien RD (1970) J. Neurochem. 17. p 1287
98. Eldefrawi ME, Eldefrawi AT, O'Brien RD (1970) J. Agr. Food 18. p 1113
99. Yamamoto I, Soeda Y, Kamimura H, Yamamoto R (1968) Agr. Biol. Chem. 32. p 1341
100. Nitta S (1934) Yakagaku Zasshi 54. p 648
101. Okaichi T, Hashimoto Y (1962) Bull. Soc. Sci. Fish 28. p 930

102. Sakai M (1964) Appl. Insec. Entomol. 8. p 73
103. Hussain M (1977) Iran. J. Agric. Res. 5. p 155
104. Sato Y (1973) Taketa Kenkyusho Ho 32. p 564
105. Witkowski W (1972) Buil. Inst. Ochr. Rosl. 54. p 299
106. Nishi K, Kodo I, Tan N (1979) J. Pestic. Sci. 4. p 35
107. Yamamoto E (1979) Kyushu Byogaichu Kenkyukai Ho 25. p 141
108. Kao CH, Cheng EY, Lin DF (1986) Chung-hua Nung Yeh Yen Chiu 35. p 107
109. Lakocy A (1975) Dokl. Soobshch. VIII Mezdunar. Kongr. Zashsc. Rast, Moscow 3. p 377
110. Horn A (1985) Mezogazd Kemizalasa Konf. Nehezvegzip. Kut. Intez. Veszeprem. Hung. 2. p 172
111. Hurny J, Ziebe S (1980) Organika 11. p 121
112. Wegorek W, Pruszynski S (1979) Pr. Nauk. Inst. Ochr. Rosl. 21. p 119
113. Chu YI, Lin D Shi, Mu T (1976) Chi Wu Pao Hu Hsueh Hui Hui K an 18. p 377
114. Benedek P, Komlodi J, Herczeg G, Dely G (1976) Novenytermeles 25. p 129
115. Ivanov SG, Zhukovskii SG (1980) Khim. Sel'sk. Khoz 18. p 15
116. Wilkaniec Z (1979) Rocz. Akad. Poln. Poznaniu 111, p 185
117. Rucner Z, Maceljski M (1983) Zast. Bilja 34. p 129
118. Szczepanska K, Kowalska T (1975) Dokl. Soobshch. VIII. Mezhudunar. Kongr. Zashch. Rast. Moscow 6. p 243
119. Nishi K, Funakoshi K, Tan N, Hattori M (1979) J. Pest. Sci. 4 p 29
120. Obayashi H, Kono Y, Kawakami T, Sato Y (1983) J. Takeda. Res. Lab. 42. p 78
121. Asahi Y, Yoshida T (1977) Chem. Pharm. Bull. 25. p 2211
122. Husain M, Begum N (1984) Indian J. Agric. Sci. 54. p 693
123. Anon. (1979) Kongetsu no Noyaku 23. p 22
124. Haruki T, Inone H (1978) Kitanippon Byogaichu Ken Kyu Kai Tokubotsu Hokoku 9. p 21
125. Fuse H (1978) Kitanippon Byogaichu Ken Kyu Kai Tokubotsu Hokoku 9. p 15
126. Ibrar-ul-Hay C, Manyoor-ul-Hay E (1981) Sind Univ. Res. J. Sci. Ser. 13. p 39
127. Ray PS, Vidyachandra B, Chandrasekhar HT (1980) Curr. Res. (Univ. Agric. Sci. Bangalore) 9. p 70
128. Makhdoom SMA, Majid A, DAR IA (1976) Agric. Pak. 27. p 23
129. Chu YI, Lin DS, Mu T (1976) Chih. wu Pao HU Hsueh Hui Hui K'an 18. p 369
130. Sakai M, Kato M, Hayiwara H, Konishi K (1976) Patent JP 51/42177
131. Berg W, Knutti HJ (1975) Proc. Br. Insect. Fungic. Conf. 2. p 683
132. Balasubramanian M, Michael RK (1976) Madras. Agric. J. 63. p 288
133. Rao VLV, Rao BHK, Reddy PS, Rao NV (1984) Indian J. Agric. Sci. 54. p 209
134. Rao CS, Rao NV, Rasvi SA (1985) Pesticides 19. p 20
135. Chiu SF (1985) Insect. Sci. Appl. 6. p 1
136. Mitsudera H, Konishi K, Sato Y (1984) TAKEDA Ltd. EP-Pat. 102062. JP-Pat. 82149633 p 53
137. Rajemani S, Kulshrestha JP, Pasalu IC, Dani RC (1984) Indian J. Plant Prot. 12. p 35
138. Gopee TG (1981) Int. Pest. Control 23. p 154
139. Garbar LI (1981) Nauk Pr. Ukr. SIL's' kogospod Akad. 3. p 87
140. Mateeva A, Avdzhinski G (1981) Nauchi Tr.-Vissh Selshostop. Int. 'V. Kolarov' Plovdiv 76. p 125
141. Abelentseva GM, Sedykh AS, Popov PV, Zkuraoleva VI (1978) Khin sel'sh Khoz 16. p 40
142. Kukhta NS (1974) Tr. Beoruss. S.-kh. Akad. 125. p 43
143. Przybylski Z (1972) Biol. Inst. Ochr. Rosl. 54. p 453
144. Perju T, Rozalia Z, Ghidra V, Cornea O (1981) Bul. Inst. Agron. Chuj-Napoca, Ser. Agric. 35. p 121
145. Swiech J, Byrdy S, Lakota S (1978) Mater. Ses. nauk. Inst. Ochr. Rosl. [Poznan] 18. p 475
146. Studzinski A, Micolajewicz M (1975) Herba Pol. 21. p 61
147. Witkowski W, Coesielski F, Mrowczyski M (1983) Pr. Nauk Inst. Ochr. Rosl. 25. p 127
148. Hurkova J (1973) SB Uuti Ochr. Rosl. 9. p 253
149. Nachev N (1976) Rastit. Zasht 24. p 14
150. Tribel sa, Ponemarenko LI (1976) Khim. Sel'sk Khoz. 14. p 14
151. Dochkova B, Stavechev A (1976) Rastit. Zasht. 24. p 17
152. Trenchev G, Pavlov A (1984) Gradinar. Lozar. Nauka 21. p 28

153. Pavlov A, Trenchev G (1981) Rastenievud Nauki 18. p 110
154. Pavlov A, Trenchev G (1981) Rastenievud Nauki 18. p 99
155. Radwatin HSA, Rizk GA (1984) Bull. Entomol. Soc. Egypt Econ. 13. p 25
156. Hussain M, Askari A (1976) J. Econ. Entomol. 69. p 343
157. Khan S, Chughtai GH, Qamar-ul-Islam E (1984) Pak. J. Agric. Res. 5. p. 40
158. Li HS, Wu KC, Liu TH, Won HT (1979) T' aiwan Nung Yeh Shuang K'an 15. p 28
159. Labanowska BH, Pala EM, Suski ZW (1978) Fruit Sci. Rep. 5. p 37
160. Alexandri AA (1971) An. Inst. Cercet. Prot. Plant Acad. Stiiute Agr. Silvice 3. p 507
161. Tsankov G, Mancheva E (1986) Gorskostop Nauka 23. p 62
162. Sakai M, Kato M, Sato Y (1967) Jap. J. Appl. Entomol. Zool. 11. p 125
163. Van de Vries M, Price JF, De Jong J (1986) med. Fac. Landbouww. Rijksuniv. Gent 51. p 879
164. Krihning J, Becker H, Beck I, Kleemann U, Kalzendorf I (1987) Wirkstoffkartei PSM/MBP, Ed. VEB Chemiekombinat Bitterfeld, Org. Chemie and PSM
165. Wei H-CH, Chang CH-T (1977) Chung-kuo Nung Yeh K'o Hsueh 14. p 67
166. Li HK (1982) Act. Entomol. Sin. 25. p 24
167. Hedley ME, Preston AF, Cross DJ, Butchner JA (1979) Int. Biodeterior. Bull. 15. p 9
168. Toth B (1986) Novenyvedelm [Budapest] 22. p 82
169. Benedek P (1985) Herba Hung. 24. p 111
170. Sakai M (1984) Jpn. Pest. Inf. 45. p 21
171. El-Bahrawi A, Saad ASA, Soliman SS (1979) Med. Fac. Landbouww. Rijksuniv. Gent 44. p 31
172. Wu JF, Huang ZH, Won RZ (1987) Act. Entomol. Sin. 25. p 42
173. Ree HI, Hong HK, Shin JC, Lee JS, Cho HW (1978) Kungnip Pogon Yon' Guwonpo 15. p 275
174. Deguchi T, Narashashi T, Haas HG (1971) Pest. Bioch. 1. p 196
175. Sakai M (1976) Bochu-Kagaku 32. p 21
176. Bettini S, D'Ajello V, Maroli M (1973) Pest. Bioch. 3. p 119
177. Sattelle DB, Callec JJ (1977) Proc. XXVIII. Int. Congr. Physiol. [Paris] 28. p 662
178. Sattelle DB (1977) Trans. Bioch. Soc. 5. p 849
179. Shaker N, Eldefrawi AT (1981) Pestic. Bioch. Physiol. 15. p 14
180. Sakai M (1966) Appl. Ent. Zool. 1. p 73
181. Sakai M, Sato Y (1971) Abstr [2nd.] Int. Cong. Pest. Chem., Tel Aviv 2. p 13
182. Fukoto TR Chemtech. 15. p 362
183. Jacobson N, Peterson LE (1983) Pest. Sci. 14. p 90
184. Kato A, Hashimoto Y (1979) in: Cavallini D, Gaull G, Zappia V (eds) Nat. Sulfur Compd. 3rd. Meeting. Plenum Press, New York, p 361
185. Konishi K (1970) Agr. Biol. Chem. 34. p 926
186. Hagiwara H, Numata M, Konishi K (1965) Chem. Pharm. Bull. 13. p 253
187. Numata M, Hagiwara H (1968) Chem. Pharm. Bull. 16. p 311
188. Konishi K (1968) Agr. Biol. Chem. 32. p 1199
189. Sato Y, Konishi K (1974) TAKEDA Ltd. JP-Pat 28-971
190. Konishi K Agr. Biol. Chem. 34. p 935
191. Konishi K (1974) TAKEDA Ltd. JP-Pat 48-27857 [73-27875]
192. Suzuki S (1970) Nayaku Seisan Gijutsu 21. p 40
193. Konishi K (1974) TAKEDA Ltd 31. p 970
194. Anon. (1966) TAKEDA Ltd. 13. p 338
195. Konishi K, Okutani T, Soma T (1967) TAKEDA Ltd. US-Pat 3332943
196. Schneider K (1972) SANDOZ Ltd. D.O.S. 2209556
197. Schneider K (1978) SANDOZ Ltd. CH-Pat. 605721
198. Sakai M (1983) Abstr. V. Int. Congr. Pestic. Chem., Kyoto Japan
199. Kuhnen F (1972) SANDOZ Ltd. D.O.S. 217286
200. Sakai M (1969) Rev. Plant Protec. Res. 2. p 17
201. Sakai M (1964) Jap. J. Appl. Ent. Zool. 8. p 324
202. Kuhnen F (1964) SANDOZ Ltd. CH-Pat. 544037
203. Schelling HP, Kuhnen F (1970) SANDOZ Ltd. D.A.S. 2039666
204. Kihara K, Kurorda M (1984) TAKEDA Ltd. EP-Pat. 0140335

205. Konishi K (1970) Agr. Biol. Chem. 34. p 1549
206. Anon. (1970) TAKEDA Ltd. JP-Pat. 80127387
207. Schneider R SANDOZ Ltd. US-Pat. 3770769
208. Schneider R SANDOZ Ltd. CH-Pat. 551999
209. Schneider R SANDOZ Ltd. CH-Pat. 553206
210. Mitsudera H, Konishi K, Sato Y (1982) TAKEDA Ltd. EP-Pat. 102062, JP-Pat. 2149633
211. Konishi K (1970) TAKEDA Ltd. JP-Pat. 7018847
212. Sakai M (1969) TAKEDA Ltd. D.O.S. 1917246
213. Kihara K, Kurodo M (1984) TAKEDA Ltd. EP-Pat. 0140352
214. Schneider R (1974) SANDOZ Ltd. CH-Pat. 553206
215. Schneider R (1972) SANDOZ Ltd. D.O.S. 2147741
216. Pezzementi L, Schmidt J (1981) J. Biol. Chem. 256. p 12651
217. Baillie AC, Corbett JR, Dowsett JR, Sattelle DB, Callec J (1975) Pest. Sci. 6. p 645
218. Baillie AC, Corbett JR, Sharpe TM (1978) Pest. Sci. 9. p 1
219. Gepner JI, Hall LM, Sattelle DB (1978) Nature 276. p 188
220. Narashi T (1976) in: Wilkinson CF (ed) Insecticide Biochemistry and Physiology. Plenum Press, New York, 24. p 327
221. Eriksson P, Nordberg A (1986) 85. p 121
222. Fonesca MT, Aquilar L, Lopes C (1986) Toxicol. Appl. Pharmacol. 84. p 192
223. Bigg DCH, Purvis SR (1976) Nature 262. p 220
224. Sherby SM, Eldefrawi AT, Albuquerque EX, Sattelle DB (1984) Mol. Pharmacol. 27. p 343
225. Eldefrawi ME, Eldefrawi AT, O'Brien RD (1971) Poc. Nat. Acad. Sci. U.S.A. 68. p 1047
226. Abbassay MA, Eldefrawi ME, Eldefrawi AT (1983) Pest. Biochem. and Physiol. 19. p 299
227. Otto D, Richter R, Karabensch KH (1989) in: Otto D, Weber B (eds) Insecticides — Mechanism of Action and Resistance. Tagungsber. 274, Acad. Agricult. Sciences GDR

Chemical Manipulation of Desiccation and Defoliation and Essential Aspects for the Application and Development of New Chemical Compounds in the Future

H. Bergmann and D. Martin
Institute for Plant Protection Research of the Academy of Agricultural Sciences of GDR, Stahnsdorfer Damm 81, DDR-1532 Kleinmachnow

The chemical desiccation and defoliation of agronomic crops especially under unfavourable climatic conditions makes harvesting much easier and is linked with several further advantages. The use of desiccants and defoliants is of increasing practical importance internationally. The dosages and the limitation of application of desiccants and defoliants recently employed in agriculture together with practical aspects of their use are described in detail.

Desiccants, and depending on their applied concentration many defoliants, too, cause drastic herbicidal effects. Nevertheless, most herbicides are not suitable as desiccants or defoliants. Considering the above mentioned facts, relations between mode and mechanism of action of chemical compounds and their desiccating or defoliating effect are discussed with regard to the search for theoretical starting points for the development of new desiccants and defoliants.

Chemistry of Plant Protection 2

1 Introduction

Chemical desiccation of cultivated plants is the chemical drying of whole plants, or parts of them, in a relatively short time before harvesting, in order to kill or destroy a plant, or certain parts of organs of it. Desiccation, in preparation for harvesting, fulfills different functions in different crops and forms an integral part of the production system of a lot of crops. Desiccants act drastically and quickly. On the one hand they damage or destroy cellular structures, e.g. cell membranes, which leads to the collapse of vital metabolic systems and the plant's dying off. On the other hand specific inhibitions of the metabolism, e.g. in connection with photosynthesis, may kill plants. Finally, acceleration of ripening and aging causes rapid completion of the vegetable development and thus brings about the desired desiccant effects. In some cases, a functional disturbance by a desiccant is so dominant that it can be the actual reason for the desiccant effect. The phytotoxic effect can also be caused by the influence of several metabolic processes. As a rule, desiccants have herbicidal properties and are not real growth regulators.

Abscission, i.e. the loss of vegetable organs, is normally accomodated to environmental and weather conditions by natural growth regulators. Chemical compounds can interfere with the hormonal control of the abscission. Chemicals inducing defoliation can be considered as wound agents e.g. by local necroses on leaves, facilitating abscission e.g. by inducing the production of wound and stress ethylene; reduction of synthesis and transport of auxin; support of water loss and the synthesis of abscisic acid; and aging and dying of cells [1].

The following example explains the relationship between herbicide-like damage, abscission of leaves, and desiccation. A low dose of sodium chlorate promoted ethylene production and abscission, while a higher dose desiccated the plants and reduced ethylene production [2]. These relationships support the conclusion that defoliants at high concentrations will often desiccate and desiccants will often defoliate if applied at low rates, and that most of the defoliants and desiccants are contact herbicides [1].

This contribution deals with the most important aspects of the use and development of desiccants and defoliants, the latter exclusively in view of practical use and not on the basis of hormonal control of growth and development of plants.

2 The Need for and Application of Chemical Desiccation and Defoliation Methods in Agronomic Crops

2.1 Chemical Desiccation and Defoliation in Agronomic Crops

More than 50 years ago the use of chemical desiccants in agronomic crops was begun. In the 1930s *copper sulfate* and sodex for killing leaves and controlling late blight (*Phytophthora infestans* L.) in *potato* [3] and even *sulfuric acid* were used [4]. The 1940s saw the application of *calcium cyanamide* fertilizer for defoliating *cotton* to facilitate the use of mechanical harvesters [5]. The enormous increase in the use of desiccants and defoliants over the past 40 years is closely connected with rationalization and the more intensive cultivation of crops in agriculture, so that a number of inorganic and organic chemical compounds for use as desiccants and defoliants have been developed by the chemical industry for practical use.

The most important advantages of desiccants and defoliants are:

1) They facilitate the use and increase the efficiency of mechanical harvesters,
2) it is possible to fix the harvest date independent of or with respect to weather conditions; this facilitates better farm management organization and reduces losses caused by weather,
3) introduction of the "short-season" culture systems to minimize the insect control problems,
4) reduction of harvest and storage losses by control of fungal and bacterial pathogenic agents, as well as viral infections, this includes protection of the harvest products (e.g. potatoes) against pathogenic agents,
5) weed control,
6) improvement in the quality of harvested products.

2.2 Influence of Chemical Desiccation and Defoliation on Quality and Quantity of Harvested Products

Chemical desiccation and defoliation have a decisive influence on quality and quantity of harvested products in modern agriculture and is, therefore, an important factor in the production of food and feeding stuffs, as well as for providing plant material for industrial processing.

Desiccative efficiency and effects on yield parameters are closely connected with the physiological state of the plant. For that reason the desiccation modalities such as date, speed and mode of action of a chemical compound are of great importance. Some examples of application in the following sections show the advantages of chemical desiccants or defoliants.

Cotton

The purpose of desiccation and defoliation of cotton (*Gossypium hirsutum L.*) is the removal of leaves prior to harvesting. Resulting acceleration of ripening often entails a quicker opening of the bolls and a positive influence on the production of cellulose and, therefore, higher seed and lint yields [6–12]. In addition, the chemicals control harmful insects and weeds, which reduces harvest losses, improves the quality of the raw cotton and facilitates the use of mechanical harvesters [13,14]. Improved ginning yields after application of desiccants have also been noticed [15]. The suppression of regrowth is another advantage [13,16]. Finally, this measure reduces the amount of energy required to dry grain [17].

Legumes

Leaves of soy bean and other legumes (*Vicia faba L.*, *Phaseolus vulgaris L.*, *Pisum sativum L.*, *Vigna unguiculata L.* and *Lupinus sp.*) have to be removed by desiccation or defoliation a short time before harvesting, at which timing of the desiccation plays a special role. To avoid premature blocking of primary substance production or translocation of nutrients out of the leaves into the reproductive organs, either fast acting chemicals at a later time or slowly acting chemicals at an earlier time have to be used. This is so in order that the translocation of the nutrients out of the leaves into the pods and seeds, during the first stage of the chemical action can take place or is even facilitated. Premature desiccation reduces not only the yield [18] but also the quality [19]. Further advantages are like those for other crops: acceleration of ripening, optimization of various conditions, a decrease of harvest losses, a decrease of water content of seeds and less infestation with seed-born diseases, a decrease of the expenditure of energy for drying seeds, and advancement of the harvest date [18, 20–25].

Potato

The chemical killing of haulms and leaves of potato (*Solanum tuberosum L.*) some weeks before harvesting has some advantages, which exert, directly or indirectly, a great influence on quality and quantity of the tubers. The biomass reduction of haulms, leaves, and weeds allows the soil to dry out faster which increases the speed of the harvesting machines and decreases dirtiness of tubers [26].

Moreover, biochemical and physiological processes of the formation of a potato tuber are considerably influenced, because they are connected with manifold changes in substantial composition and metabolism. That is why the timing of desiccation has an influence on the quality of tubers. In the case of early desiccation, e.g. to control *Phytophthora infestans* or to avoid migration of viruses into the tubers of seed potatoes, metabolic activity and water content of the tubers are relatively high and this allows high water loss during storage without damage [27]. Additionally, the relation between the content of reducing sugars and starch within a tuber is high and the content of dry substance increases with ripening, i.e. the content of reducing sugars is high and that diminishes storing stability and culinary quality of the tubers [27–32].

The formation of tuber skin is also influenced by the timing of desiccation [33]. Late desiccation increases tuber skin thickness which decreases injury during harvesting and reduces storage losses due to infestation of the tubers by fungal and bacterial diseases, and viruses [34–38].

Desiccation causes, directly or indirectly, the elimination of rots and diseases by preventing the translocation of diseases from the upper parts of a plant to the tuber and in this way reduces storage and quality losses [38–41].

Crops for Seed Production

By desiccation of trefoil (*Trifolium sp.*), alfalfa (*Medicago sp.*), beets (*Beta vulgaris L.*), cabbage (*Brassica sp.*), cucumber (*Cucumis sativus L.*) and other crops, especially during bad weather conditions in the ripening phase, seeds and fruits are supposed to ripen and dry faster and more constantly. The moisture content of the seeds is decreased [24, 42–45], and the germinability of the seeds is positively influenced in some cases [43–46]. On the whole, conditions for mechanized harvesting are improved so that harvest losses as a result of bad weather conditions can be avoided [40, 44, 46–49].

Desiccant treatment reduced the fruit rot disease incidence and the severity of the cucumber disease caused by *Rhizoctonia solani L.* Application of desiccants at first flower reduces plant canopy and leads to a reduction of disease [50].

Sunflower

The harvest date of sunflower (*Helianthus annuus L.*) is brought forward with chemically induced plant and seed drying by some days in some cases and nearly 4 weeks in others [51]. This measure increases the yield and the quality of seed and oil. By decreasing the moisture content of the inflorescence, white and gray mold are controlled [21,52–56]. Application before physiological maturity reduces yield, oil content, protein content in the meal, and to a lesser extend also the fatty acid component [51].

Rice

Pre-harvest desiccation of rice (*Oryza sativa L.*) improves timeliness of harvest and allows more efficient and faster combining of the grain [57] by decreasing moisture content in rice grain and straw, by increasing seed retention to harvest time and the percentage of full high-quality seed and decreased ill-filled chaff seed [57–62].

Grain Sorghum and Seed Corn

Desiccation of grain sorghum (*Sorghum sp.*) and seed corn (*Zea mays L.*) is carried out in some cases when the weather is damp during harvesting. This brings about advantages similar to those for the crops already mentioned [63,64].

Rape

The upper parts of the shoots of rape (*Brassica napus L.*), including the leaves, have to be desiccated 1 to 2 weeks before harvesting. This induces a steady ripening of pods and seeds. Differing rates of ripening because of twofold growth after frost damage in moist places and rapid growth of weeds in plants laid down in the dust have to be avoided. The optimization of combining conditions, including the prevention of pod splitting, reduces harvest losses. The water content of the seeds is decreased. This hinders the occurrence of mold in harvested products and saves the expenditure of energy for drying [15,65].

Sugarcane

The elimination of trash in sugarcane (*Saccharum officinarum L.*) before milling operations is widely recognized. When leaves accompany the stalks during the milling processes the recovery of sucrose is decreased. In addition to the extra costs of harvesting and transportation due to trash, the extra fibre increases the amount of milling necessary without producing more sugar, and the impurities in the trash contribute to low juice purity. The presence of trash also leads to greater insect infestation, larger rat populations, and reduced effectiveness of preemergence herbicide treatments [66]. The preharvest elimination of trash by traditional means of burning can be replaced by means of chemical desiccation [67–70]. A further aspect of the application of chemical compounds in sugarcane is the acceleration of ripening, resulting in higher sugar yield [66].

Spring Wheat

The advantages of chemical desiccants for spring wheat (*Triticum aestivum L.*), used at the beginning of waxy ripeness, are described. Desiccation accelerates ripening, increases the size of embryos, stimulates the differentiation of the meristematic tissue, stimulated the formation of conductive tissue, and stimulated organ development. Furthermore, the desiccation retarded losses of sugars and acidsoluble organic phosphor by inhibiting respiration of the harvested grain in the initial period of storage [71–75].

Onion

Desiccants produce preharvest desiccation of the leaves of onion (*Allium cepa L.*) and retard sprouting during storage [76–78]. In some cases, however, depending on the desiccant, dewatering of the bulbs may be retarded and therefore postharvest infestation with gray mold neck rot caused by *Botrytis allii* and sprouting occurs [77].

Flax

The measure of desiccation influences some agronomic and quality characteristics of flax (*Linum usitatissimum L.*), particularly hastening the drying of plant which eliminates the need for windrowing. Furthermore, the desiccants cause separation of the fiber bundles from their surrounding tissue. This resembles the release of fiber bundles, which results from conventional postharvest retting [79–82].

Hay

Experiments to decrease the moisture content of fodder grasses (e.g. perennial ryegrass, *Lolium perenne L.*) and other fodder crops (e.g. red clover, *Trifolium pratense L.*; alfalfa, *Medicago sativa L.*) by chemical preharvest desiccation have been made [83–85]. Chemical desiccation of clover and alfalfa decreased the fiber content and increased the digestibility of the hay [85]. The application of chemical desiccants to alfalfa and clover after cutting led to more radip drying than mechanically conditioned or control swats [86,87].

2.3 Evaluation of Application in Crops and the Economic Aspects

The application of desiccants and defoliants is closely linked with rationalization in agriculture and the increasingly intensive cultivation it involves. Essential aspects are that mechanical harvestating technology can be used more easily, there is a decrease in transport costs, a decrease of harvesting and storage losses, and the quality of the crop yields is increased.

How much desiccants and defoliants are used is influenced by weather conditions and varies from crop to crop. There is, however, desiccation and defoliation in many cases independent of weather conditions in order to avoid the spreading of diseases during storage of crop yields.

In the USA, about 80% of cotton crops are desiccated [88]. Production of potatoes, in many European countries, is connected with the use of desiccants to a high extent in some countries up to 90%. The control of rots and virus diseases of tubers by this method during storage is predominant in the production of seed potatoes [89]. Desiccation is highly necessary in some crops, e.g. sunflower, rape, rice, sorghum.

However, it is not applied, depending on weather conditions, every year. Crops for seed production, e.g. trefoil, lucerne, beets, cabbage, cucumber are regularly desiccated on a wide scale. The preharvest elimination of trash in sugarcane by the traditional means of burning is being increasingly replaced by chemical desiccation.

The advantages of using desiccants and defoliants already mentioned see pp 200–203 are connected with considerably economical effects. The profitable use of chemical compounds as harvest aids depends on the ratio of costs for the total management system for one crop (including costs for desiccants or defoliants) to the proceeds of harvest products. This ratio differs greatly for individual crops and is influenced by marketing situations. In spite of good results Reglone and Gramoxone for example are not applied, or only to a limited extent, because of economical and toxicological reasons. On the other hand it has to be considered that cheaper methods, on account of high persistance in soil, can lead to disadvantageous effects in following crops. Such damage, for example, reduces germination, occasionally inhibits growth, bleaching effects are hard to calculate with reference to a negative influence on the yields.

The economically rational integration of desiccants and defoliants into the management of a crop consequently demands calculation of the advantages and disadvantages caused by using chemical compounds. For that purpose the following have to be considered:

1) Facilitation of the use of mechanical harvest technology, increasing the speed of harvesting machines, increasing the harvest performance,
2) decreasing harvest losses caused by weather,
3) decreasing the harvest and storage losses by control of diseases and pests,
4) iscreasing the quality of harvest products by control of weeds, diseases, and pests, as well as by avoiding regrowth,
5) lowering of energy for transport of dirt and trash and for drying of harvest products.

All these reasons favor harvest aid practices. That means desiccants and defoliants can be integral components of economically efficient agricultural production.

3 Chemical Compounds for Desiccation and Defoliation

3.1 Chemical Dessicants and Defoliants Including Synergistic Combinations in Practical Use

The following basic demands were made on desiccants and defoliants:

1) High desiccative and defoliating efficiency or high acceleration of ripening of plants or parts of them,
2) low mammalian toxicity,
3) low persistance in soil,
4) low environmental pollution,
5) no disadvantages to the quality of harvested products and progeny,
6) weed killing,
7) prevention of regrowth,

8) no corrosion of plant protection equipment,
9) integration of the desiccants and defoliants as one component of an economically efficient agricultural production.

Table 1 gives an overview of chemical desiccants and defoliants in practical use (chemical structure, common names, trade names, acute oral LD_{50} for rats in mg/kg, crops which can be treated with the compounds). The LD_{50} data were mainly taken from Wegler [90,91] or Worthing and Walker [92].

3.1.1 Inorganic Compounds

The use of inorganic compounds for desiccation and defoliation is decreasing. Reasons for this are high standard application rates and unfavourable chemical and toxicological properties. Only a few investigations with respect to the biochemical mode of action have been made [93]. Despite these facts, the compounds are widely used in the USA, in India, and several European countries because of low costs.

3.1.1.1 Chlorates

Sodium Chlorate

This compound is soluble in water and a potent oxidising agent which reacts with organic materials. It quickly penetrates into the vegetable tissue and is reduced by the enzyme nitrate reductase to yield the highly reactive and acutely phytotoxic chlorite ion. The reduction product chlorite leads to the rapid inactivation of nitrate reductase and possibly other related enzymes and in this fashion has a phytotoxic action [94].

Sodium chlorate has been used since 1900 as a total herbicide. It is used for desiccation and defoliation of soy bean and other legumes, rice, cotton and potato [57,89,93,95,96]. Desiccative action starts relatively slowly. Standard application rates at the lower limit of the registered amount of the chemical per hectare can cause, e.g. in potato, depending on the physiological state of development of the variety at the moment of application and on weather conditions, regrowth and thereby diminish the quality of the tubers. High amounts of standard application rates can damage progeny because of high persistence of the compound and make the crop susceptible to stem-end-rot [97] and facilitate infestation of the tubers with *Phytophthora infestans* [98].

Formulations on the market contain fire depressants in form of borates, phosphates, carbonates or sulfates. Trade names are Agrosan, Atratol, Biokassol, De-Fol-Ate, Dervan, Drexel Defol, Harvest-Aid, Klorex, Tumbleaf and others more [412]. The acute oral LD_{50} for rats amounts to 1200 mg/kg. The standard application rates amount to 15 to 30 kg/ha. They are applied as aqueous solutions.

Described synergistic combinations of sodium chlorate with other compounds are formulated to avoid damage to progeny by decreasing the proportion of sodium chlorate and to improve desiccative efficiency. Described chemical compounds for binary or ternary mixtures with sodium chlorate are sodium carbonate together with urea, and a surfactant [99]; 7-oxabicyclo(2,2,1)heptane-2,3-di-carboxylic acid [1]; ammonium peroxydisulfate, 2,2-dichloropropionic acid, *p*-nitrophenol and/or chloroacetic acid [100-102].

Table 1. Chemical desiccants and defoliants in practical use

Chemical structure	1 Common name 2 Trade name 3 Toxicity[a]	1 Introduced by (year)[b] 2 Producer[b]	Use in	Refs.
Inorganic compounds				
Sodium chlorate $NaClO_3$	2 Agrosan, Atratol, Biokassol, De-Fol-Ate, Dervan, Drexel Defol, Harvest-Aid, Klorex, Tumbleaf and some more 3 1200	1 ca. 1900	Cotton Legumes Potato Rice	93, 95) 96, 387) 89, 93, 95, 387) 57)
Magnesium chlorate $Mg(ClO_3)_2$	2 Defoal, De-Fol-Ate, Desecol, E-Z-Off, Magron, MC Defoliant, Ortho MC	2 EIA	Corn Cotton Legumes Potato Sorghum Sunflower Weyrich jointweed	64) 8, 11, 12, 15, 108, 109, 387) 25, 44, 45, 96, 110) 89) 63) 21, 52–55, 111) 111)
Calcium chlorate chloride $Ca(ClO_3)Cl$			Cotton Potato	8, 12) 117)
Ammonium sulfate $(NH_4)_2SO_4$			Corn Legumes Potato Sorghum Sunflower	64) 20, 47) 121, 129) 63) 52)
Copper sulfate $CuSO_4$	1 Bordeaux mixture	1 (1885)	Potato Red clover	3, 119, 120) 47)
Ammonium nitrate NH_4NO_3			Corn Legumes Potato Sorghum Sunflower	64) 20, 47) 119, 121) 63) 52)

Sodium chloride NaCl			Cotton Potato Rice Sorghum	[15] [120, 123] [60, 122] [63]
Potassium chloride KCl			Cotton Potato Rice Sorghum	[15] [120, 123] [60, 122] [63]
Magnesium chloride $MgCl_2$			Cotton Potato Rice Sorghum	[15] [120, 123] [60, 122] [63]
Potassium carbonate K_2CO_3			Grasses Lucerne Red clover	[124–126]
Sodium carbonate/ Sodium hydrogen carbonate $Na_2CO_3/NaHCO_3$			Grasses Lucerne Red clover	[124–126]
Phosphoric acid H_3PO_4			Potato	[121, 123, 128–130]
Sulfuric acid H_2SO_4			Potato	[89, 93, 119, 121, 123, 128–133, 387]
Sodium arsenite $NaAsO_2$	2 Arcadian, Amorta 3 10–50	1 ṄCD (1901) 2 ACP, TLM	Cotton Potato	[90, 93] [89, 90]
Arsenic acid $H_3AsO_4 \times 1/2\ H_2O$	2 Desiccant L-10 Zotox 3 48–100	2 PCP	Cotton	[1, 7, 91]
Calcium cyanamide CaN-CN	2 Tsaks 3 1400	1 ACC (1905) 2 DAI, DGG	Cotton Potato	[91, 93] [91, 93, 129, 135, 138–143]

Table 1. (continued)

Chemical structure	1 Common name 2 Trade name 3 Toxicity[a]	1 Introduced by (year)[b] 2 Producer[b]	Use in	Refs.
Organic compounds				
Phenols and derivatives				
4,6-Dinitro-o-cresol 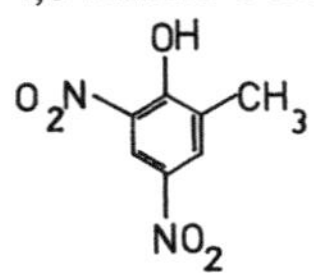	1 DNOC 2 Antinonnin, Sinox, Trifanex and some more 3 25–40	1 TEC (1932) 2 AHM, PHP, TLM	Legumes Potato	92, 93) 89, 92, 93)
2-*sec*-Butyl-4,6-dinitrophenol 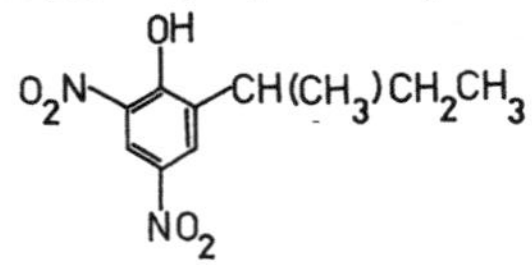	1 Dinoseb, DNBP 2 Basanite, Chemox General, Dynamyte and some more 3 58	1 DOC (1945) 2 AHM, DCC, TLM, UCP	Hop Legumes Potato	92, 387) 92, 387) 89, 92)
2-*tert*-Butyl-4,6-dinitrophenol 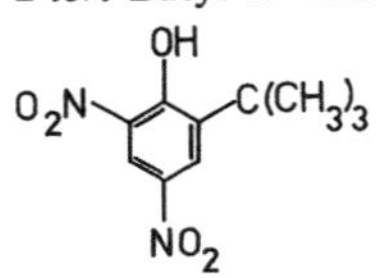	1 Dinoterb 2 Extar forte, Herbogil, Stirpar forte, Tolkan 3 25[c]	1 PEP (1967) 2 RPF	Potato	89, 387)
2-*sec*-Amyl-4,6-dinitrophenol 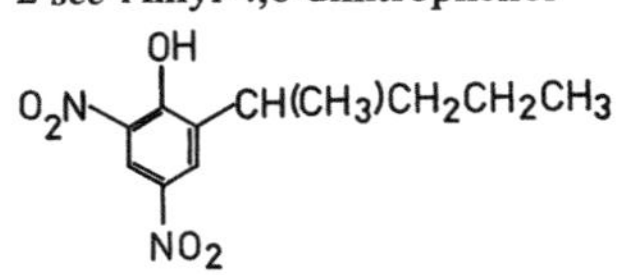	1 Dinosam, DNAP	1 FCP (1966)	Potato	89)

Sodium pentachlorophenol 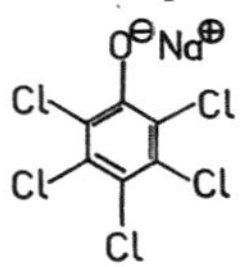	1 PCP 2 Penchlorol, Pentacon, Penwar, Sinituho 3 210	1 DOC (1936)	Potato	89, 92, 93)
3,5-Dibromo-4-hydroxybenzonitrile 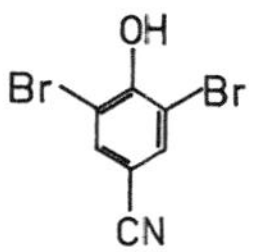	1 Bromoxynil 2 Briotril, Brittox, Brominal, Bromotril, Bronate, Buctril 3 190	1 MUB, API (1963) 2 CFP, MAB, MUB, RPA	Potato	89, 92)
Carbonic acid derivatives 3-(3-Chloro-4-methoxyphenyl)-1,1-dimethyl urea 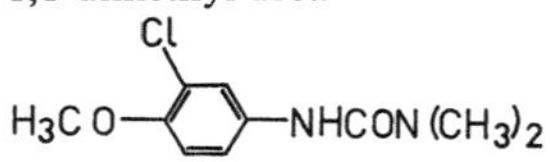	1 Metoxuron 2 Deftor, Dosaflo, Dosanex, Purivel, Surelex 3 3200	1 SAN (1968) 2 SAN	Flax Hemp Potato Tomato	92) 92, 89, 92, 173 – 175, 387) 92, 387)
Carboxylic acid derivatives Formic acid HCO_2H			Grass Potato	83, 84) 130)
Chloroacetic acid (sodium salt) $ClCH_2CO_2Na$	1 MCA 2 Monoxone 3 650	1 ICI (1956)	Hop Potato	387) 89, 91
Trichloroacetic acid (sodium salt) Cl_3COO_2Na	1 TCA 2 Konesta, NaTA, Varitox 3 3200–5000	1 DNP, DOC (1947) 2 HOE, MUB	Potato	89)
Propionic acid $CH_3CH_2CO_2H$			Grass	83)

Table 1. (continued)

Chemical structure	1 Common name 2 Trade name 3 Toxicity[a]	1 Introduced by (year)[b] 2 Producer[b]	Use in	Refs.
2,2-Dichloropropionic acid (sodium salt) $CH_3CCl_2CO_2Na$	1 Dalapon 2 Ded-Weed, Devipon, Dowpon, Gramevin, Revenge, Unipon 3 7570–9300	1 DOC (1953) 2 BAS, FPP, IAG	Potato	89)
cis-3-Chloroacrylic aeid (sodium salt) $ClCH{=}CHCO_2Na$	1 Chloroacrylic acid 2 Prep	1 UNC (1963)	Potato	197)
2,3,6-Trichlorophenylacetic acid 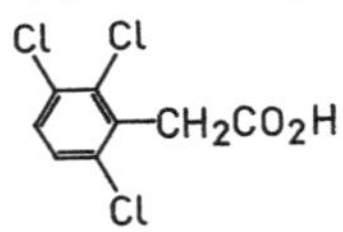	1 Chlorfenac 2 Fenac, Trifene 3 576–1780	1 API (1965)		90, 91)
4-Chloro-*o*-tolyl-oxyacetic acid 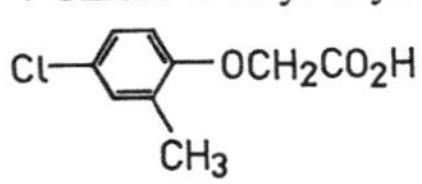	1 MCPA, Metaxon 2 Agroxone, Agritox, Chiptox, Ded-Weed, Empal, Kilsem and others more 3 700	1 ICI (1945) 2 AHM, ASA, AZC, BAG, BAS, CLA, GIM, IAG, ICI, NCI, SYS, UCP	Potato Sugarcane	202, 66)
2,4-Dichlorophenoxyacetic acid 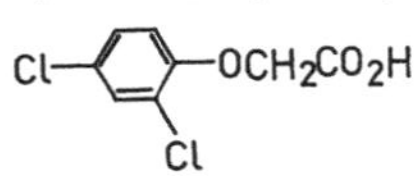	1 2,4-D 2 Acme Amine, Agrotect, Amoxone, Aqua-Kleen, Chloroxone and many others more 3 375	1 API (1944) 2 AHM, ASA, AZC, BAG, BAS, CHB, CKB, CLA, COS, CRC, FAP, FPP, GIM, IAG, NCI, PBI, RPA, RPF, SPB, UCC	Cotton Potato Red clover	15) 89) 47)

2,4,5-Trichlorophenoxyacetic acid 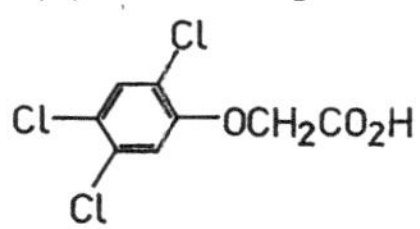	1 2,4,5-T 2 Dacamine, Ded-Weed, Farmco Fence, Rider, Forron, Inverton 245, Line Rider, T-Nox, U-46 3 300–1700	1 API (1944)	Potato	202)
7-Oxabicyclo(2,2,1)heptane-2,3-dicarboxylic acid 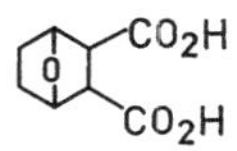	1 Endothal 2 Accelerate, Aquathol K, Des-i-cate, Endothal Turf Herbicide, Herbicide 273, Hydrothol 191, Hydrothol 191 Granular 3 51	1 SCC (1951) 2 PCP	Cotton Crops for seed production Legumes Sugarcane	92, 93) 92, 387) 23) 66)
Quarternary ammonium compounds				
1,1′-Ethylene-2,2′-bipyridylium dibromide 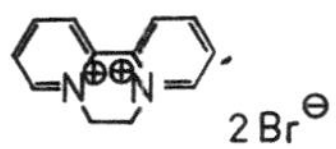	1 Diquat, Deiquat 2 Aquacide, Dextrone, Reglone, Reglox, Weedtrine-D 3 231	1 ICI (1955) 2 ICI	Crops for seed production Flax Legumes Onion Potato Rape Rice Sorghum Sugarcane Sunflower Weyrich jointweed	43–47, 49, 64, 230, 231, 233) 79) 19, 229) 76) 39, 41, 89, 176, 226, 227) 235) 57, 122) 63) 66, 70) 21, 52, 53, 55, 228) 111)
1,1′-Dimethyl-4,4′-bipyridylium dichloride or sulfate 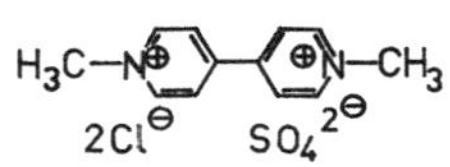	1 Paraquat 2 Actor, Dextrone, Esgram, Gramoxone and many others more 3 150	1 ICI (1960) 2 COC, CRC, HSA, ICA, ICI, POS, SCT, SSP, VCS	Cotton Crops for seed production Legumes Potato Rice Sugarcane	6, 7, 10, 12, 13, 15) 232, 245) 1, 18, 22, 96) 1, 243, 244, 387) 1, 57–59, 62, 122) 1, 66, 70)

Table 1. (continued)

Chemical structure	1 Common name 2 Trade name 3 Toxicity[a]	1 Introduced by (year)[b] 2 Producer[b]	Use in	Refs.
1,1′-Dimethyl-4,4′-bipyridylium dimethylphosphate 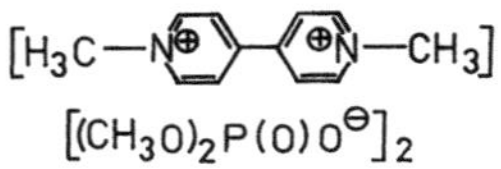	1 DDP		Cotton Crops for seed production Legumes Potato Sunflower	249) 44, 233, 249, 250) 249) 89) 21)
Heterocyclic compounds 1*H*-1,2,4-Triazol-3-ylamine 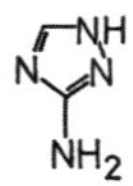	1 Amitrole 2 Amerol, Amizol, Azolan, Cytrol, Weedazol and others more 3 1100–24600	1 APC (1954) 2 BAG, CCU, CFP, FPI, MAB, RPA	Cotton	91, 259)
2-Ethylamino-4-isopropylamino-6-methylthio-1,3,5-triazine 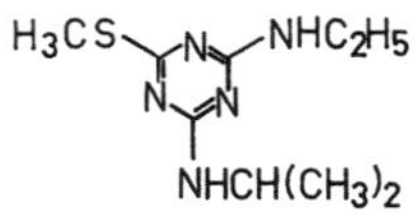	1 Ametryne 2 Ametrex, Doruplant, Evik, G-34162, Gesapax, Trinatox D 3 1110	1 GSA (1962) 2 CGC, MAB	Potato Soybean	268) 18)
2-*n*-Butylmercapto-benzothiazole 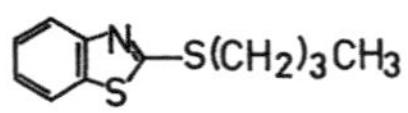	1 Butylcaptax		Cotton	270)
1-Phenyl-3-(1,2,3-thiadiazol-5-yl)urea 	1 Thidiazuron 2 Dropp 3 >4000	1 SCH (1976) 2 NAC, SCH	Cotton	16, 92, 271, 272)

2,3-Dihydro-5,6-dimethyl-1,4-dithi-ine-1,1,4,4-tetraoxide 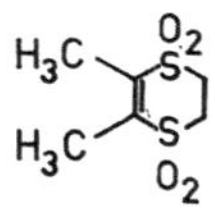	1 Dimethipin 2 Harvade 5 F 3 1180	1 UCC (1974) 2 UCC	Cotton Legumes Oilcrops Potato Rice Sunflower	17, 92, 292–296) 291) 56, 292, 296) 1, 89, 174, 176, 296–298, 387) 92, 296) 56, 92, 292, 296)
Phosphorous and arsenic compounds				
Tributylphosphorotrithioite $(nC_4H_9S)_3P$	1 Merphos 2 Folex 3 1272	1 VCC (1960) 2 RPA	Cotton	1, 13, 91–93, 387)
S,S,S-Tributylphorphorotrithioate $(nC_4H_9S)_3PO$	2 Def, De-Green, E-Z-off D, Folex 6EC, Fos-Fall A, Ortho Phosphat Defoliant 3 325	1 ETC, CHC (1956) 2 MBC	Cotton	1, 11–14, 91–93, 108, 299, 387)
2-Chloroethylphosphonic acid $ClCH_2CH_2PO(OH)_2$	1 Ethephon 2 Camposan, Cerone, Ethrel, Flordimex, Florel, Prep 3 4229	1 API (1968) 2 CKB, RPA	Cotton Crops for seed production Fruits Legumes Onion Potato Sugarcane Vine	1, 6, 92) 45, 231) 92) 23) 78) 89, 302) 66) 231)
N-(Phosphonomethyl)-glycine $HOCOCH_2NHCH_2PO(OH)_2$	1 Glyphosate 2 Glifinox, Glycel, Rodeo, Rondo, Roundup 3 5600	1 MCP (1971) 2 CRC, EIA, ETL, MCP, POS, QES	Crops for seed production Flax Legumes Rice Sugarcane	231) 80–82) 22, 96, 314, 387) 57) 66)

Table 1. (continued)

Chemical structure	1 Common name 2 Trade name 3 Toxicity[a]	1 Introduced by (year)[b] 2 Producer[b]	Use in	Refs.
Ammonium-DL-homoalanin-4-yl-(methyl)-phosphinate NH_4^+ $^-OCOCHNH_2CH_2CH_2PO(OH)CH_3$	1 Glufosinate-Ammonium 2 Basta, Buster, Finale, Ignite 3 1620–2000	1 HOE (1981) 2 HOE, HRA	Crops for seed production Legumes Potato Rape Sunflower	320, 321) 319, 320, 322) 319, 320, 322) 319, 320) 320)
Di-*n*-butyl-1-*n*-butylaminocyclo-hexylphosphonate 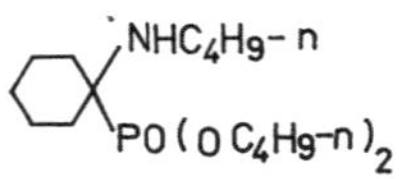	1 Buminafos 2 Trakephon 3 5110	1 CKB (1972) 2 CKB	Cotton Crops for seed production Onion Potato Sugarcane	328) 326) 76, 77) 89, 169, 176, 326) 328)
Hydroxydimethylarsine oxide (sodium salt) $(CH_3)_2AsONa$	1 Dimethylarsinic acid 2 Ansar, Bolate, Bolls-Eye, Bophy, Dilic, Phytar 3 1350	1 ACC (1958) CCC (1962) 2 IAG, MCF, PAL	Cotton	7, 9, 15, 91 – 93, 330)

[a] Acute oral LD_{50} for rats in mg/kg;
[b] List of the firms mentioned in Table 1 as discoverers and producers of the compounds concerned [412)];
[c] Acute oral LD_{50} for mice in mg/kg.

[b]ACC	American Cyanamid Company, USA
ACH	Ansul Chemical Corporation, USA
ACP	Agtrol Chemical Corporation, USA
AHM	A.H. Marks a. Co. Ltd., England
AIM	All India Medical Corp., India
APC	American Chemical Paint Company, USA
API	Amchem Products Incorp., USA (now: Union Carbide Agricultural Products Co Inc., USA)
ASA	Atanor S.A.M., Argentina
AZC	Akzo Zout Chemie Nederland bv, The Netherlands
BAG	Bayer AG, Federal Republic of Germany
BAS	BASF AG, Federal Republic of Germany

Table 1. (continued)

CCC	Crystal Chemical Corp., USA	KCC	Kerr-McGee Chemical Corp., USA
CCU	Custom Chemicides, USA	MAB	Makteshim-Agan, Israel
CFP	Compagnie Francaise de Produits Industrieis, France	MBC	Mobay Corp., USA
CGC	Ciba Geigy Corp., Switzerland	MCF	Monterey Chemical Co, USA
CHB	Chemolimpex, Hungary	MCP	Monsanto Co, USA
CHC	Chemagro Corp., USA	MUB	May and Baker Ltd., England
CKB	VEB Chemiekombinat Bitterfeld, German Democratic Republic	NAC	Nor-Am Chemical Co, USA
CLA	Chemie Linz AG, Austria	NCD	Niagara Chemical Division, USA
COC	Complets Chemical Industrial Co, Ltd., Taiwan	NCI	Nissan Chemical Industries, Ltd., Japan
COS	Compania Quimica S.A., Argentina	PAL	Pamol Ltd., Israel
CRS	Crystal Chemical Inter-America, USA	PBI	PBI-Gordon Corp., USA
CSA	Caffaro S.p.A., Italy	PCP	Pennwalt Corp., USA
DAI	Degesch America, Inc., USA	PEP	Pepro (now: Rhone-Poulenc Agrochimie S.A., France)
DCC	Drexel Chemical Co, USA	PHP	Pennwalt Holland B.V., The Netherlands
DGG	Degesch GmbH, Federal Republic of Germany	POS	Productos OSA, Argentina
DNP	E.I. Du Pont de Nemours Co Inc., USA	QES	Quimica Estrella S.A.C., I.e.I., Argentina
DOC	Dow Chemical Co, USA	RPA	Rhone-Poulenc Ag Co, USA
EIA	Energia E. Industrias Aragonesas, Spain	RPF	Rhone-Poulenc Agrochemie S.A., France
EIL	Exel Industries, Ltd., India	SAN	Sandoz AG, Switzerland
ETC	Ethyl Corp., USA	SCC	Sharples Chemical Corp., USA (now: Pennwalt Corp., USA)
FAP	Farmco Australia Pty Ltd., Australia	SCH	Schering AG, Federal Republic of Germany
FCP	Fabrik van Chemische Producten, The Netherlands	SCT	Shinung Corp., Taiwan
FPI	Fair Products, Inc., USA	SPB	Sintesul, Brasil
FPP	Fermenta Plant Protection Co, USA	SSP	Sundat (S) Pte., Ltd., Singapore
GIM	Gilmore, Inc., USA	SYS	VEB Synthesewerk Schwarzheide, German Democratic Republic
GSA	J.R. Geigy S.A., Switzerland (now: Ciba Geigy Corp., Switzerland)	TEC	Truffaut et Cie, France
HOE	Hoechst AG, Federal Republic of Germany	TLM	Tifa Ltd., USA
HRA	Hoechst-Roussel Agri-Vet. Co, USA	UCC	Uniroyal Chemical Co, Inc., USA
HSA	Hightex S.A., Spain	UCP	Universal Crop Protection Ltd., England
IAG	Inter AG, USA	UNC	Union Carbide Agricultural Products Co, Inc., USA
ICA	ICI Americas, Inc., USA	VCC	Virginia Carolina Chemical Corp., USA
ICI	ICI, England	VCS	Visplant-Chemiren S.R.L., Italy
JRS	J.R. Simplot Co, USA	WEC	Willbur-Ellis Co, USA

Magnesium Chlorate

It has chemical properties similar to sodium chlorate and a similar mode of action. Treatment of pea leads to a higher content of malonic dialdehyde, accumulation of hydroperoxide, which causes lipid peroxidation [103]. Treatment of cotton causes translocation of ^{14}C-labelled photosynthetic products to the fruits [104, 105] and of ^{32}P out of the leaves into the seeds and fibers [106]. This effect is neutralized by nitrogen fertilizers [107].

Magnesium chlorate is used for desiccation and defoliation of cotton [8, 11, 12, 15, 108, 109], soy bean [25, 96], broad bean [110], sunflower [21, 52–55, 111], sorghum [63], potato [89] as well as for the seed production of corn [64], trefoil [44, 45] and Weyrich jointweed (*Polygonum Weyrichii*) [111].

Trade names are Defoal, De-Fol-Ate, Desecol, E-Z-Off, Magron, MC Defoliant, Ortho MC [412]. The acute oral LD_{50} for rats amounts to 5250 mg/kg. The standard application rates per hectare amount to 20 to 30 kg.

The following chemical compounds for advantageous combination with magnesium chlorate are described: ammophos [112], nitroammophos [113], potassium dihydrogen phosphate [114] and urea [115], S,S,S-tributyl phosphorothioate [116], 1,1′-ethylene-2,2′-bipyridylium dibromide [48], bipyridylphosphate [42].

Calcium Chlorate Chloride

Until several years ago calcium chlorate chloride was also used to a small extent for defoliating cotton [8, 12] and desiccating potato [117].

3.1.1.2 Sulfates, Nitrates, Chlorides and Carbonates

Copper sulfate inhibits several enzymes and denaturates cellular proteins [118]. Copper sulfate solutions were used as early as 1932 for desiccation of potato [3]. At present, it is probably used for killing haulms and leaves of potato [119, 120] or desiccation of trefoil [47] in individual cases. The standard application rate ranges from 500 to 1000 L of 5 per cent aqueous solution per hectare.

The application of ammonium sulfate and ammonium nitrate for preharvest desiccation of potato [119, 121] as well as for desiccating grain sorghum [63], trefoil [47], seed corn [64], sunflower [52] and lupine [20] is described. The ammonium ion is quickly absorbed by the plant, where it changes the pH value of the cell sap and denaturates cell proteins [118].

Ammonium nitrate hastens ripening but fails to control gray rot [52], caused burning effects, and decreased germination [20]. To reach sufficient desiccative effects, high standard application rates, 60 to 200 kg/ha [63, 64], or 10 per cent solutions have to be used [47].

Sodium, potassium and magnesium chloride can be used for desiccating rice [60, 122], cotton [15], grain sorghum [63] and potato [120, 123]. As a rule, 500 to 1000 L of 5 to 20 per cent aqueous solutions are applied. Quicker ripening and decreased grain moisture are achieved only by applying the higher concentrated solutions [122, 123].

Possibly, potassium carbonate, or a mixture of sodium carbonate and sodium hydrogen carbonate, or mixtures with carbonate, silicate, phosphate can be used for drying grass, lucerne or red clover to make hay [124–126].

3.1.1.3 *Sulfuric and Phosphoric Acid*

The herbicidal effect of sulfuric acid has been known since 1874 [127]. Sulfuric acid drastically devolves water from vegetable tissue, destroys the chloroplasts, denaturates proteins, and leads to the rapid destruction of vegetable tissue [118].

Sulfuric acid was used in 1935 for eliminating haulms and leaves of potato [119]. The literature describes the application of sulfuric acid and phosphoric acid for desiccation of potato [121, 123, 128–130]. The application of phosphoric acid is of little importance. Good desiccative effects are achieved with 12 to 15 per cent sulfuric acid, a typical application rate being 135 L/ha in 1000 L water [131, 132]. However, a smaller amount of sulfuric acid per hectare, especially in combination with previous mechnical elimination of haulm and leaves, may give good results [133]. Disadvantages, e.g. the need for special application machinery because of the strong corrosive action, irregular distribution of the acid because of high viscosity at 15 to 18 °C and great harm to users and environment, have obviously decreased its use. To increase the control of weeds a mixture of sulfuric acid and potassium permanganate is recommended [134].

3.1.1.4 *Arsenic Compounds*

Among inorganic arsenic compounds sodium arsenite and arsenic acid are used in practice.

Sodium arsenite inhibits the respiratory metabolism of plants and disturbs carbohydrate metabolism [118]. This compound was introduced in 1901 as a herbicide [90] and used for defoliating cotton and desiccating potato [89, 90, 93]. Sodium arsenite may cause, like sodium chlorate, damage to progeny and stem-end-rot [135, 136]. Trade names are Arcadian and Amorta. Sodium arsenite is highly poisonous. The acute oral LD_{50} for rats amounts to 10 to 50 mg/kg. The standard application rate per hectare for desiccation of potato is 20 kg. The use as a herbicide has been prohibited in most countries. In the Netherlands it is used for desiccating seed potatoes. The compound is often used in combination with arsenic trioxide [91].

Arsenic acid is less effective than arsenite and is used, also together with 7-oxabicyclo(2,2,1)heptane-2,3-dicarboxylic acid, for defoliating and desiccating cotton [1, 7, 91]. Trade names are Desiccant L-10 and Zotox [412]. The acute oral LD_{50} for rats amounts to 48 to 100 mg/kg. Because of its high toxicity and environmental pollution, its use has been prohibited in a lot of countries and is on the decrease.

3.1.1.5 *Calcium Cyanamide*

Calcium cyanamide decomposes, depending on weather conditions, in soil relatively quickly to urea [91] or ammonia [137] and has therefore on the one hand fertilizing and on the other hand herbicidal properties. The compound was introduced in 1905 as a fertilizer and herbicide [91, 93] and has been used over the past decades very often for defoliating cotton and desiccating potato [91, 93, 129, 135, 138–143]. The effect depends to a high degree on weather conditions and is therefore very variable. A decrease in the culinary quality of the potato tubers is not excluded [135]. It is also used for defoliating vine by applying 2.45 per cent solution immediately after the normal harvest season in late August. This procedure induces bud opening and the second flash of shoot growth, making possible the harvesting of two crops per year under tropical and semitropical conditions [144].

The trade name in the USSR is Tsaks. Calcium cyanamide is no longer registered in European countries. The acute oral LD_{50} for rats amounts to 1400 mg/kg. Economic aspects have reduced its use [139]. Calcium cyanamide had been often used as defoliant in combination with disodium hexafluorosilicate [145].

3.1.2 Organic Compounds

3.1.2.1 Phenols and Derivatives

Phenol compounds belong to the oldest herbicides. As a rule phytotoxicity is increased by substituents, e.g. NO_2, Cl, which increase the acid character. Low alkyl groups in position 6 increase the phytotoxicity too, but probably by improving penetration into the plant cells [1,146]. The compounds are not transported or only to a small amount within the plant.

Dinitroalkylphenols and pentachlorophenol are uncouplers of oxidative phosphorylation [147]. The compounds contain a labile phenolic hydrogen in the molecule. They may therefore transport protons through membranes and degrade an existing pH-gradient. The result would be a sharp drop of the rate of ATP-formation with no effect on electron transport. On the other hand they also retard the photosynthetic electron transport. Studies with chloroplasts from s-triazine resistant weeds have confirmed that the binding site exists for nitrophenols overlaps with the binding site of other photosystem II inhibitors [413]. Because of the dual action on electron transport and phosphorylation, these compounds have also been termed "inhibitory uncouplers" [148,149]. The action on photosynthesis and respiration may contribute to the phytotoxicity under field conditions.

Phenols also denaturate proteins and retard the enzymatic synthesis of glycerol lipids. This changes the structure of cell membranes [150].

The following 4 compounds of dinitrophenols are used practically as desiccants and defoliants:

4,6-Dinitro-o-cresol (Common Name: DNOC)

This compound was the first organic herbicide and was introduced in 1932 [151]. In its capacity as an insecticide it was the first synthetic one used as early as 1892 [91]. It is used for preharvest desiccation of potato and leguminous seed crops [89,92,93].

Besides being sold under the names Antinonnin, Sinox and Trifanex a number of other liquid formulations, containing mineral oils or ammonium sulfate, are used [92,152,412]. The acute oral LD_{50} for rats amounts to 25 to 40 mg/kg.

2-sec-Butyl-4,6-dinitrophenol (Common Name: Dinoseb)

The compound was developed in 1945 as a herbicide [146] and is used for preharvest desiccation of potato, leguminous seed crops and hops [89,92,93].

Besides the trade names Basanite, Chemox General and Dynamyte a number of other liquid formulations, containing mineral oils, exist [412]. The acute oral LD_{50} for rats amounts to 58 mg/kg.

Desiccative action could be increased synergistically by ammonium sulfate or 2-chloroethylphosphonic acid [153] respectively 2-ethylamino-4-isopropylamino-6-methylthio-1,3,5-triazine [154], in the course of which split application was more effective than single application for the latter.

2-tert-Butyl-4,6-Dinitrophenol (Common Name: Dinoterb)

The herbicide was described and introduced in 1967 [155]. It is used for preharvest desiccation of potato, but only in a small number of countries. On the whole, its use is decreasing.

Trade names are Extar forte, Stirpar forte, Herbogil and Tolkan [89, 412]. Dinoterb is available as amine salts or as mixtures with other herbicides. The acute oral LD_{50} for mice amounts to 25 mg/kg.

Potato desiccation by a mixture of dinoterb and metoxuron is faster than desiccation by the compounds applied singly [6].

2-sec-Amyl-4,6-Dinitrophenol (Common Names: Dinosam, DNAP)

Mode of action and application compares with dinoseb. The compound was first described as a herbicide in 1966 [157]. Dinosam has the lowest practical importance as desiccant of the described for dinitrophenol compounds.

Further derivatives of phenol are:

Sodium Pentachlorophenol (Common Name: PCP)

Over the last decades pentachlorophenol had gained practical importance in the form of the sodium salt. The compound was introduced in 1936 as a timber preservative and later as a general disinfectant, herbicide, and desiccant (or defoliant), especially in potato [92].

Trade names are Penchlorol, Pentacon, Penwar and Sinituho [412]. The acute oral LD_{50} for rats amounts to 210 mg/kg. At present it is hardly used as a desiccant.

3,5-Dibromo-4-Hydroxybenzonitrile (Common Name: Bromoxynil)

The herbicide bromoxynil was described and introduced in 1963 [92]. Salts and esters of bromoxynil are contact herbicides with some systemic activity. Halogenated hydroxybenzonitrile herbicides are inhibitors of the photosynthetic electron transport at the photosystem II site [148, 160] and inhibitors [161] or uncouplers [162, 163] of oxidative phosphorylation and respiration. Bromoxynil is a very potent inhibitor of photosynthetic electron transport [148, 163, 164]. The action on intact plants has been found to depend on climatic conditions. The greatest damage will occur when the light intensity is low before and high after the application of the compound [165]. Phytotoxicity increase with rising temperature and decreases with growing humidity [166, 167]. The herbicide is rapidly decomposed in the soil and metabolized in plants and animals by hydrolysis of the ester and nitrile group with some debromination [92].

It is used for controlling the seedlings of broad-leaved weeds in various crops, especially in cereal crops, and for preharvest desiccation in potato [89, 92].

Trade names are Briotril, Brittox, Brominal, Bromotril, Bronate and Buctril. There are various formulations of sodium or potassium salts. Bromoxynil forms stable esters, of which the 2,6-dibromo-4-cyanophenyloctanoate, and occasionally, the corresponding heptanoate are used. The ester is readily hydrolysed to bromoxynil at $pH > 9$. The acute oral LD_{50} for rats amounts to 190 mg/kg.

With regard to increased desiccative action, the literature contains descriptions of synergistically acting combinations of bromoxynil and 2,3-dihydro-5,6-dimethyl-1,4-dithi-ine-1,1,4,4-tetraoxide [168] or bromoxynil and di-*n*-butyl-1-*n*-butylamino-cyclohexylphosphonate [169].

3.1.2.2 Carbonic Acid Derivatives

3-(3-Chloro-4-Methoxyphenyl)-1,1-Dimethylurea (Common Name: Metoxuron)

The phenol compound contains a dimethylurea substituent and is therefore considered as a urea herbicide and compared with a lot of other urea herbicides. Urea herbicides are, like certain other herbicide groups, e.g. s-triazines and hydroxybenzonitriles, inhibitors of the photosynthetic electron transport chain near photosystem II. They retard the Hill reaction [94, 170, 171]. Metoxuron is absorbed into the leaves and transported acropetally only. Degrading of 50% in soil occurs within 10 to 30 days. Main metabolic reactions are *N*-demethylation and hydrolysis of the urea moiety [89, 92]. The herbicide metoxuron was described and introduced in 1968 [92, 172].

A lot of European countries use metoxuron as a potato-haulm killer, but also for defoliating hemp, flax and tomato. It accelerates and synchronizes ripening. Therefore, regrowth and colouring of the fascicles of vessels are more seldom than by other chemicals [89, 92, 173–175]. The relatively slow action favours the formation of tuber skin and increases the quality of the tuber [89]. However, in some cases metoxuron is ineffective as preharvest desiccant in potato [176].

Trade names are Deftor, Dosaflo, Dosanex, Purivel and Sulerex [412]. The acute oral LD_{50} for rats amounts to 3200 mg/kg.

Binary metoxuron-dinoterb or metoxuron-trialkylamine mixtures synergistically desiccate potato haulm [156, 177].

3.1.2.3 Carboxylic Acids and Derivatives

Some carbocylic acids or their derivatives, often sodium salts, are used for desiccation, defoliation, or acceleration of ripening. They act as strong acids and denaturate proteins as well as retard enzymes of the biosynthesis of lipids [93]. As a rule, the herbicidal effect of carboxylic acids is increased by halogen near the carboxylic group [91]. The phytotoxic effects of the chlorinated aliphatic acid herbicides as a group are very similar in some respects all cause leaf chlorosis and abnormal growth responses [178, 179]. The compounds absorbed by roots, cause formative effects which indicate profound physiological disturbance. In addition to these systemic effects, the chloroacetic acids have a pronounced contact toxicity which is associated with the inability of their being translocated from the leaves [180, 181]. Most of these compounds are metabolized with difficulty and relatively persistant in soil. Therefore, application in agriculture is very restricted [89].

Formic Acid

Formic acid is not as corrosive as sulfuric acid and is applied, for example in Sweden, with the help of special machinery during mechanical haulm destruction in potato [130]. Furthermore, there are reports about ensiling of chemically desiccated grass. Spraying grass with 75 L concentrated formic acid per ha 3 to 4 days before harvesting increased dry weight of the harvested forage, whereas Gramoxone or Reglone injured grass, delayed regrowth and left toxic residues in the silage [83]. Finally, experiments on preharvest treatment of perennial rye grass by formic acid to decrease the moisture concentration in vegetative growth have been carried out [84].

Chloroacetic Acid (Common Name: MCA)

MCA is very soluble in water, is hardly metabolized within the plant and denaturates proteins [118]. Its herbicidal properties were first described in 1952 [182]. Sodium chloroacetate was described and introduced as a herbicide in 1956 [183].

The compound can be used for defoliation [91] and preharvest desiccation of potato [89].

The trade name is Monoxone [412]. The acute oral LD_{50} for rats of sodium chloroacetate is 650 mg/kg.

To improve the action, binary mixtures of chloroacetic acid with sodium chlorate, chloroethanephosphonic acid, basic copper 2-chloroethanephosphonate, 2-chloro-6-methyl-4-benzylphenol, 2,6-dinitro-4-trifluoromethyl-*N*,*N*-dipropylaniline, *p*-nitrophenol, amonium peroxydisulfate, sodium-2,2-dichloropropionate, 2-methylthio-4-ethylamino-6-isopropylamino-1,3,5-triazine or *N*-butyl-*N*-(1-(O,O-dibutylphosphonyl)cyclohexyl)ammonium chloride as synergistic crop desiccants, especially in potato, legumes and rape, have been suggested [184, 185].

Trichloroacetic Acid (Common Name: TCA)

TCA, or more so as the sodium or ammonium salt, is very soluble in water, absorbed by roots and acts systemically. The compound is hardly metabolized at all within the plant and denaturates proteins [118]. TCA possibly interferes with lipid synthesis and affects the formation of epicuticular waxes [186, 187]. However, TCA does not inhibit the incorporation of ^{14}C-acetate in pea seedlings [188] and also does not significantly decrease the epicuticular wax formation on pea leaves [189]. The persistance in soil is from 14 to 90 days, depending on the soil, moisture and temperature [92].

The herbicidal properties of the sodium salt were first described in 1947 [190]. The herbicide was introduced in 1949.

TCA is suitable for defoliating and preharvest desiccating of potato [89, 91]. Trade names are Konesta, NaTA and Varitox [412]. The acute oral LD_{50} for rats is 3200 to 5000 mg/kg.

Propionic Acid

Propionic acid is of little importance as a desiccant. The spraying of grass with 75 L/ha concentrated propionic acid 3 to 4 days before harvesting, increased dry weight of the harvested forage, whereas Gramoxone and Reglone show disadvantages. However, propionic acid desiccation decreased the β-carotene in the dry matter in 3 month silage for the first and also the third regrowth [83]. In the USA propionic acid is extempted from the requirement of a tolerance, when used as plant desiccant [191].

2,2-Dichloropropionic Acid (Common Name: Dalapon)

Dalapon is soluble in water and absorbed and translocated by roots and leaves. Dalapon interacts with proteins and inactivates enzyme systems. Interactions with lipid biosynthesis are though to take place [93]. The formation of epicuticular waxes is also affected [186, 187]. As a consequence, the penetration of other herbicides applied to the leaf surface may be facilitated and in cases where the transpiration of the treated plant is increased, xylem-mobile herbicides may be taken up more extensively via the roots [192].

Another observation is the partial antagonism between dalapon and panthothenate of β-alanine [186]. Panthoic acid and β-alanine both take part in the biosynthesis of coenzyme A. However dalapon at 20 mg/L does not inhibit lipid biosynthesis in Sesbania [193] and further experiments must clarify the effect of dalapon on lipid metabolism. Aqueous solutions are instable and corrosive against iron. It is readily broken down by soil microorganisms [92].

The sodium salt (dalapon-sodium) was introduced as a herbicide in 1953 [194]. Among the chlorinated aliphatic carbonic acids the 2,2-dichloropropionic acid is of greatest importance [91].

Dalapon is suitable for preharvest desiccation of potatoes [89]. Trade names are Ded-Weed, Devipon, Dowpon, Gramevin, Revenge and Unipon [412]. The acute oral LD_{50} for rats is 7570 to 9300 mg dalapon-sodium/kg.

The following synergistic mixtures suitable for preharvest desiccation of potatoes are described in the literature: Dalapon in combination with *p*-nitrophenol or ammonium peroxydisulfate or as a ternary mixture of all three compounds [195].

cis-3-Chloroacrylic Acid (Common Name: Chloroacrylic Acid)

Chloroacrylic acid is an unsaturated carbonic acid. This compound is used as sodium-3-chloroacrylate as a leaf herbicide with contact action and was introduced in 1963 [91, 196].

Chloroacrylic acid can be used for defoliating and desiccating, e.g. in potato [197].

The trade name is Prep [90]. The acute oral LD_{50} for rate is 160 to 320 mg/kg. Chloroacrylic acid is of little importance as a desiccant in practice.

2,3,6-Trichlorophenyl Acetic Acid (Common Name: Chlorfenac)

Chlorfenac is a representative of hormone-type herbicides with auxin-activity and acts similarly to the other compounds of this group described later. The compound is primarily absorbed by roots. It resist leaching and is persistent in the soil [92].

The herbicide was introduced in 1965 [198, 199] and is usually applied as the sodium salt.

Chlorfenac is suitable for preharvest desiccation of potato, but it has no great practical significance because of high manufacturing costs. Trade names are Fenac and Trifene. The acute oral LD_{50} for rats is 576 to 1780 mg/kg.

4-Chloro-o-tolyloxyacetic Acid (Common Name: MCPA, Metaxon)

MCPA is readily absorbed by leaves and roots and a systemic hormone-type selective herbicide with auxin activity [91–93]. A characteristic of most auxin-active structures is the free carboxy group. The main reason for herbicidal action of the phenoxy acids appears to be the metabolic stability. The phenoxy acids interact with other plant hormones and affect the biosynthesis of nucleic acids, proteins, and lipids and also influence respiration and photosynthesis. The interactions of compounds with the plant are very complex. In this context it has to be considered that absorption and transport depend on physiological conditions. Mode and mechanism of action have been described in detail by Fedtke [200].

MCPA was first introduced as a herbicide in 1945 and, later, by numerous other firms [91, 92, 201].

MCPA is suitable for desiccation of potatoes to increase the effect of a subsequent treatment with Reglone whereby MCPA is used for pretreatment [202]. Here, MCPA causes accelerated ripening via increased production of ethylene. The practical use of MCPA for desiccation and defoliation is strongly limited by its residue problems.

Trade names are Agroxone, Chiptox, Ded-Weed, Empal, Kilsem, Mephanac, Phenoxylene Plus, Rhomene, Rhonox, Shamrox, Weedar, Weedone, Weed-Rhap and Zelan [91,412]. The acute oral LD_{50} for rats is 700 mg/kg.

2,4-Dichlorophenoxy Acetic Acid (Common Name: 2,4-D)

2,4-D is the most important growth regulator herbicide with auxin-activity. It is a leaf herbicide with systemic effect and phloemmobile. 2,4-D causes, among other symptoms, disorders of water conservation in the plant. Differences in absorption, transport and metabolism are decisive for selectivity. Detailed descriptions on mode and mechanism of action are to be found in Fedtke [200] and Gräser [203].

The potent effects of its salts on plant growth were first described in 1942 [204] and the compound was introduced as a herbicide in 1944 [205]. There are various esters, salts and mixtures with other herbicides, which have been marketed by many companies over the past 40 years [92].

Using 2,4-D for desiccation of potatoes, preferably mixed with super phosphate, copper, or manganese salts, is based on accelerated ripening by an increase of the ethylene production in treated plants [89]. Defoliation of cotton by 2,4-D resulted in a positive influence on quality [15]. Desiccation of red clover seed plants with 5 per cent 2,4-D-amine increased the 100-seed weight, seed yield and germinability of seeds [47].

Trade names are Acme Amine, Agrotect, Amoxone, Aqua-Kleen, Chloroxone, Hedonal and many other more[1]. The acute oral LD_{50} for rats is 375 mg/kg.

2,4,5-Trichlorophenoxy Acetic Acid (Common Name: 2,4,5-T)

2,4,5-T acts similarly to 2,4-D [200,203]. It was first described as a herbicide in 1944 and introduced the same year [205].

Pre-treatment of potato haulm with 2,4,5-T increases the effect of the subsequent treatment by Reglone [202].

Trade names are Dacamine, Ded-Weed, Farmco Fence, Rider, Forron, Inverton 245, Line Rider, T-Nox and U 46 [412]. The acute oral LD_{50} for rats is 300 to 1700 mg/kg, depending on vehicle and rat strain [92]. The compound 2,3,7,8-Tetrachloro-dibenzo-*p*-dioxine contained in the technical product of 2,4,5-T has teratogenic properties and is highly toxic (acute oral LD_{50} for rats 0,002–0,045 mg/kg). For this reason the application of 2,4,5-T in agriculture is disputed and has been restricted in some countries.

7-Oxabicyclo(2,2,1)Heptane-2,3-Dicarboxylic Acid (Common Name: Endothal)

Endothal is a soil and contact herbicide which accelerates ripening. In treated plants it induces the production of ethylene and promotes ethephon induced abscission

[1] about 60 trade names are known and cited in [412]

synergistically [206]. Endothal inhibits lipid biosynthesis and causes changes in the permeability of membranes.

The herbicide was first described [208] and introduced in 1951 [91]. It is used as a desiccant in lucerne and potato and for defoliation of cotton and edible dry beans [23,92,154]. The use of Endothal advances the date of harvest of edible dry beans (*Phaseolus vulgaris* L.), by reducing the moisture content of the pods [23].

Trade names are Accelerate, Aquathol K, Des-i-cate, Endothal Turf Herbicide, Herbicide 273, Hydrothol 191 and Hydrothol 191 Granular [412]. The acute oral LD_{50} for rats is 51 mg/kg.

Endothal is primarily used as an additive with other defoliants, e.g. Def, Folex, Dropp or in combination with arsenic acid, paraquat or sodium chlorate to produce desiccation [1]. Application of ametryne (1.6 to 1.8 kg/ha) followed by endothal (0.5 to 0.6 kg/ha) 6 days later resulted in better potato vine desiccation than when the compounds were used alone in considerably higher rates [154]. The effect of endothal can be further increased by adding ammonium sulfate [209].

3.1.2.4 Quarternary Ammonium Compounds

It has been known for a longer time that quarternary ammonium compounds are cell poisons. Besides their bacterial and fungicidal effect, they often act as growth regulators and are strongly phytotoxic. It was discovered in 1954 that some quarternary bipyridylium salts where alkylation results in a new 6-ring have an especially high herbicidal effect [90,210]. To the compounds which have gained great significance as herbicides, desiccants and defoliants belong 1,1′-ethylene-2,2′-bipyridylium dibromide and 1,1′-dimethyl-4,4′-bipyridylium dimethyl dichloride or dimethyl sulfate. Coplanar configuration is required for herbicidal activity of these compounds [211]. Both compounds interfere with photosynthetic electron transport by absorbing one electron respectively and become stable free radicals with an oxidative effect, mediated by the photosynthetic pigment system I. The electron transport system therefore is not blocked as with the usual photosynthesis-inhibiting herbicides (see p 218–220), but the electrons are deviated from their normal path. The reoxidation of the free readicals induces, by means of the formation of peroxidic radicals or of hydrogen peroxide, some secondary reactions resulting in disorder or destruction of cell structures and functions, e.g. to an increase in the permeability of membranes and to a lipid peroxidation and all these effects together cause the strong phytotoxicity. The mentioned dipyridylium compounds are contact herbicides with some systemic properties. They act rapidly in full sunlight, but very little translocation out of the treated leaves takes place under these conditions. At low light intensities or in the dark these herbicides are translocated out of the leaves. Therefore, at low light intensities the herbicidal activity lasts longer and the toxic systems are more evenly spread [212,213].

The photosynthetic system is the most important system which reduces these herbicides, thereby initiating damage in the tissues but other biochemical processes in tissues other than the photosynthetic ones may also activate these herbicides and generate damage, although much more slowly and at higher concentrations [161,214–219]. A comprehensive survey of the bipyridylium herbicides has been published by Summers [220].

1,1'-Ethylene- 2,2'-Bipyridylium Dibromide (Common Name: Diquat, Deiquat)

Diquat was first described and introduced as a herbicide in 1958 [221,222]. It is rapidly absorbed by plant tissues which are green and which are killed on exposure to light. It is inactivated upon contact with the soil [174,223,224].

Diquat, because of its rapidity and its reliability is highly suitable as a desiccant and defoliant and is used especially for preharvest treatment of potato [39,41,89,176,226,227]; sunflower [21,52,53,55,228]; beans, lupine and pea [19,229]; crops for seed production, e.g. trefoil [44,45,47,230]; fodder beet [49]; cucumber [231]; Weyrich joint-weed [111]; *Silybum marianum* fruits [232]; sugar beet [43,46,233]; fodder grass [234]; grain sorghum [63]; seed corn [64]; rice [57,122]; rape [235]; onion [76]; and flax [79]. Besides the advantages produced by desiccation already described (see p 200–203) disadvantageous effects also sporadically occur. In potatoes the fast death of the haulm, especially under dryness, leads to damaged tubers and necroses or stem-end rot [89]. Furthermore a reduced resistance of tubers towards *Phytophthora infestans* after being treated with diquat has been described [98]. Treatment of onion with diquat retards dewatering of the bulbs, increases their water content after harvest, and therefore increases postharvest infestation by gray mold neck rot caused by *Botrytis allii* and also causes sprouting [76]. Preharvest treatment of beans (*Phaseolus vulgaris* L.) with diquat causes reduced yields and seed staining [236].

Trade names are Aquacide, Dextrone, Reglone, Reglox and Weedtrine-D [412]. The acute oral LD_{50} for rats is 231 mg/kg.

The combination of Reglone with other agents for decreasing the dosage of Reglone causes, on the one hand, a reduction of costs and a synergistic improvement of the effect on the other. Reglone can be suitably combined with magnesium chlorate [42,48], ethephon [48,237], urea [238,239], buminafos [176], polyvinyl acetate [24]. Simultaneous application of photosynthesis-inhibiting herbicides, e.g. bromacil, monuron, or diuron, synergistically increases the herbicidal activity and duration of action by diquat [240]. Adjuvants and wetting agents [241,242] also increase the effect of diquat.

1,1'-Dimethyl-4,4'-Bipyridylium Dichloride or Dimethyl Sulfate (Common Name: Paraquat)

Chemical properties as well as mode and mechanism of action of paraquat are very similar to diquat (see p 224–225). The herbicidal properties of the dichloride and dimethyl sulfate were first described by Brian et al. [221]. Both salts were introduced in 1960 [91–93].

Paraquat is very suitable for preharvest desiccation of cotton [6,7,10,12,13,15], soya [18,22,96], rice [1,57–59,62,122], potato [243,244] and crops for seed production, e.g. spinach [245], cucumber [232]. High efficiency is advantageous partly exceeding diquat, reliability and the increased quality of crop yields. In a few cases, however, disadvantages caused by paraquat were described, including reduced yields, lowering of germination of seeds, and the incidence of fungal diseases [18,78,96].

Trade names are Actor, Dextrone, Esgram, Gramoxone and many others more [412]. The acute oral LD_{50} for rats is 150 mg paraquat ion/kg.

Further improvement of the effect can be achieved by combination of paraquat with Def [13], Folex [13], ethephon [247], endothal [1] or synergistically with some thiocyanates [246].

Dipyridylphosphate (DPP)

DPP (1,1′-dimethyl-4,4′-bipyridylium dimethylphosphate) belongs to the few quarternary heterocyclic ammonium compounds which have gained practical significance as desiccants and defoliants with a similar mode of action to diquat and paraquat [248].

DDP is suitable for desiccation or defoliation of cotton, soya, lupine [249], sunflower [21], potato [89] and crops for seed production e.g. sugar beet [233, 249, 250], rutabaga [250], trefoil [44], and alfalfa [249].

DPP is less toxic than paraquat and the desiccation of potato haulms was better than that of paraquat in some trials [89].

To improve its efficiency, DPP was used in combination with magnesium chlorate [42].

3.1.2.5 Heterocyclic Compounds

1H-1,2,4-Triazol-3-ylamine (Common Name: Amitrole)

Amitrole is a non-selective herbicide and is rapidly absorbed by leaves and roots and systemically translocated. The compound inhibits carotenoid synthesis as well as regrowth of buds and leads to disturbance of chloroplast genesis and function. Thereby neutralizing the protection against photooxidative damage that is given by carotene when it inactivates the singlet oxygen inevitably produced during photosynthesis [251, 252]. With high light intensities bleaching of treated plants occurs. Some other described effects of amitrole, e.g. interference with amino acid and purine metabolism in intact higher plants is not sufficiently well known as yet. Amitrole persists for a long time within the plant, but in the soil, however, it has low persistence [253–256].

For the herbicidal properties, first reported in 1953 [257], the non-substituted C-5-atom is essential [258]. The herbicide was introduced in 1954 [91].

Amitrole is suitable for defolation, e.g. in cotton [91, 259].

Trade names are Amerol, Amizol, Azolan, Cytrol, Weedazol and others more [412]. The acute oral LD_{50} is 1100 to 24600 mg/kg.

Its activity is enhanced by the addition of ammonium thiocyanate, probably by blocking the conjugation of amitrole with serine or phosphoserine [260, 261]. Phenyl urea herbicides, e.g. monuron and diuron as well as triazines result, in combination with amitrole, in synergistic mixtures [262].

2-Ethylamino-4-Isopropylamino-6-Methylthio-1,3,5-Triazine (Common Name: Ametryne)

Ametryne is water soluble, is absorbed by leaves and roots and translocated systemically [91]. Among the s-triazines, ametryne belongs to the compounds which exhibit the strongest inhibitory effect on the photosynthetic electron transport, on cyclic phosphorylation and respiration. Ametryne also impedes the hypocotyl hook opening in cotton and stimulates the germination in lactuca. It is assumed that the herbicidal effect is caused, in the first line, by impeding photosynthesis whereby the existence of a photorespiratory pathway might contribute to a more rapid and more pronounced expression of phytotoxicity. However true physiological resistance by selective detoxification exists, e.g. in corn plants by the formation of glutathione conjugate in the resistant plant, but also by other mechanisms of degradation and resistance [263–266]. Ametryne is absorbed relatively quickly in the soil depending on the type of soil and living conditions for microorganisms in the soil.

Ametryne was first described as a herbicide in 1960 [267] and introduced in 1962 [91].

The compound is used for desiccation of potato [268] and soya [18].

Trade names are Ametrex, Doruplant, G-34162, Gesapax and Trinatox D [412]. The acute oral LD_{50} for rats is 1110 to 1405 mg/kg.

Combinations of ametryne with buminafos are suitable for the desiccation of potato [89]. Dinoseb, endothal, or tetrathiin produce, with ametryn, synergistically acting mixtures when used as a split application [154]. Synergistic effects of mixtures of ametryne and chloroacetic acid have also been described [184].

2-n-Butylmercapto-Benzothiazole

Herbicidal properties have been described for some thioether of mercaptobenzothiazoles [269]. 2-*n*-Butylmercapto-benzothiazole induces the release of ethylene in treated plants and is of practical significance as a defoliant of cotton in the USSR.

The trade name is Butylcaptax. It is also used in combination with Hydrel [270].

1-Phenyl-3-(1,2,3-Thiadiazol-5-yl) Urea (Common Name: Thidiazuron)

Thidiazuron is a plant growth regulator, induces the formation of ethylene and promotes the abscission of leaves. Leaves of the treated plant are dropped while they are still green. Thidiazuron was introduced as a plant growth regulator in 1976 [92].

The compound is recommended for defoliation of cotton [16,271,272]. However, in some cases the effectiveness of thidiazuron for defoliation of cotton is insufficient [11].

The trade name is Dropp. The acute oral LD_{50} for rate is >4000 mg/kg.

Combinations of thidiazuron with various chemical compounds, including known pesticides, result in synergistically acting mixtures [273–290].

2,3-Dihydro-5,6-Dimethyl-1,4-Dithi-ine-1,1,4,4-Tetraoxide (Common Name: Dimethipin)

Dimethipin is systemically translocated in the plant. The earliest detectable biochemical effect of dimethipin in treated plants was an inhibition of the incorporation of ^{14}C-leucine into protein. It is supposed that this initial inhibition of protein synthesis is responsible for the loss of stomatal control, associated with high rates of transpiration and subsequent desiccation [291]. Dimethipin is an ethylene-generating compound, it accelerates senescence, enhances maturation, and reduces seed moisture content at harvesting in various crops. In comparison to diquat, dimethipin acts more slowly. The slow effect positively influences formation of tuber skin in potato and leads to a decrease of damage to the tuber [89]. The content of residues is low.

The plant growth regulating properties of dimethipin were first described in 1974 [292] and introduced as a desiccant and defoliant at the same time [92].

Dimethipin is suitable for preharvest desiccation of cotton [17,92,292–296], potato [89,174,176,296–298], sunflower, rape, white mustard [56,292,296], bean [291] and rice [92,296].

The trade name is Harvade 5 F. The acute oral LD_{50} for rats is 1180 mg/kg.

A binary mixture of dimethipin and bromoxynil acts synergistically [168].

3.1.2.6 Phosphorus and Arsenic Compounds

Finding pesticidally acting phosphoric acid esters was at first connected to the development of insecticidal and relatively plant tolerant compounds. Later, herbicidal compounds also became known. The herbicidal effect is given either by herbicidal components only developing after hydrolysis of a corresponding compound or is caused by the entire molecule [91].

Tributylphosphorotrithioite (Common Name: Merphos)

Tributylphosphorotrithioite is a water soluble leaf herbicide with systemic effect. The mode of action can be compared with that of *S,S,S*-tributylphosphorotrithioate (see p 228). It was introduced in 1960 [91].

The trade name is Folex. The acute oral LD_{50} for male albino rats is 1272 mg active ingredient/kg.

Folex is suitable primarily for defoliation of cotton [13,92]. Combinations of Folex with paraquat act synergistically [13].

S,S,S-Tributyl-Phosphorotrithioate

S,S,S-Tributyl-phosphorotrithioate is a water-insoluble leaf herbicide with systemic effect. The herbicide penetrates the upper epidermis directly and causes extensive damage to isolated groups of palisade cells. This isolated wounding can be expected to have a number of abscission-promoting effects including induction of wound ethylene production and changes in the content of endogenous growth regulators. In treated cotton leaves, the compound reduces ^{14}C-transport, exhibits reductions in water potential recovery values, stomatal conductance, specific weights, soluble nitrogen content and alternation in free amino acid distribution [299].

It was introduced as a defoliant in 1956 [92,93]. In the first line it is used as a defoliant in cotton [11–14,91,108,299]; but it is also used in fruits [300].

Trade names are Def, De-Green, E-Z-off D, Folex 6 EC, Fos-Fall A, Ortho Phosphat Defoliant [412]. The acute oral LD_{50} for female rats is 325 mg/kg.

Magnesium chlorate [116], thidiazuron [273], metoxuron [177], paraquat [13], or potassium 3,4-dichloroisothiazole-5-carboxylate [300] produce, with *S,S,S*-tributyl phosphorotrithioate, synergistic mixtures. Adding adjuvant Herbac, a polymeric polyhydro acid, increases the effect of Def [301].

2-Chloroethylphosphonic Acid (Common Name: Ethephon)

Ethephon is a growth regulator and acts by releasing ethylene within plant tissue. It is stable at $pH < 3$, decomposing to liberate ethylene at higher pH values. Besides its numerous other growth regulating effects, ethylene promotes ripeness and senescence, and abscission of leaves. Among the ethylene releasing substances ethephon is the most significant [92].

It was introduced as a plant growth regulator in 1968 [93]. Ethephon is suitable for preharvest ripening of fruit and vegetables and for preharvest defoliation of cotton [1,

[6,92], potato [89,302], vine [231], onion [78], and crops for seed production, e.g. trefoil [45] and cucumber [231].

Trade names are Camposan, Cerone, Ethrel, Flordimex, Florel and Prep [207,412]. The acute oral LD_{50} for rats is 4229 mg/kg. Ethephon can be used advantageously for preharvest defoliation of various crops in combination with potassium iodide [23], dinoseb [153], diquat [236,237] and paraquat [247]. There were also described synergistic effects of binary mixtures of ethephon and chloroacetic acid [184] or *N*-methyl-2-pyrrolidone [303].

N-(phosphonomethyl)glycin (Common Name: Glyphosate)

Glyphosate is a non-selective herbicide which is absorbed by foliage and translocated throughout plants. The agent is stable in higher plants, but rapid metabolism occurs in the soil [92]. Glyphosate is an inhibitor of the enzyme 5-enolpyruvylshikimate-3-phosphate synthase and inhibits therefore the biosynthesis of aromatic amino acids. Aromatic amino acids fulfill two equally important roles in plant metabolism: forming amino acids and serving as starting compounds for phenylpropanoid metabolism. The latter role is of extreme quantitative importance in the case of phenylalanine, since the corresponding phenylpropanoid components include many pigments and the cell wall polymer lignin. It was found that the activity of the key enzyme of phenylpropanoid metabolism, phenylalanine amonia lyase (PAL) is strongly increased by glyphosate [304]. Ammonia originating at desamination of phenylalanine by PAL might be the cause of the phytotoxic effect of glyphosate [305,306]. Further investigations assume that inhibition of the biosynthesis of aromatic amino acids is not the only target for glyphosate. Glyphosate may also reduce carotenoid and chlorophyll synthesis [307] and transpiration [308] possibly as a consequence of the inhibited protein synthesis. Furthermore, the disturbance of mitotic chromosomes of plant cells by glyphosate is described [309,310]. On an ultrastructural level, disruption of the chloroplast envelopes, swelling and a progressive destruction of the endoplasmatic reticulum have been observed [311,312]. Another indirect effect is the increased production of ethylene and the increase of cellulase activities in the abscission zone of seedlings [313].

Glyphosate was first described as a herbicide and introduced in 1971 [92,93].

It is suitable for preharvest desiccation of soya [22,96], cow pea (*Vigna unguiculata*) [314], rice [57], flax [80–82], and cucumber for seed production [231].

Trade names are Glifinox, Glycel, Rodeo, Rondo and Roundup for the mono (isopropylammonium) salt [412]. The acute oral LD_{50} for rats is 5600 mg/kg.

Adding ammonium nitrate or phospholipids lead to a noticeable improvement of the effect of glyphosate [315]. Binary mixtures of glyphosate and ammonium chloride or sulfate act synergistically [316,317]. There are however, findings according to which ammonium sulfate does not increase the effect of glyphosate [318].

Ammonium-DL-Homoalanin-4-yl(Methyl)Phosphinate (Common Name: Glufosinate-Ammonium)

Glufosinate-Ammonium is a non-selective contact herbicide. It is partially systemic, i.e. it is translocated within leaves but not from them [92]. On the other hand there are some evidences for glufosinat-ammonium in the potato tubers after leaf application which indicates phloem transport [414]. A short time after application glufosinate-ammonium inhibits photosynthesis and the activity of glutamine synthetase. Thereby accumulation of ammonia in the plant tissue occurs and cell membranes are damaged. As production of ammonia is closely correlated with the photosynthetic electron transport the effect of glufosinate-ammonium is better in full day light than in shadow or in darkness [319,320]. Glufosinate-ammonium is a stable chemical compound, but is, however, rapidly decomposed in the soil mainly by microbial activity, whereby primarily 3-methyl-phosphinicopropionic acid and finally phosphoric acid, carbon dioxide and water are formed [92,319–321].

The herbicide glufosinate-ammonium was described and introduced in 1981 [322].

The suitability of glufosinate-ammonium for preharvest desiccation in potato, rape, legumes for seed production and sunflowers is currently being tested [320]. For chemical desiccation of cucumber, as a method for rapid screening of breeding lines and for simulated once-over harvest, glufosinate-ammonium had already been succesfully used [321].

Trade names are Basta, Buster, Finale and Ignite [412]. The acute oral LD_{50} for amounts to 1620 to 2000 mg glufosinate-ammonium/kg [323].

Di-n-butyl-1-n-Butylamino-Cyclohexyl-Phosphonate (Common Name: Buminafos)

Buminafos is a non selective contact herbicide. It is absorbed by leaves and not translocated significantly within the plant [324–326]. The biochemical mode of action is unknown.

Buminafos was first described as a herbicide in 1971 [327] and introduced as defoliant and desiccant in 1972 [328].

Buminafos is suitable for desiccation and defoliation of cotton [328], potato [89,326], onion [76,77], sugar cane [328] and various crops for seed production [326].

The trade name is Trakephon. The acute oral LD_{50} for rats is 5110 mg/kg.

Buminafos and diquat are an advantageous binary mixture [176]. Synergistic effects of mixtures of buminafos and chloroacetic acid [185], bromoxynil [169] or azides and *p*-nitrophenol [329] are described.

Hydroxydimethylarsine Oxide (Common Name: Dimethylarsinic Acid)

The monosodium salt of dimethylarsinic acid is a non-selective contact herbicide. By application of dimethylarsinic acid as a foliar spray the compound is absorbed via the leaves and was found in roots, leaves and reproductive structures [330]. It becomes virtually inactive on contact with the soil [92].

The herbicidal properties were first described in 1959 [331] and the herbicide was introduced in 1962 [92].

Dimethylarsinic acid is very suitable for defoliation and desiccation of cotton [7,9,15,91–93,330].

Trade names are Ansar, Bolate, Bolls-Eye, Bophy, Dilic and Phytar [412]. The acute oral LD_{50} is 1350 mg/kg.

3.2 Chemical Compounds Proposed as Desiccants and Defoliants

The patent literature describes numerous desiccating and defoliating chemical compounds. It is very difficult to evaluate these compounds for agricultural use, because

Table 2. Chemical compounds as potential desiccants and defoliants

Chemical name	Refs.
Inorganic compounds	
Silicates, Carbonates, Phosphates	125)
Phosphoric acid	332, 333)
Sulfuric acid	334, 335)
Peroxodisulfuric acid, Peroxodisulfate	336)
Organic compounds	
Phenols	
2,5-Dichloro-4-nitrophenols and derivatives	337)
2,4,5-Trichloro-6-nitrophenol	338)
2-*tert*-Butyl-6-chloro-5-methyl-4-nitrophenol	339)
4-Benzyl-2-chloro-6-methylphenol or *o*-Cresol and other chlorination products of *o*-Cresol	340)
Tetrahydronaphth-1-yl-2-phenols	341)
Diphenylether	
substituted Diphenylethers	342)
2,6-Dichloro-3′-fluoro-4-fluoromethyl-6′-nitro-diphenylether	343)
2-Nitro-5-(4-trifluoromethylphenoxy)toluene	344)
Phenoxybenzaldehydeacetales	345)
N-Aroxyphenyl-*O*-methyl hydroxylamines	346)
Phenoxyphenylazolylalkylethers	347)
Carbonic acid derivatives	
3-chloro-4-methoxyphenyl-N,N-dimethyl urea	415)
Phosphonomethylurea derivatives	348)
substituted Phenylsulfonylureas	349)
1,3-Bis(3,4-dihalo-5-oxafuran-2-yl)ureas	350, 351)
Phenoxyacylurethanes	352)
Dicyclohexylcarbodiimide	416)
Imidodithiocarbonates	353)
Arylisothiocyanates	354)
Carboxylic acids and derivatives	
Dihalophenoxypropionic acids and derivatives	355)
Phenoxyphenylpropionic acid derivatives	356)
2-Thioureidopropionic acid derivatives	357)
Chloroformamidines	358)
N,N-Dimethylformamidine	359)
N-Acylmethylchloroacetanilides	360)
Quarternary ammonium compounds	
Benzylammonium chlorides	361)
Trialkyl benzylammonium chlorides	362)
Trialkyl benzylammonium dithiocarbamates	363)
Trialkyl ammonium thiocyanates	364)
N,N′-Bis-bipyridylium methylcarboxylates	365)
4-Pyridinium quinazoline derivatives	366)
Heterocyclic compounds	
Pyrrolidin-2-one	367)
1-Phenyl-2-oxo-pyrrolidine-4-carboxylic acid	368)
5-Amino-4-heterocyclyl-1-phenylpyrazoles	369)
1,2,3-Triazolecarboxylic acid amides	370)
1,2,3-Thiadiazole-5-carboxamide derivatives	371)

Table 2. (continued)

Chemical name	Refs.
1,2,3-Thiadiazol-5-yl urea derivatives	372, 373)
1,2,3-Thiadiazolin-2-ide derivatives	374)
1,2,3-Thiadiazol-3-in-5-ylidene urea derivatives	375)
Quinolines	376)
2,4-Diamino-6-halogenylalkylpyrimidine	417)
Thienylquinoxalines	377)
1,2,4-Triazin-5-on derivatives	378)
3-Aryl-5,6-dihydro-1,4,2-oxathiazines	379)
Phosphorus compounds	
Aminomethylphosphonic acid	380)
Phosphorylaminophosphonic acid esters and amides	418)
Bis(α-methoxy-β,β,β-trichloroethyl)-β-bromoethane phosphonate	381)
Trichlorophosphazocarbacyl derivatives	382)
Miscellaneous	
3,5-Dichloro-2-hydroxybenzosulfonamide	383)
Petrolium oil	384)
Petrolium sulfonic acids	385)
Water, sodium chloride, blood meal and ascorbic acid	386)

both chemical and biological modes of action have not been investigated sufficiently. Efficiency under field conditions for various agronomic crops cannot therefore be evaluated at present. Finally economic integration of these compounds into an efficient agricultural management system cannot be evaluated. Until now they have not been registered for practical use.

Table 2 summarizes chemical compounds appearing in the literature published over the last few years.

4 The Development of New Chemical Compounds for Desiccation and Defoliation

4.1 Disadvantages and Limitation of Chemical Compounds Recently Applied in Agriculture

In addition to the described chemical compounds (see Sect. 3.1) a number of desiccants and defoliants for various crops is available. They possess, however, a variety of disadvantages which limits their use:

1) For sufficient effect and efficiency under unfavourable weather conditions, high standard application rates are required, as in the case of inorganic compounds. This in addition to the unfavourable chemical and toxicological properties of these compounds often means the danger of extensive damage to the environment. Furthermore, high transport costs arise.

2) High acute toxicity, e.g. dinitrophenoles, and quarternary ammonium compounds, causes for danger to the user and a general danger when compounds drift. This leads increasingly to restrictions in the use of these compounds.
3) High persistence in soil, e.g. inorganic chlorates but also of phenoxyacetic acid derivatives, leads to damage to progeny and limits the use of these agents.
4) In some cases the use of desiccants, e.g. sodium chlorate, Reglone, and Gramoxone leads to disadvantageous effects in harvest products.
5) The use of acids, e.g. formic or sulfuric acid, requires special application techniques because of their corrosive properties and additional precautions to avoid endangering users and the environment.
6) Finally the application of some compounds e.g. Reglone, Gramoxone and ethylene developing preparations is being limited because of economical reasons.

4.2 Mode and Mechanism of Action of Chemical Compounds Suitable as Desiccants and Defoliants

Most of the conventional herbicides inhibit photosynthesis, but only a few of these herbicides are suitable as desiccants or defoliants. Evidently the uncoupling of oxidative phosphorylation or the inhibition of photosynthetic electron transport and the thereby reduced CO_2 assimilation does not correlate with necessarily high desiccation or defoliation. On the other hand some dinitrophenols, bromoxynil and metoxuron act as inhibitory uncouplers, that means they are uncouplers as well as inhibitors of electron transport and not only photosynthesis but also respiration is slowed. It seems that such complex effects lead to a rapid disorder of essential metabolism and thereby to the destruction of the plant.

Of special importance as desiccants or defoliants are photosystem I interfering compounds, e.g. diquat and paraquat, forming reactive free radicals in the plant thereby leading to peroxidation of lipids (see p 224). The lipid peroxidation leads to rapid destruction of essentual membrane structures and functions and, as a consequence, to a drastic desiccative effect. The formation of superoxide radicals after treatment with paraquat is decisive for the phytotoxic effect, as application of copper chelate, which has a superoxide dismutative effect, reduces the damage [388]. In phytotron grown peas, 1% magnesium chlorate caused lipid peroxidation [103]. Experiments with ^{14}C-marked DNOC and dinoterb shows the destruction of lipid membranes by these compounds [389]. Diphenylether [215,390,391] and thiocarbamate herbicides [392] caused, among other effects, the formation of free radicals and lipid peroxidation. The plant possesses, however, with the enzyme system superoxide dismutase (SOD), catalase and peroxidase, a mechanism for the fast detoxification of such reactive free radicals. In paraquat-resistant biotypes of *Conyza linifolia* and *Lolium perenne* an evidently higher chloroplast SOD activity was found compared with that of susceptible types. This increase was accompanied by an increase in catalase and peroxidase was accompanied by an increase in catalase and peroxidase activities in the leaves [393]. Reactive free radicals obviously develop in the plant under the influence of different phytotoxic chemical compounds. They possibly only cause drastic damage if the enzyme system SOD, catalase and peroxidase is either overstrained or blocked. That might mean, that by specific SOD inhibition, so-called weak radical inducing components represent potentially highly efficacious desiccants or defoliants.

Rapid damage of cell membranes as a principle of good desiccative efficiency of chemical compounds is also proved by the efficiency of glufosinate-ammonium and glyphosate. Glufosinate-ammonium inhibits the enzyme glutamine synthetase as well as photosynthesis and thereby induces formation of phytotoxic concentrations of ammonia in the leaves [319]. Glyphosate leads to increased activity of phenylalanine ammonia lyase (PAL), which also causes ammonia to be produced by desamination of phenylalanine [304-306].

The first visible macroscopic effect of a foliar spray of dimethipin on kidney beans was a severe loss of leaf turgor followed by desiccation and abscission. But the higher rates of transpiration and loss of leaf turgor were not due to a massive nonspecific disruption of leaf cells, but by a loss in stomatal control [291]. It is supposed that the initial inhibition of protein synthesis by dimethipin is responsible for the loss of stomatal control, as similar effects are also reached by cycloheximide, a known protein synthesis inhibitor [394]. The finding that the desiccative effect of paraquat but also of 2,4,5-T is linked with the loss in stomatal control underlines this assumption [395]. The effects of foliar application of Def on some physiological events in cotton leaves include reduced water potential recovery values, stomatal conductance, soluble nitrogen content and an alteration in free amino acid distribution [300]. That means that in this case, too, a disturbed protein synthesis can be the cause for a disorder of the normal mechanism of transpiration and the prerequisite for a good desiccative effect. Thus inhibitors of protein synthesis are to seen as potential desiccants.

Auxin or herbicide treatments that promote abscission are correlated with enhanced ethylene production [396-398], and it could be shown that all chemicals that promoted leaf abscission also induced ethylene production [399]. The induction of wound ethylene production therefore often signals potentials defoliant activity. One of the best known chemical compounds which acts by being converted to ethylene and apparently the ultimate defoliant is ethephon. High costs, however, and inconsistency of action limit its use [1]. The commercial cotton defoliants e.g. endothal, thidiazuron and dimethipin, also induce the production of ethylene and have been less expensive than ethephon. In addition, when ethephon is applied under unfavourable conditions, such as cool weather, defoliation is often poor [400]. The action of ethylene in abscission continues only as long the gas is present, low temperatures prevent the induction of abscission by ethylene and a minimum exposure from 12 to 24 hours is needed before ethylene causes abscission [401,402]. Furthermore, defoliants may contribute to the abscission process under unfavourable conditions via effects on water loss, reduction in auxin levels, promotion of abscisic acid synthesis, and senescence [1].

Besides searching for other chemical compounds forming ethylene in the plant or inducing the evolution of ethylene, the defoliating effect can be enhanced by stabilizing the ethylene evolution activity of ethylene-releasing compounds. Examples are the use of chelating agents [403] or cyclodextrin complexation of ethephon [404].

The existing correlation between auxin induced ethylene production and supported abscission lead to the auxin-transport-abscission-hypothese, according to which inhibitors of auxin transport increase the ethylene effect on abscission. Application of inhibitors of the auxin transport e.g. 3,3-α-dihydro-2-(*p*-methoxyphenyl)-8-pyrazolo-(5,1-α)-oxo-indole-8-(1) or morphactine after ethephon treatment of cotton seedlings in deed increases the rate of leaf abscission [405]. With morphactine the effect of phenoxyalkyl carbonic acids, such as MCPA, 2,4-D and 2,4,5-T, can also be increased synergistically [406]. This means that inhibition of auxin transport can be used potentially to increase the defoliating activity of chemical compounds. Although some effects of auxins and ethylene are similar and it was originally suggested that auxins might act via the induction of excessive ethylene production, auxins and ethylene are, however, in many respects, e.g. in growth, abscission and ripening, antagonists rather than synergists [200]. It explains why the effect of ethephon is decreased by 4-chlorophenoxyacetic acid, an inductor of the auxin indole acetic acid, and why the effect is increased synergistically by antagonists of indole acetic acid, e.g. butylcaptax, 2-chlorophenoxyacetic acid or 2,4-dichloroanisole [407].

The analysis of the physiological effect of abscission-promoting compounds showed that e.g. thidiazuron in excised barley leaves or disks from leaves stimulates the activity of ribonucleic acid polymerase, delayed aging and showed thereby cytokinin-like activity [408,409]. *N*(-4-pyridyl)*N*-phenylureas with electron-withdrawing substituents at position 2 and/or 6 of the pyridine ring showed strikingly strong cytokinin activities in the tabacco callus test. Studies with these compounds in a number of plant systems show them to have a wide range of activities based on the appropriate timing and concentration in a large variety of crop plants. The compounds show considerable herbicidal activity and can either induce or prevent abscission according to the concentration used and the plant system to which they are applied [410]. However, cytokinin-like activity seems to be associated with the process of defoliation and should be considered when new defoliants are developed.

Practical use of desiccants or defoliants show combinations of chemical compounds being applied frequently. Proceeding in this way, sufficiently desiccating or defoliating effects are reached in case of the insufficient effect of one component. In addition, disadvantageous chemical properties of a component are reduced. Finally, costs of application are lowered in some cases. The combination of chemical compounds with different mechanisms of action can result in not only an addition of desiccating or defoliating effects but also synergistical ones. In the theoretical combination of chemical compounds one has to consider the mode and mechanism of action, as well as the block of metabolism by inhibition of metabolizing enzymes or enzyme systems.

Normally, desiccants have contact herbicidal properties. Addition of surfactans therefore often leads to an improvement of the efficiency of such compounds by better coverage or retention of agents on the surface of treated plant organs. In some cases the effect can be improved by influencing absorption and transport of agents. Paraquat-resistant biotypes of *Erigeron philadelphicus* and *E. canadensis* take up only 0.5% during 48 hours in comparison to sensitive types [411]. This example shows, that the increase of the uptake and translocation of agents is not achieved by changing the physical-chemical properties of chemical compounds alone but also by influencing uptake and transport mechanisms existent in the plant.

4.3 Evaluation of New Chemical Compounds for Desiccation and Defoliation

A successful method for finding new desiccants and defoliants requires a hierarchically structured screening system incorportating tests for a characterization of biochemical physiological, and complex effects on living organisms.

For the selection of potentially suitable substances, strong phytotoxic or growth regulating effects are required at low concentrations of substances. For determinating such effects callus cultures, leaf disks of etiolated and photosynthetic active plants, excised plant organs, e.g. leaves or roots, and intact seedlings are suitable.

The chemical properties and the known or supposed mechanisms of action of the selected compounds are characterized by their biochemical or physiological mode of action. This includes lipid peroxidation, degradation of leaf pigments, inhibition of activity of nitrate reductase as a consequence of uncoupled oxidative phosphorylation, activity of lipoxigenase correlating to membrane damages, activity of phenylalanine ammonium lyase, activity of glutamine synthetase disorder of the mechanism of transpiration, inhibition of protein synthesis, ethylene evolution, activity of ribonucleic acid polymerase, activity of cellulase, auxin activity, and inhibition of auxin transport. The effects described can normally only occur under in vivo conditions so that mainly intact plants or organs can be used at that test level. This also refers to the detection of synergistical effects on the basis of the inhibition of mechanisms of detoxification.

The following testing includes evaluation of the correlation between biochemical, or physiological effects and the efficiency of desiccation or defoliation. The following statements are to elucidate, when using intact plants or parts of plants under greenhouse conditions: a) dependence of the effect on concentration, b) the highest attainable desiccative or defoliative efficiency, c) chronological course of action, d) mode of action, i.e. occurrence of necroses, burnings, yellowing or aging, and e) finally statements with regard to transport of the compounds in the plant.

The evaluation of new potential desiccants and defoliants under field conditions is ultimately decisive. Essential aspects to be considered are: a) efficiency, b) chronological course of action, c) effect on weeds, d) effect on quality of harvest products, i.e. the influence of the substances on the material composition of harvest products and on the infestation by diseases, e) influence on progeny, f) costs.

Parallel to testing and evaluation of new potential desiccants and defoliants in the way described, the toxicological characterization of compounds is necessary. To that end, the following are necessary: the determination of acute toxicity on different animal organism including long-term experiments, as well as the analysis of residues in plants (especially in harvest products), in water and in soil, and including the peristence.

5 Concluding Remarks

It seems likely that trends will continue and make chemical desiccation and defoliation of many crops shortly before harvesting highly beneficial practices. Modern agriculture includes mechanized harvesting, temporary storage of products during harvesting-processing operations, high quality harvest products, management options to minimize insect and disease problems, minimized risk from weather losses, the need to minimize the use of pesticides, the need to minimize costs, the protection of food supply and the environment. Application of chemical desiccants and defoliants serves that purpose.

Highly efficacious, normally fast acting and less toxic compounds are necessary. Systemic substances seem to be especially suitable for an increase in efficiency and consistency of action. Furthermore, the use of desiccants and defoliants may be improved by adjustment in solvents, surfactants and spray equipment to maximize coverage and penetration through waxy cuticles. Finally, substances are required that can be integrated economically into the agricultural management system of the corresponding crop.

6 References

1. Morgan PW (1985) Chemical manipulation of abscission and desiccation. In: Hilton JE (ed) Agricultural chemicals of the future. Rowman and Allanheld, Totowa, p 61
2. Hall WC, Truchelot GB, Leinweber CL, Herrero FA (1957) Physiol Plant 10: 306–317
3. Poeteren Nvan (1932) Versl Pl Ziekt Dienst Wageningen 66
4. McDowell RK (1935) Scot J Agric 18: 243
5. Hall EE, Harrell FM (1942) cit in: Morgan PW (1985) Chemical manipulation of abscission and desiccation. In: Hilton JE (ed) Agricultural chemicals of the future. Rowman and Allanheld, Totowa, p 62
6. Singh K, Tripathi HP (1977) Indian J Agron 22: 127–129
7. Cathey GW (1979) Agron J 71: 505–508
8. Imamaliev AJ, Turakhodzhaev TJ, Barjetas PK (1980) Khlopkovodstvo 8: 23–24
9. Brar ZS, Singh M (1983) Indian J Ecol 10: 254–259
10. Nagwekar SN, Pai SD, Singh V, Kaairon MS (1985) Haryana Agric Univ J Res 15: 309–312
11. Zufarova M, Lozovaya V, Marapov L, Okhotnik S (1985) Khlopkovodstvo 10: 31–33
12. Turakhodzhajev TJ, Arslanov N (1986) Khlopkovodstvo 6: 22–23
13. Del C Laca-Buendia JP, Ferreira L (1979) Planta Daninha 2: 102–106
14. Yusbashyan O, Babajev T (1980) Khlopkovodstvo 7: 26–27
15. Brar ZS, Singh M (1985) Indian J Ecol 12: 85–90
16. Hopkins AR, Moore RF (1980) J Econ Entomol 73: 768–770
17. Romander L (1981) Agrichem Age 25: 58–59
18. Whigham DK, Stoller EW (1979) Agron J 71: 630–633
19. Betts MF, Morrison IN (1980) Can J Plant Sci 60: 1115–1121
20. Buzmakov VV, Leontev SF (1979) Zernovoe Khoz 9: 39–40
21. Vasiliev DS, Degtyarenko VA (1980). In: Pukhalskii AV (ed) Semenovod Tekhnol Vozdelyvaniya Tekh Kult. Izd Kolos, Moscow, p 194
22. Andreoli C, Ebeltoft DC (1979) Pesqui Agropecu Bras 14: 135–140
23. Forbes JJ, Pratley JE (1983) Aust J Exp Agric Anim Husb 23: 426–428
24. Lipinskii FB, Zavetskii MY (1983) Khim Selsk Khoz 6: 41–43
25. Oskarev VI, Gureeva M, Mikhaleva VP (1984) Khim Selsk Khoz 3: 49–50
26. Scholz B (1982) Kartoffelbau 7: 232–236
27. Brazda G (1971) Feldwirtschaft 12: 311–313
28. Grassert V, Vogel J, Heerkloss B (1978) EAPR-Proc 7th Triennial Conf Warsaw, p 314
29. Talburt WF, Smith O (1967) Potato processing, Westport, Connecticut
30. Grison C (1981) La pomme de terre francaise 404: 129
31. Henrikson JB, Ostergaard SP (1984) T Planteavl 88: 571
32. Grassert V, Vogel J (1981) Nahrung 25: 321–327

33. Wilcockson SJ, Griffith RL, Allen EJ (1980) Ann Appl Biol 96: 349–353
34. Ullrich G (1966) Wiss Z Humboldt Univ Berlin, Math Naturwiss Reihe 15: 211
35. Brazda G (1972) Dissertation, Akad Landwirtschaftswiss. DDR, Berlin
36. Gall H, Truckenbrodt M (1974) Feldwirtschaft 15: 351–354
37. Bouman A, Bouma J (1983) Agric Engineer 38: 25
38. Ambrosov AL, Bardyshev MA, Zagurskaya LE, Shchutskaya AL (1979) Dokl Akad Nauk BSSR 23: 746–748
39. Barskaya TA, Prokofeva LP (1978) Zashch Rast 8: 29–30
40. Kurilov VI, Blagirova TD (1978) Zashch Rast 10: 40
41. Bogdanov AI, Koval YV, Buryi VS, Bidnenko LI, Derevyanko SI, Bondar PI, Lobov VP (1979) Fiziol Biokhim Kult Rast 11: 351–357
42. Dolya VS, Pastukh NA (1980) Khim Selsk Khoz 18: 30–31
43. Dolya VS, Bidnenko LI, Buryi VS (1980) Zashch Rast 8: 26
44. Selevicius A, Jundulas J (1980) Khim Selsk Khoz 18: 45–46
45. Izyumov LV, Titova OV (1983) Mezhvuz Sb Nauchn Tr 204: 33–37
46. Dolya VS, Pastukh NA (1981) Khim Selsk Khoz 11: 38–39
47. Ilynshchenko EI, Zakharchenko EN (1982) Sib Vestn Skh Nauki 1: 37–40
48. Kolesnikov VA, Ermolaeva OI (1984) Khim Selsk Khoz 10: 24–26
49. Petrov AV, Fomichev AM (1984) Khim Selsk Khoz 3: 47–49
50. Halterlein AJ, Sciumbato GL, Barrentine WL (1981) HortScience 16: 189–190
51. Gubbels GH, Dedio W (1985) Can J Plant Sci 65: 841–848
52. Bozhko MF (1978) Maslo-Zhir Prom st 12: 9–12
53. Degtyarenko VA, Golovin AV (1979) Zashch Rast 8: 38
54. Kovalchuk VS (1979) Zashch Rast 9: 25–26
55. Anonymous (1982) Zashch Rast 6: 32–34
56. Witkowski W, Ciesielski F, Mrowczynski M, Musnicki C (1985) Mater Ses Nauk Inst Ochr Rosl 25: 337–350
57. Eastin EF (1980) Crop Sci 20: 389–391
58. Eastin EF (1978) Crop Sci 18: 1068–1070
59. Ramanujam T, Nagarajan M, Sethuraman S, Pillaiyar P (1979) Pesticides 13: 40–44
60. Sasidharan NK, Oommen SK, Thomas MJ (1984) Agric Res J Kerala 22: 177–178
61. Karivaratharaju TV, Ramakrishnan V (1985) Oryza 22: 214–218
62. Venkateswarlu B, Rao S, Gajbhiye VT, Agnihotri N (1986) Indian J Plant Physiol 29: 1–13
63. Shcherbakov VY (1982) Khim Selsk Khoz 6: 40–42
64. Naumenko AI, Kirpa NY (1980) Khim Selsk Khoz 18: 38–40
65. Karaman J, Gal T (1985) Mezogazd Kemizalasa Conf, p 152; Chem Abstr (1986) 105: 20377t
66. Nickell LG (1983) Sugarcane. In: Nickell LG (ed) Plant growth regulating chemicals. CRC Press Inc, Boca Raton, Vol. 1, p 185
67. Evans H, Bates JF (1963) Proc Int Soc Sugarcane Technol 11: 298–300
68. Chen JCD, Liu PPD (1965) Sugar J 27: 22–25
69. Arvier AC (1970) Exp Agric 6: 309
70. Alexander AG, Montalvo-Zapata R (1971) Int Sugar J 73: 261
71. Gracheva EG (1981) Sb Nauch Tr 196: 37–40
72. Nikolskaya NA (1981) Sb Nauch Tr 196: 47–51
73. Zhukova GA (1981) Sb Nauch Tr 196: 41–47
74. Nikolskaya NA (1983) Sb Nauch Tr 204: 24–28
75. Titova OV (1983) Sb Nauch Tr 204: 3–15
76. Pospisilova J, Janyska A (1978) Rost Vyroba 24: 879–888
77. Pospisilova J, Janyska A (1978) Rost Vyroba 24: 869–878
78. Gorecki R (1980) Ogrodnictvo 17: 152–153
79. Gubbels GH, Kenashuk ED (1981) Can J Plant Sci 61: 575–581
80. Fraser TW, Courtney AD, Harvey BMR (1982) Ann Appl Biol 101: 533–538
81. Harvey BMR, Crothers SH, Courtney AD (1985) Rec Agric Res 33: 87–94
82. Harvey BMR, Spence L, Crothers SH (1985) Ann Appl Biol 107: 263–270
83. Michna G (1980) Rocz Nauk Zootechn 7: 295–305

84. Tetlow RM (1983) Grass Forage Sci 38: 27–32
85. Cermak B, Vokral M (1977) Agrichemia 17: 255–258
86. Tullberg JN, Minson DJ (1978) J Agric Sci 91: 557–561
87. Pitt RE (1985) J Agric Sci 105: 223–230
88. Nickell LG (1985) Plant growth regulators: in retrospect and in prospect. Seminar plant growth regulators Agric, Tokyo, p 9
89. Brazda G, Ebert HD (1986) Fortschrittsber Landwirts Nahrungsgüterwirts Akad Landwirtschaftswiss DDR 24: 1–56
90. Wegler R, Eue L (1970) Herbizide. In: Wegler R (ed) Chemie der Pflanzenschutz- und Schaedlingsbekaempfungsmittel. Springer, Berlin Heidelberg New York, Vol 2, p 163
91. Wegler R, Eue L (1977) Herbizide. In: Wegler R (ed) Chemie der Pflanzenschutz- und Schaedlingsbekaempfungsmittel. Springer, Berlin Heidelberg New York, Vol 5, p 1
92. Worthing CR, Walker SB (1983) The pesticide manual, Lavenham Press Limited, London
93. Jaeger G (1977) Unkrautbekaempfungmittel (Herbizide). In: Buechel KH (ed) Pflanzenschutz und Schaedlingsbekaempfung. Georg-Thieme, Stuttgart, p 155
94. Fedtke C (1982) Biochemistry and physiology of herbicide action, Springer, Berlin Heidelberg New York, p 190
95. Schumacher JC (1952) US Pat 2 733 982; Chem Abstr (1956) 50: 8980c
96. Cerkauskas RF, Dhingra OD, Sinclair JB, Foor SR, Stephen R (1982) Weed Sci 30: 484–490
97. Nielsen NK (1966) EAPR-Proc 3rd Triennial Conf Zuerich, p 168
98. Henriksen JB (1978) T Planteavl 82: 227–229
99. Johnson BN, Cartsunis LP (1978) US Pat 4 125 396; Chem Abstr (1979) 90: 82151v
100. Bergmann H, Banasiak L, Burth E, Kochmann W, Kramer W, Lyr H, Schoenfeld H, Steinke W, Zanke D, Lang S, Thust U, Wolter G (1984) DD Pat 211 944; Chem Abstr (1984) 102: 19627f
101. Bergmann H, Burth E, Kochmann W, Kramer W, Lyr H, Radzuhn B, Siering G, Steinke W, Thust U, Wolter G, Zanke D (1985) DD Pat 221 058; Chem Abstr (1985) 103: 155835v
102. Bergmann H, Banasiak L, Lyr H, Radzuhn B, Zanke D, Kochmann W, Kramer W, Steinke W, Thust U, Wolter G (1985) DD Pat 225 900; Chem Abstr (1986) 104: 143982y
103. Ovcharov PK, Shevchenko NV (1982) Agrokhimiya 2: 110–112
104. Barjetas P, Faiziev S, Imamaliev AI (1979) Usb Biol Zh 3: 27–30
105. Barjetas P, Faiziev S, Imamaliev AI (1979) Fiziol Rast 26: 161–166
106. Barjetas P, Faiziev S, Imamaliev AI (1978) Usb Biol Zh 5: 21–25
107. Aleev BG, Zakirov TS (1978) Khim Selsk Khoz 16: 54–55
108. Barjetas P, Turakhodzhaev TI, Arslanov N, Faiziev S (1979) Khlopkovodstvo 8: 35
109. Imamaliev AI, Turakhodzhaev TI, Barjetas P (1979) Sov Agric Sci 11: 37–39
110. Botnar LS, Nol Z (1982) Effect of biologically active substances on physiological functions of broad beans. In: Ilvitskii VA (ed) Fiziol-Biokhim Genet Issled Rast. Shtiintsa, Kishinev, p 3
111. Filatov VI, Lashin AI (1985) Timiryazev SKH Akad 1: 56–60
112. Nabiev MN, Imaliev AI, Tukhtaev S, Shammasov RE, Igamberdyev LR, Rubo VM (1979) SU Pat 641 947; Chem Abstr (1979) 104: 116445m
113. Nabiev MN, Sadykov AS, Tukhtaev S, Shammasov RE, Bogdanov SA, Kiselev A, Musaev NY (1985) SU Pat 1 184 500; Chem Abstr (1986) 104: 68202p
114. Kostrova GL, Mulkidzhanyan YI, Teodoranskii VS (1984) SU Pat 1 123 614: Chem Abstr (1985) 102: 91438n
115. Libershtein II, Lungu GI (1981) SU Pat 880 386; Chem Abstr (1982) 96: 138007w
116. Abbasov YM, Nagiev DA (1978) Khim Selsk Khoz 16: 18–19
117. Volovik AS, Trifimec LN, Shnejder J (1984) Zashch Rast 6: 42–44
118. Neubert P, Kulitzscher H, Laue W (1967) Fortschrittsber Landwirts Nahrungsgüterwirts Akad Landwirtschaftswiss DDR 5: 1–95
119. Vasiliev M (1974) Rast Nauk 11: 98–104
120. Muchametova A (1978) Kartofel Ovochi 6: 10
121. Murphy HJ (1968) Am Potato J 45: 472
122. Nagajaran M, Sethuraman S, Pillaiyar P (1981) Malays Agric J 53: 58–63

123. Andrjusina NA, Zemojc AA, Kljukvina JV (1976) VNII Informazii i Tekhniko-Ekonomitsheskich issl po Selsk Khoz, Moscow
124. Anonymous (1984) US Environm Protec Agency Fed Regist 49: 17759–17760
125. Fenn GS (1982) EUR Pat 63920; Chem Abstr (1983) 98: 121367g
126. Pitt RE (1985) J Agric Sci 105: 223–230
127. Schmitt N (1969) Z Pflanzenkrankh Pflanzenschutz 76: 282–290
128. Ahlborg UG (1981) Matpotatis-vaextskydd och kemiska medel, Mark u Vaexter, Uppsala
129. Hoymann WG (1947) Am Potato J 24: 110
130. Larson K (1982/1983) Ann Rep Swed Inst Agric Engineer: 1–35
131. Anonymous (1972) Ministry Agric Fish Food London Potatoes Bull 94: 1–118
132. Saunders AR (1981) Agric North Ireland 56: 19
133. Bach H, Loelke H (1985) Feldwirtschaft 26: 119–120
134. Brazda G, Holst J, Bach H, Loelke H, Uhlig V (1985) DD Pat 224480; Chem Abstr. (1985) 103: 208926i
135. Blaszyk P (1959) Kartoffelbau 10: 131–132
136. Headford DWR (1966) EAPR-Proc 3rd Triennial Conf Zuerich, p 291
137. Remy H (1955) Grundriss der anorganischen Chemie, Geest u Portig KG, Leipzig
138. Dambroth M (1967) Dissertation, Justus v Liebig Univ Giessen, Giessen
139. Maykuhs F (1979) Kartoffelbau 30: 240
140. Maykuhs F (1981) Kartoffelbau 32: 206–207
141. Maykuhs F (1981) Kartoffelbau 32: 310–311
142. Maykuhs F (1984) Kartoffelbau 35: 317
143. Nabiev MN, Shammasov RE, Tukhtaev S, Aleev BG (1979) Khlopkovodstvo 8: 17–18
144. Lin CH, Lin JH, Chang LR, Lin HS (1985) Am J Enol Vitic 36: 114–117
145. Marcovitch (1945) J Econ Entomol 38: 395
146. Crafts AS (1945) Science 101: 417
147. Kerr MW, Wain LR (1964) Ann Appl Biol 54: 441
148. Moreland DE (1969) Inhibitors of chloroplast electron transport: structure-activity relations. In: Metzner H (ed) Progress in photosynthesis research. Laupp Press, Tübingen, p 1693
149. Alsop WR, Moreland DE (1975) Pestic Biochem Physiol 5: 163–170
150. John S, Hilton JL (1973) Weed Sci 21: 477–480
151. Truffaut G, Pastac J (1932) FR Pat 751 855; Chem Zbl (1934) II: 831
152. Lecorre P (1984) La pommes de terre francaise 422: 157
153. Mutch DR, Penner D, Roggenbuck F, Chase RW (1984) Am Potato J 61: 577–586
154. Sieczka JB (1980) Proc Ann Meet Northeast Weed Sci Soc 34: 222–228
155. Poignant P, Crisinel P (1967) C R J Etud 4th Herbic Conf Columna, p 196
156. Anonymous (1979) BE Pat 874 352; Chem Abstr (1979) 91: 205646r
157. Van den Boogart K (1966) DE Pat 1 214 014
158. Wain RL (1963) Nature 200: 28
159. Carpenter K, Heywood BJ (1963) Nature 200: 28
160. Reimer S, Link K, Trebst A (1979) Z Naturforsch 34c: 419
161. Foy CL, Penner W (1965) Weeds 13: 226
162. Van Oorschot JLP (1970) Pest Sci 1: 33
163. Moreland DE, Blackmon WJ (1970) Weed Sci 18: 419–426
164. Katoh S (1972) Plant Cell Physiol 13: 273
165. Van Oorschot JLP (1974) Weeds Res 14: 81
166. Houseworth LD, Tweedy BG (1971) Weed Sci 19; 732–735
167. Figuerola LF, Furtick WR (1972) Weed Sci 20: 60–63
168. Blanco HG, Lobato dos Santos CA, Shiba S (1985) Biologico 51: 57–66
169. Bergmann H, Lyr H, Radzuhn B, Zanke D, Kochmann W, Kramer W, Steinke W, Wolter G, Schluckwerder H, Lezius W, Ende F, Sieberhein K, Wildgrube W (1988) DD Pat 257 379
170. Renger G (1973) Biochem Biophys Acta 314: 113
171. Trebst A (1980) Inhibitors of electron flow: tolls for the functional and structural localization of carriers and energy transformation sites. In: San Pietro A (ed) Methods in enzymology. Academic Press, New York London, Vol 69, p 675
172. Berg W (1968) Pflanzenkrank Pflanzenpathol Pflanzenschutz 75: 233

173. Horvath S (1978) Erfahrungen über die Anwendung von Purivel zur Krautabtötung von Kartoffeln, Univ Keszthely, Keszthely
174. Rozkosova V (1978) Agrochemia 18: 268–270
175. Kravtsov AA, Golyshin NM, Samersov VF, Posunko LA (1981) Zashch Rast 5: 53–55
176. Rozkosova V, Foltyn J (1984) Agrochemia 24: 282–286
177. Eder F, Helfersberger H, Leber JP (1979) DE Pat 2 914 164; Chem Abstr (1980) 92: 35994d
178. Woodford EK, Holly K, McCready CC (1958) Ann Rev Plant Physiol 9: 311
179. Van Overbeek J (1964). In: Audus LJ (ed) The physiology and biochemistry of herbicides. Academic Press, New York, p 392
180. Freed VH, Montgomery VH (1956) Res Prog Rep West Weed Control Conf, p 98
181. Blanchard FA, Muelder WW, Smith GN (1960) J Agric Food Chem 8: 124
182. Hitchcock AE, Zimmermann PW (1952) US Pat 2 622 976; Chem Zbl (1954) 2257
183. Breese TC, Wheeler AFJ (1956) Proc 3rd Brit Weed Control Conf, p 759
184. Bergmann H, Burth E, Kochmann W, Kramer W, Lyr H, Radzuhn B, Siering G, Steinke W, Thust U, Wolter G, Zanke D (1985) DD Pat 221 058; Chem Abstr (1985) 103: 155835v
185. Banasiak L, Bergmann H, Guenther E, Kochmann W, Lyr H, Radzuhn B, Roethling T, Steinke W, Wolter G (1985) DD Pat 225 899; Chem Abstr (1986) 104: 104440a
186. Hilton JL, Jansen LL, Hull HM (1965) Ann Rev Plant Physiol 14: 353–384
187. Holmsen JL, Bidlack HD (1971) Proc Weed Sci Soc Am, No 204
188. Harwood JL, Stumpf PK (1971) Arch Biochem Biophys 142: 281
189. Still GG, Davis DG, Zander GL (1970) Plant Physiol 46: 307–314
190. Barrons KC, Watson AJ (1947) Dow Chem Co Res Rep, North Cent Weed Control Conf, p 43, p 284
191. Anonymous (1985) US Environm Protec Agency Fed Regist 50: 28944–28945
192. Fedtke C (1982) Biochemistry and physiology of herbicide action, Springer, Berlin Heidelberg New York, p 142
193. Mann JD, Pu M (1968) Weed Sci 16: 197
194. Barrons KC (1951) US Pat 2 642 354; Chem Abstr (1953) 47: 8308e
195. Bergmann H, Banasiak L, Kochmann W, Kramer W, Lyr H, Schoenfeld H, Siering G, Steinke W, Zanke D, Lang S, Thust U, Wolter G (1984) DD Pat 211 940; Chem Abstr (1984) 102: 41586p
196. Herrett RA, Kurtz AN (1963) Science 141: 1192
197. Burgova LV, Rudnev GK, Christich AI (1973) Zh Prikl Chim 46: 1529
198. Tischler N (1958) US Pat 2 977 212; Chem Abstr (1961) 55: 13756g
199. Newcomer JS, Weil ED, Dorfman D (1965) US Pat 3 299 120; Chem Abstr (1967) 67: 31855v
200. Fedtke C (1982) Biochemistry and physiology of herbicide action, Springer, Berlin Heidelberg New York, p. 159
201. Slade RE, Templeman WG, Sexton WA (1945) Nature 155: 497
202. Budnik J, Jedrzejowska E, Walczak W (1974) Biul Inst Ziemniaka 14: 29–43
203. Graeser H (1978) Biochemie und physiologie der phytoeffektoren, VEB Deutscher Verlag der Wissenschaften, Berlin, p 331
204. Zimmermann PW, Hitchcock AE (1942) Contrib Boyce Thompson Inst 12: 321
205. Hamner CL, Tuckey HB (1944) Science 100: 154
206. Davis JT, Sterrett JP, Leather GR (1972) HortScience 7: 478–480
207. Wagenbreth D (1981) Fortschrittsber Landwirts Nahrungsgüterwirts Akad Landwirtschaftswiss DDR 19: 1–48
208. Tischler N, Bell EP (1951) Proc 5th Northeast Weed Control Conf, p 51
209. Tischler N, Quimba GP, Bejuki WM (1951) Proc 5th Northeast Weed Control Conf, p. 35
210. Crafts AS (1961) The chemistry and mode of action of herbicides, Interscience Publishers Inc, New York London, p 214
211. Baldwin BC (1969) On the mode of action of bipyridylium herbicides. In: Metzner H (ed) Progress in photosynthesis research, Laupp Press, Tübingen, Vol 3, p 1737

212. Moreland DE (1967) Ann Rev Plant Physiol 18: 365–386
213. Headford DWR (1970) Pest Sci 1: 41
214. Schneider J, Guenther G (1974) Biol Zbl 93: 459–466
215. Prendeville GN, Warren GF (1977) Weed Res 17: 251–258
216. Harris N, Dodge AD (1972) Planta 104: 213–219
217. Davis DG, Shimabukuro RH (1973) Abstr Weed Sci Soc Am Meet, p 71
218. Sutton DL, Foy CL (1971) Bot Gaz 132: 299
219. Bell SL, Schwarz OJ, Hughes KW (1976) Can J Genet Cytol 18: 93
220. Summers LA (1980) The bipyridinium herbicides, Academic Press, London Toronto Sydney San Francisco
221. Brian RC, Homer RF, Stubbs J, Jones RL (1958) Nature 181: 446
222. Fielden RJ, Homer RF, Jones RL (1957) GB Pat 785 732; Chem Abstr (1958) 52: 6707g
223. Dodge AD (1971) Endeavour 30: 130–135
224. Boon WR (1967) Endeavour 26: 27–32
225. Sanderson JB, Ivany JA, White RP (1984) Am Potato J 61: 691
226. Rioux R (1983) Phytoprotection 64: 27–30
227. Roztropovicz S, Przeobrazenski J, Dzienia (1978) Biul Inst Ziemniaka 22: 77–95
228. Gimesi A (1983) Novenyvedelem 19: 497–501
229. Gubbels GH (1982) Can J Plant Sci 62: 555–560
230. Rybak H (1983) Zesz Probl Postepow Nauk Roln 253: 79–87
231. Wehner TC, Monaco TJ, Bonanno AR (1984) HortScience 19: 671–673
232. Zalecki R, Górna A (1985) Herba Pol 29: 217–222
233. Goryachikh AS (1982) Khim Selsk Khoz 6: 42–43
234. Tucholska H, Bartz J (1983) Zesz Probl Postepow Nauk Roln 282: 185–196
235. Karaman J, Gal T (1985) Mezogazd Kemizalasa Conf, p 152; Chem Abstr (1986) 105: 20377t
236. Hole CC, Hardwick RC (1978) Ann Appl Biol 88: 421–427
237. Fuerste K, Schmidt J, Woehlert W, Kochmann W, Steinke W, Kramer W, Kuehnel F (1983) DD Pat 159 140; Chem Abstr (1983) 99: 117869a
238. Bach H, Uhlig V, Seltmann U, Schinkel W, Hermann R, Herles K (1982) DD Pat 157 753; Chem Abstr (1983) 99: 1829h
239. Libershtein II, Lungu GI (1982) SU Pat 936 859; Chem Abstr (1982) 97: 158039g
240. Fedtke C (1982) Biochemistry and physiology of herbicide action, Springer, Berlin Heidelberg New York, p 89
241. Murphy HJ, Morrow LS (1979) Proc Northeast Weed Sci Soc 33: 53–58
242. Rasocha V (1978) Agrochemia 18: 145–149
243. Ahrens JF (1975) Proc Northeast Weed Sci Soc 29: 276
244. Struckmeyer BE, Binning LK (1983) Am Potato J 60: 41
245. Domanski R, Peplinska M (1980) Acta Physiol Plant 2: 209–216
246. Stephenson LW (1981) EUR Pat 38 948; Chem Abstr (1982) 96: 47577r
247. Salinas R, Garciduenas MR (1981) Turrialba 31: 263–267
248. Melnikov NN, Chaskin BA, Sablina IV (1968) Zh Obshch Khim 38: 1558
249. Stonov LD (1976) Zashch Rast 10: 21–22
250. Rykin VK, Kakhadse SP, Kviloriya GI (1978) Khim Selsk Khoz 16: 15–17
251. Krinski NJ (1978) Carotenoid pigments: multible mechanism for coping with the stress of photosensitized oxidations. In: Shilo M (ed) Strategies of microbial life in extreme environments. Verlag Chemie, Weinheim New York, p 163
252. Oequist G, Samuelsson G, Bishop NJ (1980) Physiol Plant 50: 63–70
253. Muzik TJ (1965) Weed Res 5: 207
254. Ercegovich CD, Frear DE (1964) J Agric Food Chem 12: 26
255. McRae IC, Alexander M (1965) J Agric Food Chem 13: 72
256. Riempa P (1962) Weed Res 2: 41
257. Behrens R (1953) Proc North Cent Weed Control Conf, p 61
258. Gomaa EAA (1968) Acta Phytopathol 3: 175–180
259. Miller CS, Hall WC (1957) Weeds 5: 218
260. Carter MC (1965) Physiol Plant 18: 1054–1058
261. Smith LW, Bayer DE, Foy CL (1967) Abstr Weed Soc Am, p 61

262. Graeser H (1978) Biochemie und physiologie der phytoeffektoren, VEB Deutscher Verlag der Wissenschaften, Berlin, p 185
263. Shimabukuro RH, Swanson HR (1969) J Agric Food Chem 17: 199
264. Guddewar MB, Dautermann WC (1979) Phytochemistry 18: 735–740
265. Van Oorschot JLP (1979) Z Naturforsch 35c: 900–904
266. Aigle N (1980) La Defense des Vegetaux 34: 323
267. Gysin H, Knuesli F (1960) Adv Pest Control Res 3: 289
268. Murphy HJ, Morrow LS (1982) Ann Rep Main Potato Com 1981/1982 278: 9
269. Rozkova NK, Bezumnova AN (1968) Khim Geterozikl Soed 2: 376
270. Kadyrov M, Loj NP, Umarov AA (1982) Khim Selsk Khoz 10: 30–32
271. Cruz LSP, Leiderman L, Crossi JMM (1977) Biologico 43: 179–186
272. Imamaliev AI, Turaev M, Akbarov K, Khudaibergenov K (1978) Khlopkovodstvo 40: 81–83
273. Rusch R, Arndt F (1978) DE Pat 2 646 712; Chem Abstr (1978) 89: 101907n
274. Rusch R, Krueger HR (1979) DE Pat 2 745 968; Chem Abstr (1979) 91: 85277x
275. Rusch R, Taylor K (1982) DE Pat 3 116 007; Chem Abstr (1983) 98: 48680x
276. Rusch R, Taylor K (1982) DE Pat 3 116 008; Chem Abstr (1983) 98: 12936w
277. Rusch R, Taylor K (1982) DE Pat 3116009; Chem Abstr (1983) 98: 12397x
278. Rusch R, Taylor K (1982) DE Pat 3116010; Chem Abstr (1983) 98: 12938y
279. Rusch R, Taylor K (1982) DE Pat 3116011; Chem Abstr (1983) 98: 12935v
280. Rusch R, Taylor K (1982) DE Pat 3116012; Chem Abstr (1983) 98: 12928v
281. Rusch R, Taylor K (1982) DE Pat 3116013; Chem Abstr (1983) 98: 12934u
282. Rusch R, Taylor K (1982) DE Pat 3116015; Chem Abstr (1983) 98: 12943w
283. Rusch R, Taylor K (1982) DE Pat 3116016; Chem Abstr (1983) 98: 29691k
284. Rusch R, Taylor K (1982) DE Pat 3116017; Chem Abstr (1983) 98: 12939z
285. Rusch R, Taylor K (1982) DE Pat 3116019; Chem Abstr (1983) 98: 12940t
286. Rusch R, Taylor K (1982) DE Pat 3116020; Chem Abstr (1983) 98: 12941u
287. Rusch R, Taylor K (1982) DE Pat 3116021; Chem Abstr (1983) 98: 12942v
288. Rusch R, Taylor K (1983) DE Pat 3128420; Chem Abstr (1983) 98: 102709d
289. Krueger PJ (1985) US Pat 4552582; Chem Abstr (1986) 104: 104445f
290. Zubkova NF, Grusinskaya NA, Shekhtman LM, Kasyanenko NA, Shakhramov AAC (1986) SU Pat 1066061; Chem Abstr (1986) 105: 166905k
291. Metzger JD, Keng J (1984) J Plant Growth Regul 3: 141–156
292. Ames RB, Brewer AD, McIntire WS (1974) Proc Beltwide Cotton Produc Res Conf, p 61
293. Hegman AR (1982) Proc Beltwide Cotton Produc Mechan Conf, p 104
294. Neidermyer RW, Brewer AD, Judge FD (1974) Proc 12th Brit Weed Control Conf, p 104
295. Sarpe N, Pipie F, Popa G, Florea A, Ciotae V, Pascu M, Dascalu I (1975) Prod Veget Cereale Plant Tech 27: 23
296. Ames RB, Blem AR, Pryzbylek JM, Walz AM, Jackson D (1982) Proc Brit Crop Protec Conf, Weeds, p 563
297. Bell AR, Ames RB, Neidermyer RW, Brewer AD (1974) Proc North Centr Weed Control Conf, p 69
298. Cremaschi D, Vender C (1977) Inst Sperimentale Couture Indust 9: 297
299. Cathey GW, Elmore CD, McMichael BL (1981) Physiol Plant 51: 140–144
300. Kukalenko SS, Volodkovich SD, Martynenko VI (1985) Zashch Rast 11: 46–54
301. Haahn JW, Peugh DD, Skloss LJ (1978) Proc South Weed Sci Soc 31: 83
302. Chrominsky A, Rozej B (1976) Potato Res 1: 123
303. McCarthy RF, Kliegmann JM (1979) DE Pat 2840063; Chem Abstr (1979) 91: 118672c
304. Fedtke C (1982) Biochemistry and physiology of herbicide action, Springer, Berlin Heidelberg New York, p 185
305. Hoagland RE, Duke SO (1982) Am Chem Soc Ser 181: 175–205
306. Bergmann H, Opitz C, Radzuhn B, Sammler P, Schmidt A (1987) Biol Rundschau 25: 81–92
307. Abu-Irmaileh BE, Jordan LS (1978) Weed Sci 26: 700–703
308. Shaner DL (1978) Weed Sci 26: 513–516
309. Boyle WS, Evans JO (1974) J Heredity 65: 250

310. Goltenboth F (1980) Weed Abstr 29: 427
311. Campell WF, Evans JO, Reed SC (1976) Weed Sci 24: 22–25
312. Uotila M, Evjen K, Iversen TH (1980) Weed Res 20: 153–158
313. Abu-Irmaileh BE, Jordan LS, Kumamoto J (1979) Weed Sci 27: 103–106
314. Cole AW, Cerdeira AL (1982) HortScience 17: 244–246
315. Gimesi A (1986) Novenytermeles 35: 319–324
316. Lutman PJW, Richardson WG (1978) Weed Res 18: 65–70
317. Bakuniak E, Borucka B, Kubiak Z (1984) PL Pat 121281; Chem Abstr (1984) 101: 105799z
318. Hampton JG, Hebblethwaite PD (1982) Grass Forage Sci 37: 243–248
319. Koecher H (1984) Mitt Biol Bundesanst Landwirts Forstwirts Berlin-Dahlem 223: 303–305
320. Anonymous (1986) Basta, Technical information Hoechst AG
321. Lindhoud WM (1984) Meded Fac Landbouw Riksuniv Gent 49: 1085–1090
322. Schwerdtle F, Bieringer H, Finke K (1981) Z Pflanzenkrank Pflanzenschutz Sonderheft 9: 431–440
323. Langelueddeke P, Reuss HU, Ceroni C (1982) Meded Fac Landbouw Riksuniv Gent 47: 95–104
324. Kramer W (1973) Nachrichtenbl Pflanzenschutz-Dienst DDR 27: 213–214
325. Richter S (1975) Biochem Physiol Pflanz 167: 173–181
326. Kramer W, Loettge W, Wolter G, Sorge N (1982) Nachrichtenbl Pflanzenschutz DDR 36: 241–245
327. Guenther E, Loettge W (1971) DD Pat 80586
328. Guenther E, Loettge W, Kochmann W, Kramer W (1972) DD Pat 94280
329. Bergmann H, Burth U, Kochmann W, Lyr H, Steinke W, Zanke D (1983) DD Pat 159734; Chem Abstr (1983) 99: 153864d
330. Marcus-Wyner L, Rains DW (1982) Environm Qual 11: 715
331. Mowev RG, Cornman JF (1959) Proc Northeast Weed Control Conf 13: 62
332. Takala MI (1978) DE Pat 2820359; Chem Abstr (1979) 90: 49658t
333. Kosar J (1978) CS Pat 172584; Chem Abstr (1978) 89: 192511z
334. Bach H, Loelke H, Schroeder G (1981) DD Pat 146541; Chem Abstr (1981) 95: 37118g
335. Uhlig V, Bach H, Boettcher H, Loelke H, Schuman F (1984) DD Pat 213127; Chem Abstr (1985) 102: 91440g
336. Matschiner H, Thiele W, Wildner K, Balej J (1981) DD Pat 150226; Chem Abstr (1982) 97: 26388q
337. Kochmann W, Naumann K, Roethling T, Hoelzel H, Loettge W, Wolter G, Guenther E (1980) DD Pat 140405; Chem Abstr (1980) 93: 90197s
338. Barth P, Guenther E, Hoelzel H, Kochmann W, Kramer W, Loettge W, Naumann K, Roethling T, Wolter G (1978) DD Pat 131448; Chem Abstr (1979) 90: 67725s
339. Knops HJ, Luerssen K, Eue L (1979) DE Pat 2815237; Chem Abstr (1980) 92: 53390
340. Everest-Todd S (1985) EUR Pat 155783; Chem Abstr (1985) 103: 191491p
341. Hoelzel H, Creuzburg D, Stohr P, Kaestner A, Biernig H, Kochmann W, Luthardt H, Roethling T (1986) DD Pat 235413; Chem Abstr (1986) 105: 185889z
342. Dickore K, Klauke E, Luerssen K (1979) DE Pat 2745006; Chem Abstr (1979) 91: 56662
343. Klusacek H, Klauke E, Luerssen K, Roessler P (1979) DE Pat 2745005; Chem Abstr (1979) 91: 56616b
344. Anonymous (1983) DE Pat 3141861; Chem Abstr (1983) 99: 53377d
345. Foerster H, Eue L, Schmidt R, Luerssen K (1982) DE Pat 3118371; Chem Abstr (1983) 98: 143395s
346. Duerr D (1986) EUR Pat 166694; Chem Abstr (1986) 104: 206897x
347. Foerster H, Eue L, Schmidt R, Luerssen K (1982) DE Pat 3122174; Chem Abstr (1983) 98: 143411u
348. Large GB, Buren LL (1985) RO Pat 88229; Chem Abstr (1986) 105: 74403s
349. Stetter J, Wroblowsky HJ, Schmidt R, Santel HJ, Haenssler G, Luerssen K (1985) DE Pat 3420769; Chem Abstr (1986) 104: 88598a
350. Jaehnisch K, Seeboth H, Pinon D, Callejas DR (1984) DD Pat 216236; Chem Abstr (1985) 103: 87765v

351. Jaehnisch K, Seeboth H, Pinon D, Callejas DR (1984) DD Pat 216462; Chem Abstr (1985) 103: 22441s
352. Schlee HG, Luerssen K (1980) DE Pat 2851766; Chem Abstr (1980) 93: 204308m
353. Walek W, Creuzburg D, Kochmann W, Lang S, Pallas M, Schulze F, Steinke W, Winkler J, Wozniak H (1985) DD Pat 226179; Chem Abstr (1986) 104: 64220v
354. Schmid W (1978) SU Pat 587835; Chem Abstr (1978) 88: 132030
355. Barton JED, Collins DJ, Headford DWR (1980) GB Pat 2024213; Chem Abstr (1980) 93: 2256g
356. Kirste R, Wegler R, Eue L, Santel HJ, Schmidt R (1985) DE Pat 3419952; Chem Abstr (1986) 104: 168165x
357. Ackermann H, Krueger S, Lexow B, Creuzburg D, Stohr P (1987) DD Pat 243663
358. Michel HJ, Schwarz G, Grabarse M, Land S, Wozniak H, Borchert F, Einicke W, Oertel M, Thieme H, Jasche K (1985) DD Pat 223053; Chem Abstr (1985) 103: 174071j
359. Pedersen EB (1979) Synthesis 7: 546
360. Regel E, Eue L, Schmidt R (1979) DD Pat 137878; Bekanntmach Amt Erfind Patentwesen DDR (1979) 20: 61
361. Roethling T, Creuzburg D, Hornuf A, Lang S, Naumann K, Hoelzel H, Kochmann W, Kranz L, Steinke W, Thust U, Wozniak H, Creuzburg A (1983) DD Pat 215228; Chem Abstr (1985) 103: 2158b
362. Roethling T, Creuzburg D, Lang S, Naumann K, Hoelzel H, Hoh G, Kochmann W, Steinke W, Thust U, Hornuf A, Guenther E, Kranz L, Schiewald W (1984) DD Pat 215688; Chem Abstr (1985) 103: 49772n
363. Roethling T, Creuzburg D, Lang S, Naumann K, Steinke W, Kochmann W, Hoelzel H, Hornuf A, Thust U, Toepfer R, Wozniak H (1984) DD Pat 215686; Chem Abstr (1985) 103: 33501n
364. Makhamatkhanov MM, Avezov K, Maksudova FN (1978) SU Pat 631512; Chem Abstr (1979) 90: 17686j
365. Parham MR, White BG (1979) US Pat 4171967; Chem Abstr (1980) 92: 35989f
366. Holyoke CP (1985) US Pat 4502880; Chem Abstr (1986) 105: 42833r
367. Anonymous (1979) FR Pat 2400013; Chem Abstr (1980) 92: 163836r
368. Bellus D, Fory W, Tobler H (1977) DD Pat 128633
369. Stetter JG, Reinhold LM, Schallner O, Eue L, Schmidt R, Santel HJ, Luerssen K (1986) DE Pat 3423101; Chem Abstr (1986) 104: 186405e
370. Krueger HR, Arndt F, Baumert D, Joppien H, Pierch EA, Rusch R (1980) DE Pat 2834879; Chem Abstr (1980) 93: 71785r
371. Krueger HR, Arndt F, Baumert D, Rusch R (1980) DE Pat 2913977; Chem Abstr (1981) 94: 121547h
372. Krueger HR, Arndt F, Rusch R (1983) DE Pat 3139505; Chem Abstr (1983) 99: 53761z
373. Krueger HR, Arndt F, Rusch R (1983) DE Pat 3139506; Chem Abstr (1983) 99: 38473p
374. Arndt F, Rusch R (1978) DE Pat 2719810; Chem Abstr (1979) 90: 84467n
375. Krueger HR, Kraehmer H, Rusch R, Sjut V (1984) DE Pat 3319008; Chem Abstr (1985) 102: 78895q
376. Kempter G, Sarodnick G, Creuzburg D, Klepel M, Kranz L, Kuehne S, Lehmann H (1978) DD Pat 132752; Chem Abstr (1979) 91: 103732y
377. Clark MT (1980) GB Pat 1574429; Chem Abstr (1981) 94: 203841p
378. Pissiotas G, Meyer W, Schwarze W, Klenk H, Leuchtenberger W (1980) DD Pat 139990; Bekanntmach Amt Erfind Patentwesen DDR (1980) 21: 61
379. Brouwer WG, Blem AR, Bell AR, Davis RA (1984) EUR Pat 104940; Chem Abstr (1984) 101: 38485d
380. Luerssen K, Coelln R (1978) DE Pat 2719721; Chem Abstr (1979) 90: 49656r
381. Hesse B, Kochmann W, Loettge W, Moll R, Pallas M (1978) DD Pat 132632; Chem Abstr (1979) 91: 103731x
382. Shiranov DF, Khomitskii BP, Rudavskii VP (1978) Fiziol Akt Veshchest 10: 18
383. Berg D, Reiser R, Luerssen K (1984) DE Pat 3315593; Chem Abstr (1985) 102: 57833c
384. Phillips CAL, Archer HW (1980) US Pat 4192668; Chem Abstr (1980) 92: 210202t
385. Isaeva FG, Eyubov RE (1978) SU Pat 592400; Chem Abstr (1978) 88: 147507w
386. Caroti JA (1980) FR Pat 2436566; Chem Abstr (1980) 93: 90201p

387. Kidd H, Harley D, Kennedy JM (1986) European directory of agrochemical products, Unwin Brothers Limited, Old Woking (Herbicides vol 2)
388. Youngman RJ, Dodge AD (1979) Z Naturforsch 34c: 1032–1035
389. Pont V, Collet GF (1984) Weed Res 24: 239–247
390. Orr GL, Hess FD (1982) Plant Physiol 69: 502–507
391. Vanstone DE, Stobbe EH (1977) Weed Sci 25: 352–354
392. Ruzo O, Casida JE (1985) J Agric Food Chem 33: 272–276
393. Harvey BMR, Harper DB (1982) Tolerance to bipyridilium herbicides. In: Le Baron HM, Gressel J (eds) Herbicide resistance in plants. J Wiley, New York, p 215
394. Thimann KV, Sattler S (1979) Proc Nat Acad Sci USA 76: 2770–2773
395. Rao JM, Swamy PM, Das VSR (1977) Weed Sci 25: 469–472
396. Abeles FB, Rubinstein B (1964) Plant Physiol 39: 963–964
397. Baur JR, Morgan PW (1969) Plant Physiol 44: 831–838
398. Hallaway M, Osborne DJ (1969) Science 163: 1067–1068
399. Cooper WC, Rasmussen GK, Rogers BJ, Reece PC, Henry WH (1968) Plant Physiol 43: 1560–1576
400. Morgan PW (1973) Proc Beltwide Cotton Prod Res Conf, p 95
401. Beyer EM (1973) Plant Physiol 52: 1–5
402. Beyer EM, Morgan PW (1971) Plant Physiol 48: 208–212
403. Cooke AR, Esposito JE, Fritz CD (1977) S Afric Pat 7602179; Chem Abstr (1978) 88: 17325z
404. Tetenyi M, Szejtli J (1984) Acta Agron Acad Sci Hung 39: 345–348
405. Morgan PW, Durham JI (1973) Planta 110: 91–93
406. Skrabka G, Gschis E (1988) Acta Agrobot Warszawa (in press)
407. Zubkova NF, Stonov LD, Bokarev KS (1977) Khim primen pestits prep. In: Kondratev YA, Promonenkov VK (eds). NIITEKHIM, Moskow, p 50; Chem Abstr (1980) 93: 126958g
408. Zubkova NF, Bukashkina ZV, Sharenkova KA, Kulaeva ON (1983) Dokl Akad Nauk SSSR 272: 1011–1013
409. Kulaeva OM, Baskakov YA, Borisova NN, Kuznetzov VV, Tsibulya LV, Shapovalov AA (1982) Fiziol Rast 29: 266–273
410. Okamoto T, Shudo K, Takahashi S, Kawachi E, Isogai Y (1981) Chem Pharm Bull 29: 3748–3751
411. Tanaka Y, Chisaka H, Saka H (1986) Physiol Plant 66: 605–608
412. Sine C (1988) Farm chemicals handbook, Meister Publishing Co, Willonghby (vol 74)
413. Fedtke C (1982) Biochemistry and physiology of herbicide action, Springer, Berlin Heidelberg New York, p 49
414. Schmutzler K (1988) Hoechst AG, personal communication
415. Angelova J, Hadjiivanova Z, Baneva P (1987) Abstr Posters Communic 31th Int Congr Pure and Appl Chem Sofia 2: 44
416. Michel HJ, Dittrich B, Grabarse M, Kochmann W, Kranz L, Lang S, Schulze F, Steinke W, Wozniak H (1987) DD Pat 242557
417. Schwamborn M, Klauke E, Schmidt R, Santel HJ, Strang RH (1988) DD Pat 257380; Bekanntmach Amt Erfind Patentwesen DDR (1988) 29: 24
418. Chodak AA, Tichonina NA, Porshnev UN (1987) Izvest Akad Nauk USSR, Ser Chim (1987) 11: 2547–2550

Subject Index

MIX
Papier aus verantwortungsvollen Quellen
Paper from responsible sources
FSC® C105338

If you have any concerns about our products,
you can contact us on
ProductSafety@springernature.com

In case Publisher is established outside the EU,
the EU authorized representative is:
Springer Nature Customer Service Center GmbH
Europaplatz 3, 69115 Heidelberg, Germany

Printed by Libri Plureos GmbH
in Hamburg, Germany